T0342482

Three-Dimensional Geometry and Topology

Princeton Mathematical Series

EDITORS: LUIS A. CAFFARELLI, JOHN N. MATHER, *and* ELIAS M. STEIN

1. The Classical Groups *by Hermann Weyl*
3. An Introduction to Differential Geometry *by Luther Pfahler Eisenhart*
4. Dimension Theory *by W. Hurewicz and H. Wallman*
8. Theory of Lie Groups: I *by C. Chevalley*
9. Mathematical Methods of Statistics *by Harald Cramér*
10. Several Complex Variables *by S. Bochner and W. T. Martin*
11. Introduction to Topology *by S. Lefschetz*
12. Algebraic Geometry and Topology *edited by R. H. Fox, D. C. Spencer, and A.W. Tucker*
14. The Topology of Fibre Bundles *by Norman Steenrod*
15. Foundations of Algebraic Topology *by Samuel Eilenberg and Norman Steenrod*
16. Functionals of Finite Riemann Surfaces *by Menahem Schiffer and Donald C. Spencer*
17. Introduction to Mathematical Logic, Vol. I *by Alonzo Church*
19. Homological Algebra *by H. Cartan and S. Eilenberg*
20. The Convolution Transform *by I. I. Hirschman and D. V. Widder*
21. Geometric Integration Theory *by H. Whitney*
22. Qualitative Theory of Differential Equations *by V. V. Nemytskii and V. V. Stepanov*
23. Topological Analysis *by Gordon T. Whyburn* (revised 1964)
24. Analytic Functions *by Ahlfors, Behnke, Bers, Grauert et al.*
25. Continuous Geometry *by John von Neumann*
26. Riemann Surfaces *by L. Ahlfors and L. Sario*
27. Differential and Combinatorial Topology *edited by S. S. Cairns*
28. Convex Analysis *by R. T. Rockafellar*
29. Global Analysis *edited by D. C. Spencer and S. Iyanaga*
30. Singular Integrals and Differentiability Properties of Functions *by E. M. Stein*
31. Problems in Analysis *edited by R. C. Gunning*
32. Introduction to Fourier Analysis on Euclidean Spaces *by E. M. Stein and G. Weiss*
33. Étale Cohomology *by J. S. Milne*
34. Pseudodifferential Operators *by Michael E. Taylor*
35. Three-Dimensional Geometry and Topology, Volume 1 *by William P. Thurston. Edited by Silvio Levy*
36. Representation Theory of Semisimple Groups: An Overview Based on Examples *by Anthony W. Knapp*
37. Foundations of Algebraic Analysis *by Masaki Kashiwara, Takahiro Kawai, and Tatsuo Kimura. Translated by Goro Kato*
38. Spin Geometry *by H. Blaine Lawson, Jr., and Marie-Louise Michelsohn*
39. Topology of 4-Manifolds *by Michael H. Freedman and Frank Quinn*
40. Hypo-Analytic Structures: Local Theory *by François Treves*
41. The Global Nonlinear Stability of the Minkowski Space *by Demetrios Christodoulou and Sergiu Klainerman*
42. Essays on Fourier Analysis in Honor of Elias M. Stein *edited by C. Fefferman, R. Fefferman, and S. Wainger*
43. Harmonic Analysis: Real-Variable Methods, Orthogonality, and Oscillatory Integrals *by Elias M. Stein*
44. Topics in Ergodic Theory *by Ya. G. Sinai*
45. Cohomological Induction and Unitary Representations *by Anthony W. Knapp and David A. Vogan, Jr.*

Three-Dimensional Geometry and Topology

VOLUME 1

William P. Thurston

EDITED BY SILVIO LEVY

PRINCETON UNIVERSITY PRESS
PRINCETON, NEW JERSEY
1997

Published by Princeton University Press, 41 William Street,
Princeton, New Jersey 08540
In the United Kingdom: Princeton University Press, Chichester, West Sussex

Library of Congress Cataloging-in-Publication Data
Thurston, William P.
Three-dimensional geometry and topology / William P. Thurston ;
edited by Silvio Levy.
p. cm. — (Princeton mathematical series ; 35)
Includes bibliographical references and index.
ISBN 0-691-08304-5 (cl : alk. paper)
1. Geometry, Hyperbolic. 2. Three-manifolds (Topology) I. Levy, Silvio.
II. Title. III. Series.
QA685.T49 1997
516'.07—dc21 96-45578

Frontispiece: The sculpture in the photograph is a marble rendition by Helaman Ferguson of the Klein quartic, a genus-three surface with the greatest possible amount of symmetry (336 isometries). It is not possible to represent all this symmetry in space, but the twenty-four heptagons that tile the surface are in fact intrinsically identical. They come from regular heptagons in the hyperbolic plane, meeting three to a vertex; each such heptagon is made of fourteen triangles of the 2-3-7 tiling of Figures 2.10 and 2.14. If you start on any edge of the surface, proceed along it to a fork and turn right, then turn left at the next fork, and keep alternating in this way, you arrive back where you started after eight turns. For this reason the sculpture, which stands at the Mathematical Sciences Research Institute in Berkeley, is named "The Eightfold Way."

The publisher would like to acknowledge William P. Thurston and Silvio Levy for providing the camera-ready copy from which this book was printed

Princeton University Press books are printed on acid-free paper and meet the guidelines for permanence and durability of the Committee on Production Guidelines for Book Longevity of the Council on Library Resources

http://pup.princeton.edu

Printed in the United States of America

17 19 20 18 16

ISBN-13: 978-0-691-08304-9 (cloth)

Contents

Preface vii

Reader's Advisory ix

1 What Is a Manifold? 3

1.1 Polygons and Surfaces 4
1.2 Hyperbolic Surfaces 7
1.3 The Totality of Surfaces 17
1.4 Some Three-Manifolds 31

2 Hyperbolic Geometry and Its Friends 43

2.1 Negatively Curved Surfaces in Space 45
2.2 The Inversive Models 53
2.3 The Hyperboloid Model and the Klein Model 64
2.4 Some Computations in Hyperbolic Space 74
2.5 Hyperbolic Isometries 86
2.6 Complex Coordinates for Hyperbolic Three-Space . . . 98
2.7 The Geometry of the Three-Sphere 103

3 Geometric Manifolds 109

3.1 Basic Definitions 109
3.2 Triangulations and Gluings 118
3.3 Geometric Structures on Manifolds 125
3.4 The Developing Map and Completeness 139
3.5 Discrete Groups 153
3.6 Bundles and Connections 158
3.7 Contact Structures 168
3.8 The Eight Model Geometries 179
3.9 Piecewise Linear Manifolds 190
3.10 Smoothings 193

4 The Structure of Discrete Groups **209**

4.1 Groups Generated by Small Elements 209
4.2 Euclidean Manifolds and Crystallographic Groups . . 221
4.3 Three-Dimensional Euclidean Manifolds 231
4.4 Elliptic Three-Manifolds 242
4.5 The Thick-Thin Decomposition 253
4.6 Teichmüller Space 258
4.7 Three-Manifolds Modeled on Fibered Geometries . . . 277

Glossary **289**

Bibliography **295**

Index **301**

Preface

This book began with notes from a graduate course I gave at Princeton University on the geometry and topology of three-manifolds, over the period 1978–1980. The notes were duplicated and sent to people who wrote to ask for them. The mailing list grew to a size of about one thousand before a version was frozen. Much of the original draft was written by Steve Kerckhoff and Bill Floyd.

The notes were originally aimed for an audience of fairly mature mathematicians, and presented material not in the standard repertoire. A number of seminars worked through these notes. Some of the feedback from seminars and individuals convinced me that it would be worth filling in considerably more detail and background; there were several places where people tended to get stuck, sometimes for weeks. I embarked on a project of clarifying, filling in and rearranging the material before publishing it.

Far more time (and blood, sweat and tears) has elapsed since the time of the original notes than I intended or anticipated. The present text originated from several chapters of the original notes, but it has undergone deep transmutation.

The ultimate emergence of this book would not have happened without the support of the Geometry Center and the vision of its founding director, Albert Marden. In particular, from 1990 through 1992 the Geometry Center hosted five intense bookwriting workshops centered on drafts of this text, during which both the ideas and the means of communicating them gained greatly from the scrutiny of multiple eyes and the thoughts of many minds. I owe many thanks to Dick Canary, Jim Cannon, David Epstein, Bill Floyd, Steve Kerckhoff, Yair Minsky, who participated in these workshops and contributed to examining and editing the text and graphics on these occasions.

Most of all, I want to say that I have tremendously appreciated and deeply admired working with Silvio Levy, who has stuck with me and with this project through thick and through thin.

William P. Thurston
June 1996

Figure Credits

Almost all the figures in this book were created by Silvio Levy using Mathematica, with labels automatically converted to TEX by Levy's Mathfig utility. Many of these figures were redrawn from originals by the author. The following figures have a different provenance:

The cover image and Figure 2.19 are frames from the video *Not Knot*, produced by the Geometry Center and directed by Charlie Gunn and Delle Maxwell. The scenes were generated by Gunn using custom software and Renderman.

The photograph in Figure 3.14 was taken by Kenneth Olson, with the collaboration of Aribert Munzner. The color photograph showing *The Eightfold Way* was taken by Jon Ferguson. Both were kindly provided by the sculptor, Helaman Ferguson, and are reproduced here with permission.

Figures 4.13, 4.14, and 4.17 were created by the author using Xfig, and Figures 4.22 and 4.23 using Adobe Illustrator. Labels were subsequently retypeset in TEX by Levy.

Reader's Advisory

The style of exposition in this book is somewhat experimental.

The most efficient logical order for a subject is usually different from the best psychological order in which to learn it. Much mathematical writing is based too closely on the logical order of deduction in a subject, with too many definitions before, or without, the examples which motivate them, and too many answers before, or without, the questions they address. In a formal and logically ordered approach to a subject, readers have little choice but to follow along passively behind the author, in the faith that machinery being developed will eventually be used to manufacture something worth the effort.

Mathematics is a huge and highly interconnected structure. It is not linear. As one reads mathematics, one needs to have an active mind, asking questions, forming mental connections between the current topic and other ideas from other contexts, so as to develop a sense of the structure, not just familiarity with a particular tour through the structure.

The style of exposition in this book is intended to encourage the reader to pause, to look around and to explore. I hope you will take the time to construct your own mental images, to form connections with other areas of mathematics and interconnections within the subject itself.

Think of a tinkertoy set. The key is the pieces which have holes, allowing you to join them with rods to form interesting and highly interconnected structures. No interesting mathematical topic is self-contained or complete: rather, it is full of "holes," or natural questions and ideas not readily answered by techniques native to the topic. These holes often give rise to connections between the given topic and other topics that seem at first unrelated. Mathematical exposition often conceals these holes, for the sake of smoothness—

but what good is a tinkertoy set if the holes are all filled in with modeling clay?

In the present exposition, many of the "holes" or questions are explicitly labeled as exercises, questions, or problems. *Most of these are not walking-the-dog exercises* where the dog follows behind on a leash until the awaited event. You may or may not be able to answer the questions, even if you completely understand the text. Some of the questions form connections with ideas discussed more fully later on. Other questions have to do with details that otherwise would have been "left as an exercise for the reader." Still others relate the material under discussion to topics which are neither discussed nor assumed in the main text.

It is important to read through and think about the exercises, questions and problems. It should be possible to solve some of the more straightforward questions. But please don't be discouraged if you can't solve all, or even most, of the questions, any more than you are discouraged when you can't immediately answer questions which occur to you spontaneously.

There are other ways in which the order of development deviates from the order of logical deduction. For instance, manifolds and geometric structures on manifolds are discussed intuitively in the first two chapters, even though the formal definition and basic properties are only presented in Chapter 3. These definitions are somewhat heavy until one has seen some good examples. The concept of an orbifold is defined only in Chapter 5 (to appear in Volume 2), even though it is significant for the material in Chapter 4.

On the other hand, for purposes of reference, the logical order is sometimes chosen over the psychological order. For example, Chapter 2 contains a fuller treatment of hyperbolic geometry than is motivated by the examples and applications which have been given up to that point. The reader may wish to skip some of it, and refer back only as needed for later reference.

Often a beginner gives up reading a book, or parts of it, when he or she hits a morass of unknown terms and notation. Given the non-linear method of exposition used in this book, it is likely that you'll encounter unfamiliar terms that are not explained in the text. Don't let that discourage you; it may be useful to read ahead and return to the sticky passage later. Some terms are defined in the Glossary, the first occurrence of each being marked with a dagger (†).

Three-Dimensional Geometry and Topology

Chapter 1

What Is a Manifold?

Manifolds are around us in many guises.

As observers in a three-dimensional world, we are most familiar with two-manifolds: the surface of a ball or a doughnut or a pretzel, the surface of a house or a tree or a volleyball net...

Three-manifolds may seem harder to understand at first. But as actors and movers in a three-dimensional world, we can learn to imagine them as alternate universes.

Mathematically, manifolds arise most often not as physical entities in space, but indirectly: the solution space of some set of conditions, the parameter space for some family of mathematical objects, and so on. Translating such abstract descriptions, where possible, into our concrete imagery of three-dimensional space is generally a big aid to understanding.

Even when we do this, however, it is often not easy to recognize the identity of a manifold: the same topological object can have completely different concrete descriptions. Furthermore, manifolds may have inherent symmetry that is not apparent from a concrete description.

How can we know a manifold?

Exercise. Which manifold is this?

1.1. Polygons and Surfaces

The simplest and most symmetric surface, next to the sphere, is the torus, or surface of a doughnut. This surface has symmetry as a surface of revolution in space, but it has additional "hidden" symmetry as well. The torus can be described topologically by gluing together the sides of a square. If the square is reflected about its main diagonal to interchange the a and b axes, the pattern of identification is preserved.

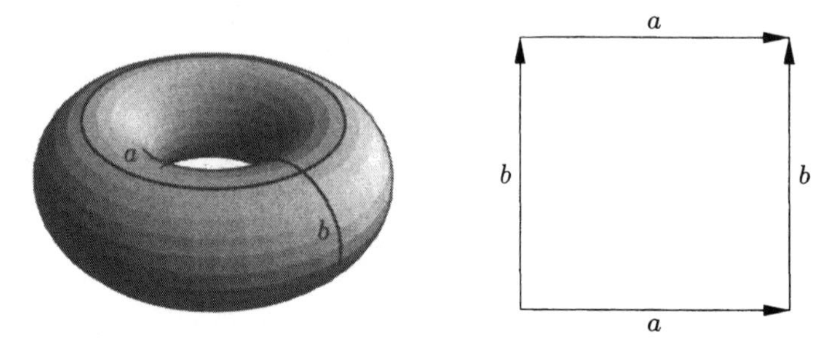

Figure 1.1. The square torus. A torus can be obtained, topologically, by gluing together parallel sides of a square. Conversely, if you cut the torus on the left along the two curves indicated, you can unroll the resulting figure into the square on the right.

Problem 1.1.1 (square torus in space). The *one-point compactification* $\widehat{\mathbf{R}^n}$ of $\mathbf{R}^n$ is the topological space obtained by adding a point ∞ to $\mathbf{R}^n$ whose neighborhoods are of the form $(\mathbf{R}^n \setminus B) \cup \infty$ for all bounded sets B.

(a) Check that the one-point compactification of $\mathbf{R}^n$ is homeomorphic to the sphere S^n.

(b) Consider an ordinary torus in $S^3 = \widehat{\mathbf{R}^3}$, and show that the interchange of curves a and b in Figure 1.1 can be achieved by moving the torus in S^3 (without necessarily preserving its geometric shape). (This question will become much easier after you read Section 2.7.)

(c) Show that this cannot be done in $\mathbf{R}^3$.

Curiously, a torus is also obtained by identifying parallel sides of a regular hexagon (Figure 1.2). This alternate description has six-fold symmetry which is not compatible with the symmetry of the previous description.

Problem 1.1.2 (reconciling the symmetries of a torus). We've seen three concrete descriptions for a torus: as a physical surface in space, as a

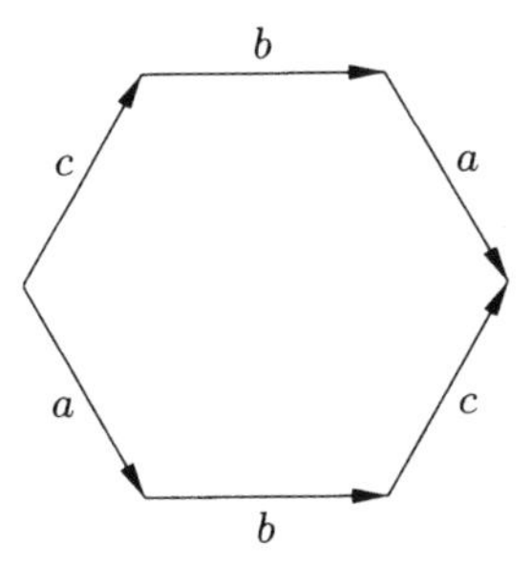

Figure 1.2. The hexagonal torus. Here is another gluing pattern of a polygon which yields a torus. This pattern reveals a different kind of symmetry from the first.

square with identifications, and as a hexagon with identifications. Can you reconcile them to your satisfaction?

(a) Check that gluing the hexagon does yield a torus. Draw the curves needed to cut a torus into a hexagon.

(b) What transformation changes a hexagon with identifications to a quadrilateral with identifications?

(c) Is it possible to †embed the torus in $\mathbf{R}^3$ or in S^3 in such a way that the six-fold symmetry extends to a symmetry of the †ambient space?

(d) One can divide the torus into seven countries in such a way that every country is in one piece and has a (non-punctual) border with every other. In other words, a political map of a torus-shaped world may require up to seven colors. Construct such a seven-colored map. Can it be done symmetrically?

These two descriptions of the torus are closely related to common patterns of †tilings of the †Euclidean plane $\mathbf{E}^2$. Take an infinite collection of identical squares, all labeled as in Figure 1.1, or of hexagons, labeled as in Figure 1.2. Begin with a single polygon, then add more polygons layer by layer, identifying edges of the new ones with similarly labeled edges of the old ones. Make sure the local picture near each vertex looks like the local picture in the original pattern, when the edges of a single polygon were identified: if you follow this rule, each new tile fits in in exactly one way. The result is a tiling of the Euclidean plane by congruent squares or hexagons.

These tilings show that the plane is a †covering space for the torus: the †covering map for the square tiling (say) is the map that identifies corresponding points in each square, taking them all to the same point on the glued-up torus. Since the plane is †simply

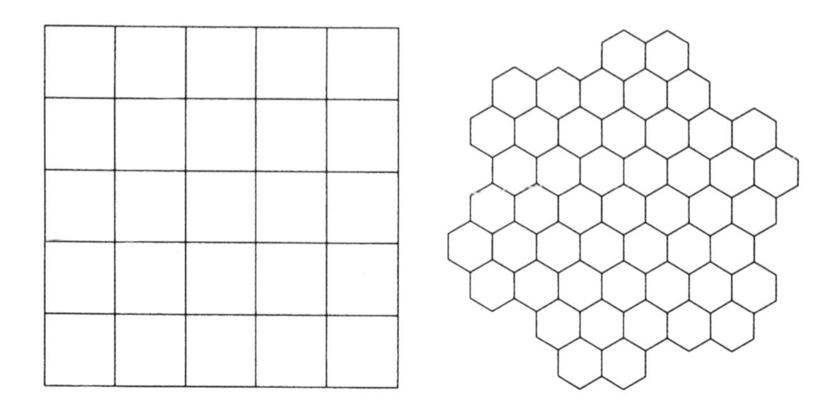

Figure 1.3. Tiling the plane with tori. These tilings of the plane arise from the two descriptions of the torus by gluing polygons. They show the universal covering space of the torus, obtained by "unrolling" the torus.

connected, it is the †universal cover of the torus. The covering map singles out a group of homeomorphisms of the plane, namely those that take any point into another point that has the same image on the torus under the covering map. For the square tiling, for instance, these †covering transformations are the translations that preserve vertices. The torus is the †quotient space of the plane by the †action of this †covering group.

Since the covering transformations are Euclidean isometries, we can give the torus a *Euclidean structure*, that is, a metric that is locally isometric to Euclidean space. This is done as follows: given a point x on the torus, we choose a neighborhood U of x small enough that the inverse image of U in the plane is made up of connected components homeomorphic to U under the covering map p. By shrinking U further, we can make sure that the diameter of these components is less than the distance separating any two of them. Then we declare p to be an isometry between any of these components and U; it doesn't matter which component we choose, because they're all isometric (Figure 1.4).

This locally Euclidean geometry on the torus is not the same as the geometry it has as a surface of revolution in space, because the former is everywhere flat, whereas the torus of revolution has positive Gaussian curvature at some places and negative at others (Section 2.1).

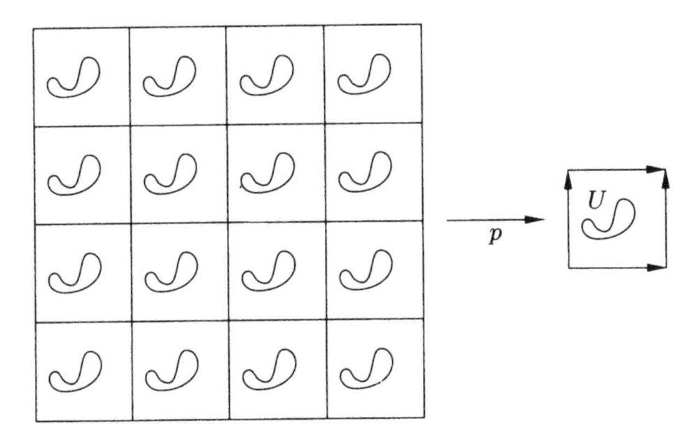

Figure 1.4. Transferring the geometry from the plane to the torus. As a quotient of the plane by a group of isometries, the torus has a locally Euclidean geometry, in which a small open set U is isometric to any component of its inverse image under the covering map p. In this geometry the images of straight lines are †geodesics on the torus. They're usually not geodesics in the geometry of the torus of revolution.

1.2. Hyperbolic Surfaces

Just like the torus, the two-holed torus or *genus-two surface* (Section 1.3) can be obtained by identifying the sides of a polygon. Most familiar is the pattern shown in Figure 1.5, in which we cut along four simple closed curves meeting in a single point, to get an octagon. The four curves can be labeled so that the resulting octagon is labeled $aba^{-1}b^{-1}cdc^{-1}d^{-1}$.

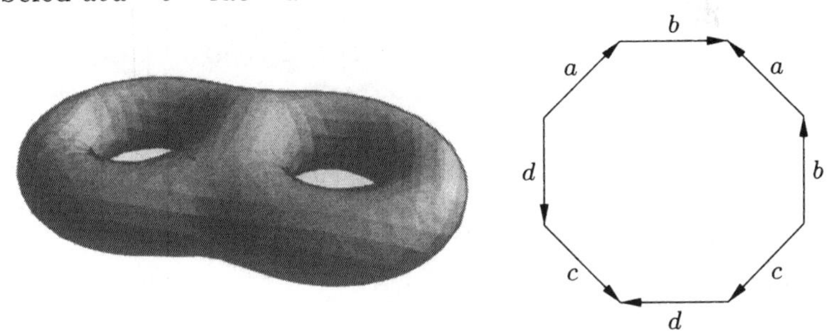

Figure 1.5. A genus-two surface. A two-holed torus, or surface of genus two, can be cut along curves until the result is topologically a polygon. Here we see the most common cutting pattern.

Are there tilings of the plane by regular octagons, coming from the gluing pattern in Figure 1.5? The answer is clearly no in the Euclidean plane: the interior angle of a regular octagon is $(8 - 2) \cdot 180^\circ/8 = 135^\circ$, so not even three octagons fit around a vertex, whereas eight would be needed.

But that's not the end of the story: if we don't insist that our plane satisfy †Euclid's parallel axiom, there is nothing to force the sum of the angles of a triangle to be 180°, and we could perhaps choose regular octagons with 45° angles, so they fit nicely eight to a vertex. We will soon describe a concrete construction to do exactly that.

Until the late eighteenth century the validity of the parallel axiom was taken for granted, and in fact much work was invested in frustrated attempts to prove its redundancy by deriving it from Euclid's other axioms and "common notions," all of which seemed much more intuitive. By the 1820s, however, three people had independently come to realize that a self-consistent geometry, with lines, planes, and angles otherwise similar to the usual ones, does not have to satisfy the parallel axiom: they were János Bolyai in Hungary, Carl Friedrich Gauss in Germany and Nikolai Ivanovich Lobachevskii in Russia. Gauss was there first, but he chose not to publish his conclusions, and Bolyai received no recognition until long after his death; and so it is that this non-Euclidean geometry became known as Lobachevskiian geometry until Felix Klein, at the turn of this century, introduced the term hyperbolic geometry, the most current today.

The denial of one of Euclid's axioms remained a profoundly disturbing idea, and although continuing work by Lobachevskii and others not only failed to lead to a contradiction but showed that hyperbolic geometry was remarkably rich, it was a matter of debate throughout most of the nineteenth century whether or not such a geometry could exist. Such doubts lasted until Eugenio Beltrami, in 1868, constructed an explicit model of hyperbolic space—something like a map of hyperbolic space in Euclidean space. Actually, Georg Friedrich Riemann seems to have reached this level of understanding much earlier, for he exhibited a metric for any space of constant curvature in his famous "Lecture on the Hypotheses That Lie at the Foundation of Geometry" (1854), and in general he wrote about lines and planes in such spaces in a manner that indicates a clear grasp of their nature. However, this understanding only entered the general mathematical consciousness with Beltrami's work.

Later other models were introduced, each with its advantages and disadvantages. These models, or maps, are helpful in the same way that maps of the earth are helpful: they are perforce distorted, but with some imagination one can develop a feeling for the true nature of the landscape by studying them.

Hyperbolic geometry will be an essential tool for us throughout this book, so it's good to get more or less familiar with it right away, at least in the two-dimensional case. Our initial study of hyperbolic geometry will be based on a particular model, but it's best to keep in mind from the start that the same geometric construct—hyperbolic space $\mathbf{H}^n$—can be represented in many different ways. We first give a characterization of hyperbolic lines (geodesics), and of certain line-preserving transformations; from this we derive many other properties of hyperbolic space, including the metric. As we go along we'll develop a "dictionary" to translate between hyperbolic objects and their representations in the model, and as we become fluent we'll start doing this translation automatically.

The hyperbolic plane $\mathbf{H}^2$ is homeomorphic to $\mathbf{R}^2$, and the *Poincaré disk model*, introduced by Henri Poincaré around the turn of this century, maps it onto the open unit disk D in the Euclidean plane. Hyperbolic straight lines, or geodesics, appear in this model as arcs of circles orthogonal to the boundary ∂D of D, and every arc orthogonal to ∂D is a hyperbolic straight line (Figure 1.6). There is one special case: any diameter of the disk is a limit of circles orthogonal to ∂D and it is also a hyperbolic straight line. For simplicity, from now on we will include diameters when talking about arcs orthogonal to ∂D.

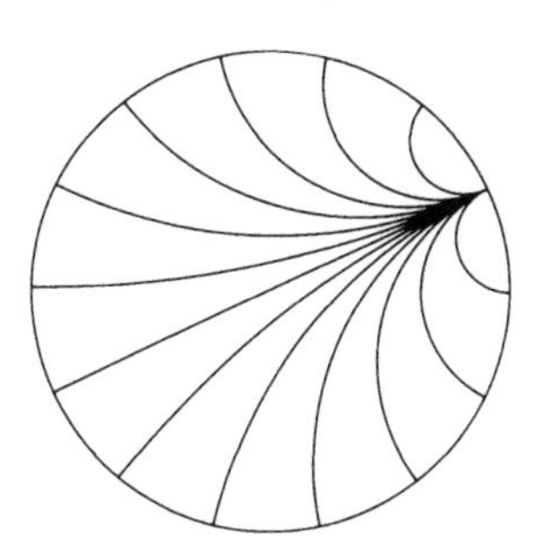

Figure 1.6. Straight lines in the Poincaré disk model. Straight lines in the Poincaré disk model appear as arcs orthogonal to the boundary of the disk or, as a special case, as diameters.

A hyperbolic reflection in one of the lines represented by a diameter of the disk translates, in our model, into a Euclidean reflection in the same diameter. How about hyperbolic reflections in lines not represented by diameters? They translate into certain Euclidean transformations, called inversions, that generalize reflections:

Definition 1.2.1 (inversion in a circle). If C is a circle in the Euclidean plane, the *inversion* i_C in C is the unique map from the complement of the center of C into itself that fixes every point of C, exchanges the interior and exterior of C and takes circles orthogonal to C to themselves.

Exercise 1.2.2 (inversions are well-defined). (a) Show the following standard result from Euclidean plane geometry: If A is a point outside a circle C and l is a line through A intersecting C at P and P', the product $AP \cdot AP'$ is independent of l and is equal to AT^2, where $\overrightarrow{AT}$ is a ray tangent to C at T. This product is the *power of A with respect to C*.

(b) Use this to show that Definition 1.2.1 makes sense. (Hint: see Figure 1.7, left.)

(c) Prove that if C has center O and radius r, the image $P' = i_C(P)$ is the point on the ray $\overrightarrow{OP}$ such that $OP \cdot OP' = r^2$.

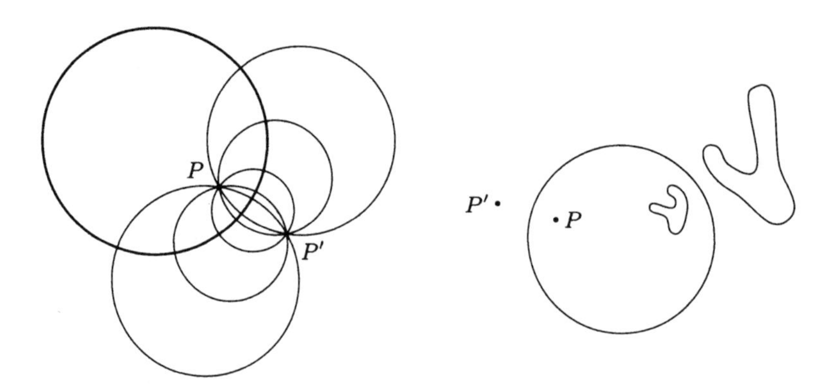

Figure 1.7. Inversion in a circle. All circles orthogonal to a given circle and passing through a given point P also pass through a point P'. We say that P' is the image of P under inversion in the circle. Inversion interchanges the interior and exterior of the circle.

Inversions have lots of neat properties, and none neater than the following, which we will use time after time:

Proposition 1.2.3 (properties of inversions). *If C is a circle in the Euclidean plane, i_C is conformal, that is, it preserves angles.*

Also, i_C takes circles not containing the center of C to circles, circles containing the center to lines, lines not containing the center to circles containing the center, and lines containing the center to themselves.

Proof of 1.2.3. Given two vectors at a point not in C we can construct a circle tangent to each vector and orthogonal to the circle of inversion, and these circles, which are preserved by the inversion, meet at the same angle at their other point of intersection. This shows conformality everywhere but on C. The case of a point on C can be handled by continuity.

Next we show that circles and lines tangent to C are taken to such circles and lines (disrespectively). For the rest of this proof, we let "circle" stand for "circle or line". For any point x on C the plane is filled by a family $\mathfrak{F}_O$ of circles orthogonal to C at x and similarly by a family $\mathfrak{F}_T$ of circles tangent to C at x (see Figure 1.8). Any

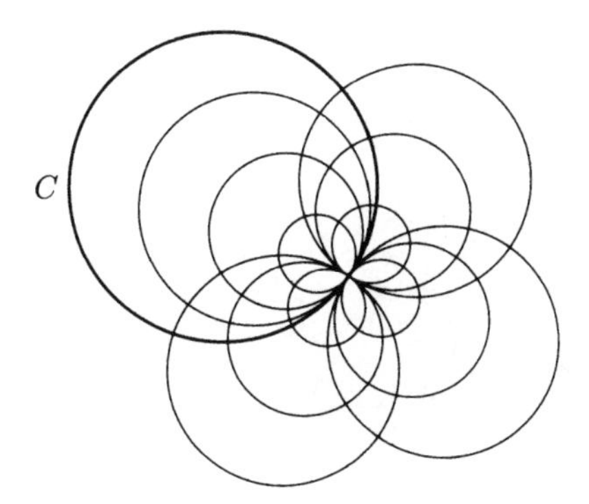

Figure 1.8. Orthogonal families of circles through a point. The circles tangent to a circle C at a given point x are the orthogonal trajectories of the family of circles orthogonal to C at x.

circle from $\mathfrak{F}_T$ meets any circle from $\mathfrak{F}_O$ perpendicularly at both their intersection points, and thus $\mathfrak{F}_T$ forms the set of orthogonal trajectories of $\mathfrak{F}_O$. We already know that $\mathfrak{F}_O$ is preserved by i_C, so since the inversion is conformal the family of orthogonal trajectories of $\mathfrak{F}_O$ is preserved, and thus any circle tangent to C is taken to another circle tangent to C.

Any circle other than a line through the center of C can be blown up or shrunk down to a circle tangent to C by means of a †homothety centered at the center of C. The circle's image under inversion suffers exactly the opposite fate, by exercise 1.2.2(c): it shrinks down or blows up by the same factor. Since the image of the tangent circle is a circle, so is the image of the original circle. Lines through the center go to themselves by 1.2.2(c). 1.2.3

Notice that circles through the center of C and straight lines are sent to each other because points closer and closer to the center are sent further and further away.

Example 1.2.4 (mechanical linkages). Around the middle of the nineteenth century, with the rapid development of the Industrial Age, there was great interest in the theory of mechanical linkages. An important problem, for a time, was to construct a mechanical linkage that would transform circular motion into straight-line motion, that is, maintain some point on the linkage in a straight line as another point described a circle. In the 1860s Lippman Lipkin and Peaucellier independently found a solution to the problem—the same solution, in fact, involving inversion in a circle. It turned out to be of little use because in practice the relatively large number of moving components (seven bars and six joints) made the linkage wobble more than simpler linkages that, mathematically speaking, only approximated straight-line motion.

Exercise 1.2.5. (a) Prove that the linkage of Figure 1.9 performs as advertised in its caption.

(b) Construct a mechanical linkage that achieves straight line motion.

Back to the Poincaré model. If a hyperbolic line appears in the model as (an arc of) a circle orthogonal to ∂D, the hyperbolic reflection in this line appears as (the restriction to D of) the Euclidean inversion in this circle. This is plausible because, by Proposition 1.2.3, such an inversion maps D into itself pointwise and preserves hyperbolic lines. We will presently see that it also preserves distances, if distances are defined the right way, so the word "reflection" is fully justified.

How then should distances be defined in the hyperbolic plane? Answering this question boils down to describing the †Riemannian metric of the hyperbolic plane, that is, to assigning each point an inner product for the tangent space at that point. In Section 2.2 we will write down a formula, but for now we can learn a whole lot just from geometric constructions. The driving idea is that hyperbolic reflections should preserve distances, that is, they should be

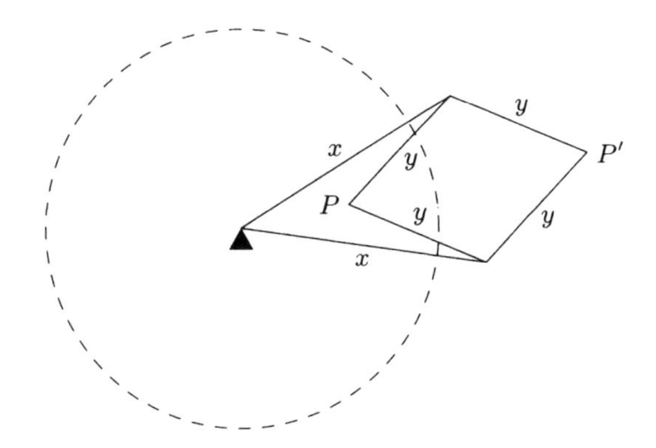

Figure 1.9. A mechanical inversor. This mechanical linkage performs an inversion in the circle of radius $r = \sqrt{(x^2 - y^2)}$. The small triangle near the center of the circle indicates an anchor point for the linkage. If point P is moved around a figure, P' moves around the image of the figure under inversion in the dotted circle.

isometries. This is enough to pin down the metric up to a constant factor.

We start with Figure 1.10. Consider two orthogonal hyperbolic lines L and M, seen in the model as Euclidean arcs of circles orthogonal to ∂D, and another Euclidan circle C that intersects ∂D in the same two points as L. A simple argument shows that C, too, is orthogonal to M, so that, by the definition of inversion, the hyperbolic reflection in M leaves both L and C invariant. If this reflection is to preserve hyperbolic distances, corresponding points of C on both sides of M must be equally distant from L. In fact, by varying M among the lines orthogonal to L, we see that *all* points of C must be equally distant from L, that is, C must be an *equidistant curve*. The banana-shaped region between L and C can be filled with segments orthogonal to L, all having the same hyperbolic length.

Now apply any hyperbolic reflection to the whole picture. The angle α at the tips of the banana doesn't change, because inversions preserve angles; neither does the width of the banana—the hyperbolic length l of the transversal segments—since we want reflections to be isometries. This means that l should be a function solely of α! Moreover, this function has a finite derivative at $\alpha = 0$, because the Euclidean length l_E of any particular transversal segment is roughly proportional to α for α small, and Euclidean and hyperbolic lengths

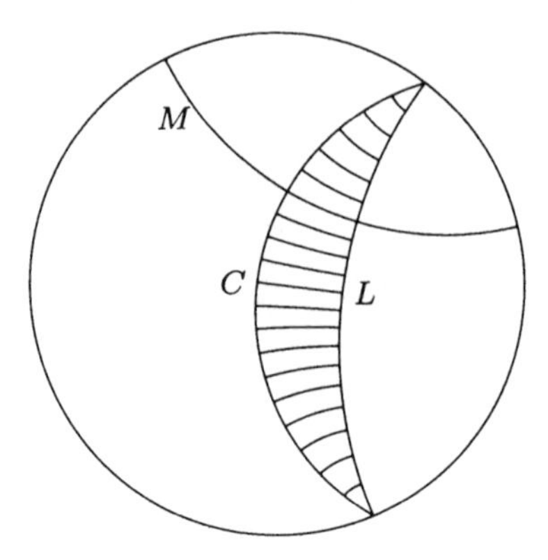

Figure 1.10. Equidistant curve to a line. All points on the arc of circle C lie at the same hyperbolic distance from the hyperbolic line L.

should be proportional to first order. We may take the derivative $dl/d\alpha$ at $\alpha = 0$ to be 1.

We're now equipped to go back and define the Riemannian metric by means of this construction (Figure 1.11). To find the length of a tangent vector v at a point x, draw the line L orthogonal to v through x, and the equidistant circle C through the tip. The length of v (for v small) is roughly the hyperbolic distance between C and L, which in turn is roughly equal to the Euclidean angle between C and L where they meet. If we want an exact value, we consider the angle α_t of the banana built on tv, for t approaching zero: the length of v is then $d\alpha_t/dt$ at $t = 0$.

Actually things are even simpler than that, because two vectors at x having the same Euclidean length also have the same hyperbolic length. This follows from the existence of a hyperbolic reflection fixing x and taking one vector to the other, and from the fact that the derivative of an inversion at a point on the the inversion circle is an orthogonal map (by the remark after Definition 1.2.1). Since Euclidean and hyperbolic vector lengths at a point are proportional, so are the inner products. In particular, the Poincaré model is conformal, because Euclidean and hyperbolic angles are equal.

We can now get back to our tiling of the hyperbolic plane using regular octagons. Remember that we need a tile with hyperbolic angles equal to 45°: but since the Poincaré model is conformal, the Euclidean angle between the arcs that form the edges will be the same. Now imagine a small octagon centered at the origin; since

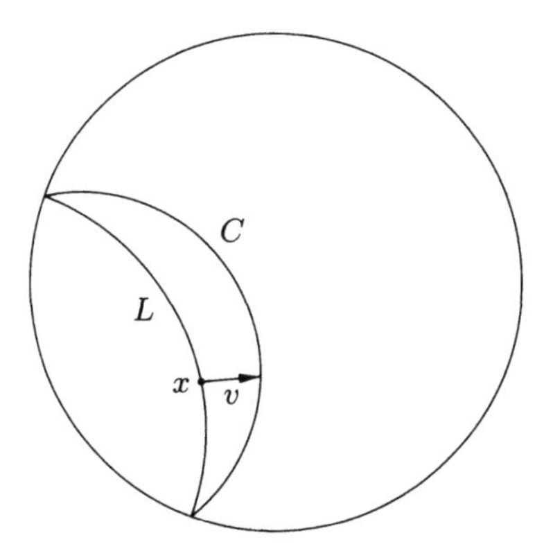

Figure 1.11. Hyperbolic versus Euclidean length. The hyperbolic and Euclidean lengths of a vector in the Poincaré model are related by a constant that depends only on how far the vector's basepoint is from the origin.

its edges (in the model) bend just a little, its angles are close to 135°. By moving the vertices away from the origin we can make the angles as small as we want. By continuity (or, more pedantically, the †intermediate value theorem), there is some octagon in between whose angles are exactly $\pi/4$ (Figure 1.12).

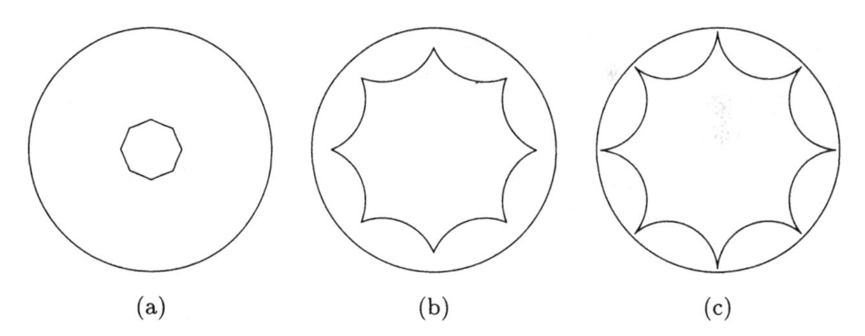

Figure 1.12. Bigger octagons in hyperbolic space have smaller angles. Between a tiny, Euclidean-like octagon with large angles (a) and a very large one with arbitrarily small angles (c) there must be one with angles exactly $\pi/4$ (b).

Once we've found the octagon we want, we take identical copies of it and place them on the hyperbolic plane respecting the identifications prescribed by Figure 1.5, to give the pattern shown in Figure 1.13(a). The copies look different depending on where they are in the model—in particular, they quickly start looking very small as we move away from the origin—but they can all be obtained from one another by hyperbolic isometries. For example, the two copies in Figure 1.13(b) are mapped to one another by a reflection in L, followed by a reflection in M.

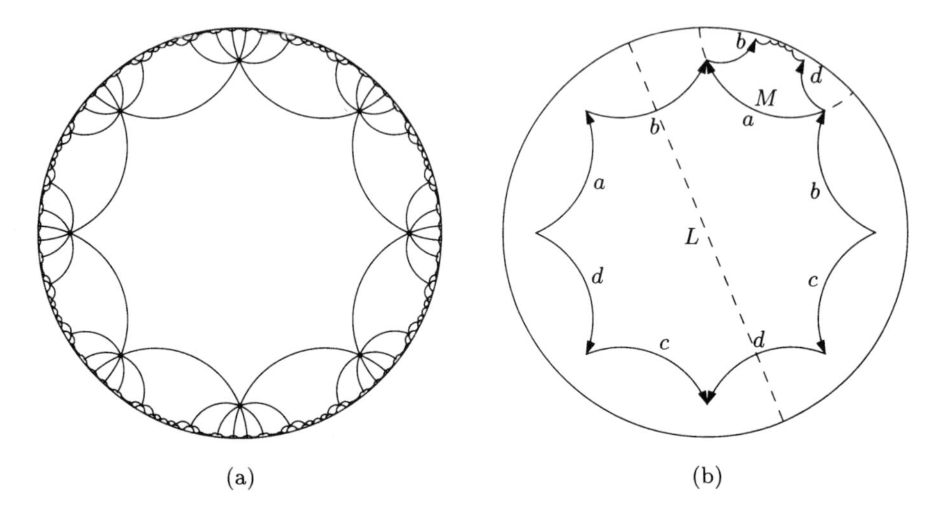

Figure 1.13. A tiling of the hyperbolic plane by regular octagons. (a) A tiling of the hyperbolic plane by identical regular octagons, seen in the Poincaré disk projection. (b) To get the small octagon from the big one, reflect in L, then in M.

This tiling of the hyperbolic plane shows that a genus-two surface can be given a *hyperbolic structure*, a geometry such that the surface looks locally like the hyperbolic plane. The construction exactly parallels the one we saw for the torus at the end of Section 1.1: again we have a covering space, the hyperbolic plane, with a Riemannian metric preserved by all the covering transformations, so we can transfer that metric to the quotient space.

It is not an accident that we were able to cover the torus with the Euclidean plane, and the genus-two surface with the hyperbolic plane: all surfaces can be given simple geometric structures, as we'll see in Section 1.3.

Problem 1.2.6 (genus-two symmetry). How much symmetry does a surface of genus two have?

(a) Show how to embed a genus-two surface in space so as to have three-fold symmetry.

(b) Show that a surface of genus two may be obtained from either a regular octagon or a regular decagon by identifying parallel sides. Draw pictures of the curves needed to cut the surface into an octagon, or a decagon.

(c) Let T_8 be the rotation of †order 8 of the octagon, and T_{10} the rotation of order 10 of the decagon. These transformations go over to homeomorphisms of order 8 and 10 of the surface of genus two. How many fixed points do T_8 and T_{10} have? How many periodic points of order less than 8 or 10?

(d) Is it possible to embed a genus-two surface in space so as to admit a symmetry of order 8? of order 10? What if you consider †immersions instead of embeddings, that is, if you allow self-intersections?

1.3. The Totality of Surfaces

Gluing edges of polygonal regions in dimension two, in the way we have been doing, always gives rise to a two-dimensional manifold. To be precise, let $F_1, \ldots, F_k$ be †oriented †polygonal regions, and suppose that the total number of boundary edges is even. Give the edges the orientations induced by the orientations of the regions. A *gluing pattern* consists of a pairing of edges and, for each such pair, a choice of $+$ or $-$ indicating whether the pair should be identified by an †orientation-preserving or †orientation-reversing homeomorphism. For example, the gluing pattern shown in Figure 1.1 pairs opposite edges and assigns to each pair the symbol $-$, since both gluing maps reverse orientation. (Notice that the arrows in the figure indicate matching directions, rather than edge orientations. See also Figure 1.14.)

Exercise 1.3.1 (homeomorphisms of an interval). Prove that two homeomorphisms of an interval to itself are †isotopic if and only if they both preserve orientation, or both reverse orientation. (Write down a formula that works.)

Exercise 1.3.2 (gluings in two dimensions). (a) Using Exercise 1.3.1, show that a gluing pattern determines a unique topological space.

(b) Show that this space is always a two-dimensional manifold.

(c) Show that the manifold is oriented if the gluing pairs edges with opposite orientations.

(d) Figure 1.14 shows three gluing patterns that identify opposite edges of a quadrilateral. What two-manifold is obtained according to each of them?

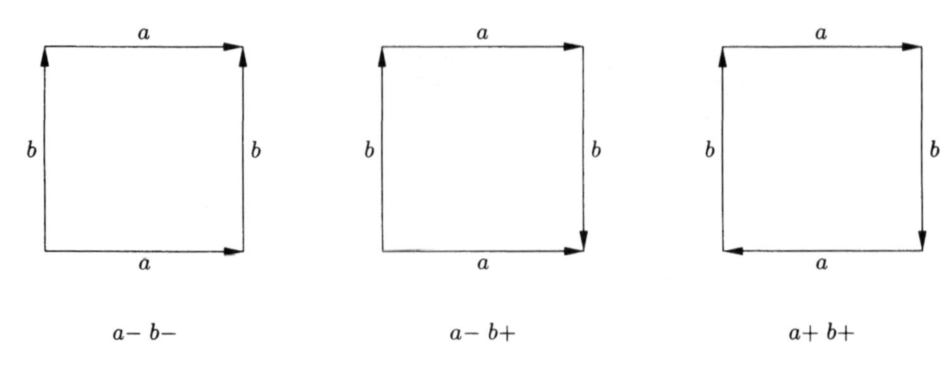

Figure 1.14. Three square gluings. Three possible gluings for a square. The signs associated with each pair of edges indicate whether the gluing map preserves or reverses orientation; the arrows convey the same information.

(e) Explain a necessary and sufficient condition for the two-manifold to be orientable, that could be read off by a computer from the gluing pattern.

It is not easy to visualize directly what surface is obtained when you glue a many-sided polygon, or several polygons, in a given pattern. However, there is an easily computed numerical invariant, the *Euler number* of a surface, that enables one to recognize surfaces quickly. If F is the number of constituent polygons (or faces) and E and V are the numbers of their edges and vertices *after* identification, the Euler number $\chi(S)$ of the glued-up surface S is given by $F - E + V$.

For instance, when a torus is formed by gluing a square as in Figure 1.1, we have one polygon, two edges (the sides of the square identified in pairs) and one vertex (all four vertices identified into one). Therefore $\chi(T^2) = 1 - 2 + 1 = 0$. For the hexagonal torus of Figure 1.2, we get three edges and two vertices, since the vertices are identified in triples; so again $\chi(T^2) = 1 - 3 + 2 = 0$. The sphere S^2 can be divided into four triangles, to form a tetrahedron with six edges and four vertices, so $\chi(S^2) = 4 - 6 + 4 = 2$; if it is divided into six squares to form a cube, the computation is $\chi(S^2) = 6 - 12 + 8 = 2$.

Cutting up a surface into polygons and their edges and vertices, as above, is an example of cell division. A *cell* is a subset $C \subset X$, where X is any Hausdorff space, homeomorphic to an open disk of some dimension, with the condition that the homeomorphism can be extended to a continuous map from the closed disk into X, called the *cell map*. A *face* is a two-cell, an *edge* is a one-cell, and a *vertex* is a zero-cell. A *cell division* of X is a partition of X into cells, in such a way that the boundary of any n-cell is contained in the union of all cells of dimension less than n.

If X is a †differentiable manifold, we will generally assume that our cell divisions are *differentiable*: this means that, for each cell C, the cell map can be realized as a differentiable map from a convex polyhedron onto the closure of C, having maximal rank everywhere. (The idea of differentiability at a point requires that the map be defined on a neighborhood of the point in $\mathbf{R}^n$. So if $X \in \mathbf{R}^n$ is not an open set, we say that a map $X \to \mathbf{R}^m$ is differentiable if it is the restriction of a †differentiable map on an open neighborhood of X.)

Differentiable cell divisions correspond closely to our intuitive idea of cutting a surface into polygons. Occasionally we will encounter cell divisions that are not of this type—for example, in Problem 1.1.1(a) a sphere is expressed as a union of a vertex and a face.

The *Euler number* of a space X having a finite cell division is defined as the sum of the numbers of even-dimensional cells, minus the sum of the numbers of odd-dimensional cells. It is natural to ask: Is the Euler number independent of the cell division? The answer is yes, and we'll prove it for differentiable surfaces.

A surface can have cells of dimension at most two, by the theorem on the †invariance of domain. Let's check what happens to the Euler number when a two-cell or a one-cell is further subdivided. If an edge is divided into two, by placing a new vertex in its middle, this adds one edge and one vertex. They contribute with opposite signs, so they cancel. If a two-cell is subdivided into two, by means of a new edge between two of its existing vertices, this adds one two-cell and one one-cell. These also contribute with opposite signs, so they cancel.

One way to show the invariance of the Euler number would be to prove that these two operations and their inverses are enough to go between any two finite cell divisions. But this approach is not very satisfactory—one can easily get lost in the technical details, and fail to see what's really going on. A more insightful idea is to relate the Euler number to something that clearly doesn't depend on any cell division: vector fields on the surface.

Let's look at a simple example first, before tackling the problem in full generality. Consider the sphere S^2, carrying a cell division that is realized as a convex polyhedron in $\mathbf{E}^3$. Arrange the polyhedron in space so that no edge is horizontal—in particular, so there is exactly one uppermost vertex U and lowermost vertex L.

Put a unit + charge at each vertex, a unit − charge at the center of each edge, and a unit + charge in the middle of each face. We will show that the charges all cancel except for those at L and at U. To do this, we displace the vertex and edge charges into a neighboring face, and then group together all the charges in each face. The direction of movement is determined by the rule that each charge moves horizontally, counterclockwise as viewed from above (Figure 1.15).

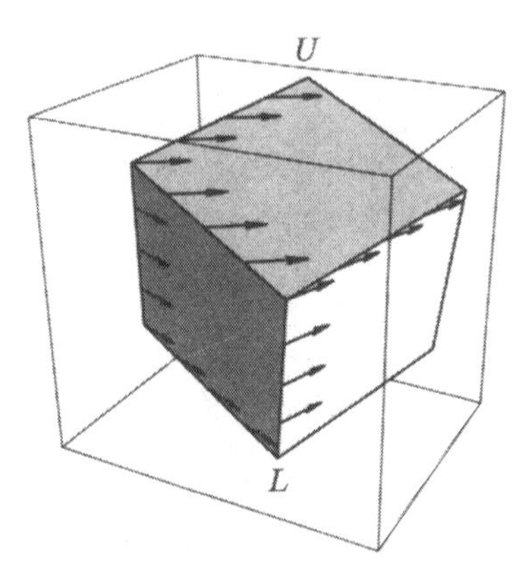

Figure 1.15. Charges on a convex polyhedron. The arrows are part of a horizontal vector field sweeping the surface of this polyhedron; the field is undefined only at the uppermost and lowermost vertices. When the + charges on vertices and the − charges on edges move according to the vector field, they cancel the + charges on the faces.

In this way, each face receives the net charge from an open interval along its boundary. This open interval is decomposed into edges and vertices, which alternate. Since the first and last are edges, there is a surplus of one −; therefore, the total charge in each face is zero. All that is left is +2, for L and for U.

We now generalize this idea to any differentiable surface with a *differentiable triangulation*. This means a differentiable cell division where the faces are modeled on triangles, in such a way that the cell map for any face is an †embedding taking each side of the model triangle onto an edge of the cell division, and the cell map for this edge is compatible with the cell map for the face (that is, they differ by an affine map between domains: see Figure 1.16).

The assumption that the surface is triangulated isn't really restrictive: if we start with a differentiable cell division that is not a triangulation, we can subdivide edges and faces so each face is

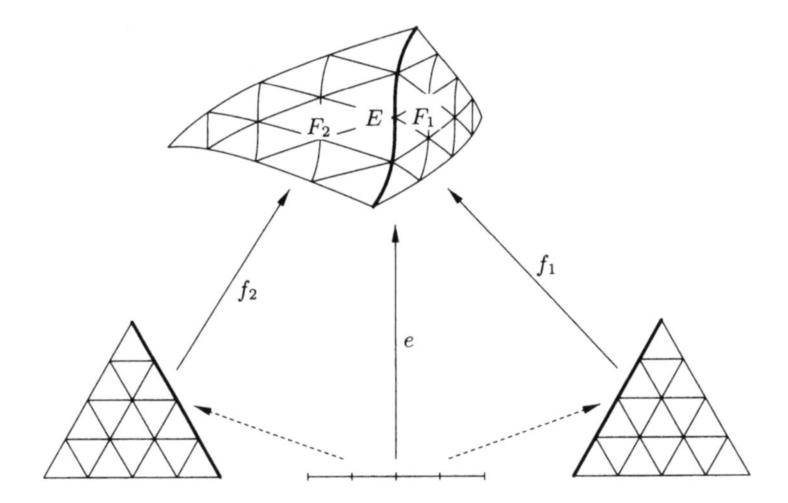

Figure 1.16. Compatibility condition for a triangulation. The cell map f_1 from the model triangle onto face F_1, composed with an affine map from the model interval to the appropriate side of the model triangle, agrees with the cell map e from the model interval onto edge E. Similarly, f_2 agrees with e after composition. This means we can refine the triangulation by subdividing the model triangle and the model interval.

modeled on a triangle with embedded sides. We have already seen that this process doesn't change the Euler number. We also have to adjust the cell maps, by a process similar to that of Exercise 1.3.1, so that edge maps are compatible with face maps.

Proposition 1.3.3 (nonvanishing vector fields). *If a differentiably triangulated closed surface admits a nowhere zero tangent vector field, its Euler number is zero.*

Proof of 1.3.3. Suppose first that the vector field is everywhere *transverse* to the triangulation, that is, nowhere tangent to an edge. In Problem 1.3.4 you're asked to show that we can arrange for that to be the case, by subdividing the triangulation and otherwise adjusting it, without changing the Euler number. By subdividing we can also make the field nearly constant within each triangular face: more precisely, for each face, in some coordinate chart, the direction of the field should change by at most ε, and the direction of the edges should change by at most ε along the edge.

Given such a transverse triangulation, we apply the idea of moving vertex and edge charges in the direction given by the vector field.

If a vertex's charge moves into a face, so do the charges for the two adjacent edges around the face. This means that in each face either exactly one edge charge gets pushed in or two edge charges and one vertex charge; the case of three or zero edge charges being pushed in is ruled out because it cannot occur for a constant field, and our field is nearly constant. In both allowable cases, the face is left with a total charge of zero, so the Euler number is zero. [1.3.3]

Problem 1.3.4 (transverse triangulation). To complete the proof of Proposition 1.3.3, we must show the triangulation can be changed so as to become transverse to the field and so the field and edge directions are nearly constant within each face.

(a) Cover the surface with a finite number of coordinate patches. By drawing equally spaced lines parallel to each edge as in Figure 1.16, subdivide the triangulation so finely that the *star* of each vertex v—that is, the union of edges and faces incident on v—lies in a single coordinate patch, and that the direction of each edge and of the field in the star of v, measured in these coordinates, changes by no more than ε.

(b) Imagine the sets of directions of the edges and of the field as intervals on the circle, of length bounded by ε. Show that you can make the intervals of directions of the edges avoid the interval of directions of the field by moving v a little bit in the appropriate direction, and extending the movement to each edge incident on v by means of a Euclidean †similarity (with respect to the patch coordinates) that keeps the other endpoint of the edge fixed.

(c) Now extend this process to all vertices simultaneously. First show that we can assume that the vertices can be colored red, green and blue, so that no two vertices of the same color are joined by an edge. (Hint: use †barycentric subdivision.) Adjust all red vertices at once, then all green vertices. This leaves all edges transversal.

The torus T^2 has nowhere zero vector fields: consider a uniform field on $\mathbf{E}^2$ and take the quotient as in Figure 1.4. So its Euler number is zero.

What about other surfaces? Most of them do not admit a nowhere zero vector field. The best we can do is to find a vector field that is zero at isolated points (see Exercise 1.3.8). The proof of Proposition 1.3.3 suggests that charges cancel in regions away from the zeros of such a field, so we now need to study its behavior near its zeros.

Let X be a vector field on a surface with an isolated zero at a point z. Working as in the proof of Problem 1.3.4, construct a small polygon containing z in its interior and having edges transverse to X. Place a + charge on each vertex, a − charge on each edge, and a + charge in the interior of the polygon, and flow the charges off the boundary of the polygon by using X. The *index of X at z*, denoted $i(X, z)$, is the sum of the charges in the interior of the polygon after the operation of the flow.

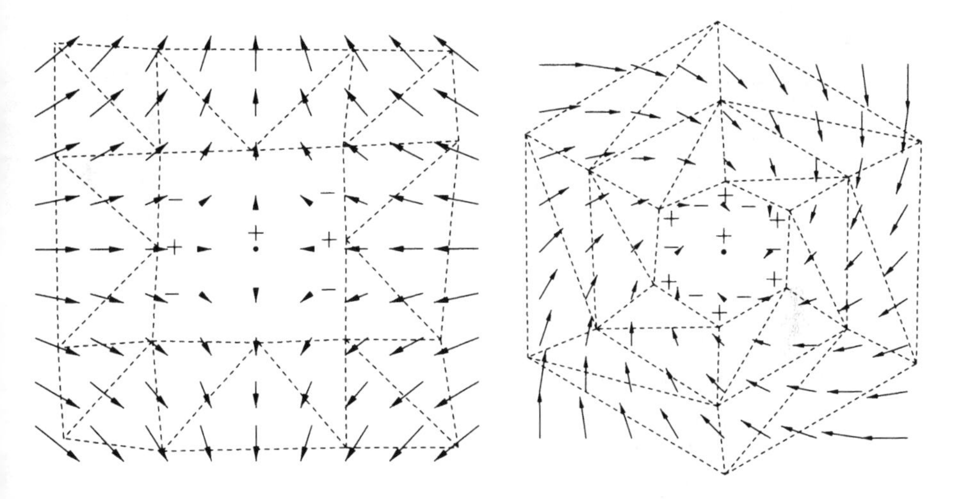

Figure 1.17. The index of a vector field. An isolated zero of a vector field leaves a small region where charges don't cancel. For the field on the left, the total charge of the polygon containing the zero—known as the index of the vector field at that point—is -1. Such a zero is called a *saddle*. For the field on the right, the index is 1; an isolated zero with index 1 is called a *sink* or a *source*, depending on whether the field points in or out.

Lemma 1.3.5 (index independence). *If X is a vector field with an isolated zero z, the index of X is independent of the polygon enclosing z; it depends only on the restriction of X to an arbitrarily small neighborhood of z.*

Proof of 1.3.5. Given one polygon containing z with edges transverse to X, we need only show that it gives the same index as another much smaller polygon with the same properties. Using Exercise 1.3.6, we subdivide the annulus between the two polygons into triangles, and jiggle the vertices to make the edges transverse to X. The flow causes some charge to enter the annulus across the outer boundary and some charge to leave across the inner boundary. The charge left on each triangle of the annulus is zero, and so the charge entering is equal to the charge leaving. 1.3.5

Exercise 1.3.6 (triangulating an annulus). A minor technical detail was suppressed above—how do we triangulate an annulus in the plane? Here we set things up so that the triangulation is easy—with the necessary background, one could instead just quote a theorem about the triangulability of surfaces.

(a) A set D is said to be *star-shaped* with respect to a point v in its interior if each ray from v to the boundary of D is contained in the interior of D. Show that a star-shaped polygon can be triangulated with v as a vertex.

(b) Given two polygons that are star-shaped with respect to v and such that one is contained in the interior of the other, triangulate the annulus between their boundaries.

(c) Suppose that we have a polygon which is star-shaped with respect to the isolated zero z of X. Show that the boundary of the polygon can be made transverse to X by jiggling vertices only in the radial direction, and hence that the polygon obtained after jiggling is still star-shaped.

(d) (Harder.) Can you work out a proof without the assumption that the polygons are star-shaped?

The simplest vector fields with isolated zeros are the linear vector fields in the plane, those where the value of the field at a point is obtained by applying a linear map to the point. Clearly the origin is a zero of any linear vector field; it is isolated if and only if the linear map has non-zero determinant.

If a vector field with isolated zeros is †homotoped in such a way that the points where it is zero do not change, the indices at the zeros must remain constant since they are integers. This implies that two linear vector fields whose determinants have the same sign must have the same index, since then they are †homotopic through linear vector fields of non-zero determinant.

Exercise 1.3.7 (index is sign of determinant). (a) Sketch enough pictures of linear vector fields that you understand the relationship between determinant and qualitative appearance.

(b) Prove that the index of a linear vector field in the plane is the sign of its determinant.

Exercise 1.3.8 (isolated zeros). Given a finite cell division of a surface, find a way to construct a differentiable vector field on the surface with a source in the middle of each two-cell, a sink at each zero-cell, and a saddle in the middle of each edge (see Figure 1.17 for definitions).

Problem 1.3.9. What are the possible values of the $i(X, z)$ when X is a general vector field? Can you find a formula for $i(X, z)$ valid for all vector fields with isolated zeros?

Proposition 1.3.10 (Poincaré index theorem). *If S is a smooth surface and X is a vector field on S with isolated zeros, the Euler number of S is*

$$\chi(S) = \sum_{z \in \text{zeros}} i(X, z).$$

Consequently, the Euler number of a surface does not depend on the cell division used to compute it—it is a topological invariant.

Proof of 1.3.10. Given a finite cell division of S, start by replacing it with a differentiable triangulation, as discussed just before Proposition 1.3.3. Subdivide and jiggle the triangulation as necessary to make all the zeros lie inside faces, no more than one to a face. Enclose each zero with a polygon contained in a face and transverse to the field, as explained in the paragraph preceding Lemma 1.3.5. Triangulate the annulus formed by taking away the polygon from the face (Exercise 1.3.6). Finally, make the rest of the triangulation transverse, again by using the technique in the proof of Problem 1.3.4.

Each polygon's contribution to the Euler number is the index of the vector field at the corresponding zero. Each triangle's contribution outside the polygons is zero. This proves the formula.

The last sentence of the proposition follows because every closed surface admits a vector field with isolated zeros (Exercise 1.3.8).

1.3.10

Challenge 1.3.11. Show that this discussion about the Euler number and the index of isolated zeros can be carried out for a smooth manifold of any dimension.

If the topology of a closed surface determines its Euler number, could the converse be true as well? This sounds unlikely—so much information in a single number! But amazingly, it's almost true: knowing the Euler number *and* whether or not the surface is orientable is enough to determine the surface. This important result has been known for more than a century; if you are not familiar with it, you are urged to work through the following steps:

Problem 1.3.12 (classification of surfaces). Consider a (connected) surface S obtained by gluing polygons (we allow digons, but not monogons).

(a) S can be obtained by gluing a single polygon.

(b) If the number of vertices in S is greater than one, try to reduce the number by shrinking one of the edges. This is always possible unless the polygon is a digon and the two vertices correspond to distinct points on S, in which case $S = S^2$. Assume from now on that S has one vertex.

(c) Let E be the number of edges of S, so the polygon has $2E$ edges and $\chi(S) = 2 - E$. If $E = 1$, we have the projective plane $\mathbf{RP}^2$, and if $E = 2$, we have the torus or Klein bottle (Figure 1.14). Assume from now on that $E \geq 3$.

(d) An *elementary move* in this context consists of cutting a polygon along a diagonal which separates paired edges e and e', and then gluing e to e'. Show that an elementary move does not change the topology of S.

(e) If S is orientable, there exist two pairs of paired edges x, x' and y, y' that separate each other.

(f) Suppose that x, x' and y, y' are pairs of paired edges that separate each other, and that they are paired with reversal of orientation. The remaining edges of the polygon form segments of length m, n, p and q. Using one elementary move, arrange that $m = n = 0$. Follow this with another elementary move, to arrange that $m = n = p = 0$.

(g) If S is orientable, it can be obtained by a gluing of the form

$$a_1 b_1 a_1^{-1} b_1^{-1} a_2 b_2 a_2^{-1} b_2^{-1} \ldots a_g b_g a_g^{-1} b_g^{-1}.$$

(In this notation, we choose for each edge an orientation that is consistent with the identification, and read the resulting word going once around the polygon.) Therefore, a closed, orientable surface is determined up to homeomorphism by its Euler number, which is any even integer ≤ 2. The number g is called the *genus* of the surface, and the surface is a g-holed torus.

(h) If S is non-orientable, it can be rearranged by one elementary move in such a way that its gluing has two adjacent edges that are paired by a gluing map that preserves orientation.

(i) The non-orientable surface obtained from a hexagon by the gluing $aabbcc$ is homeomorphic to that obtained by the gluing $aabcb^{-1}c^{-1}$.

(j) Any non-orientable surface can be obtained by a gluing of the form

$$a_1 a_1 a_2 a_2 \ldots a_g a_g.$$

The number g is called the *non-orientable genus* of the surface. Two closed non-orientable surfaces of the same Euler number are homeomorphic.

Exercise 1.3.13. How many gluing patterns are there for a $2n$-gon? How many lead to an oriented surface? How many topological types can be obtained? This shows there is a huge amount of repetition in descriptions of surfaces by gluings.

Exercise 1.3.14 (closed surface minus a disk). Let S be a closed connected surface. Prove that by removing a disk from S we get a disk with attached strips as in Figure 1.18. Can you always make all the strips untwisted? All but one?

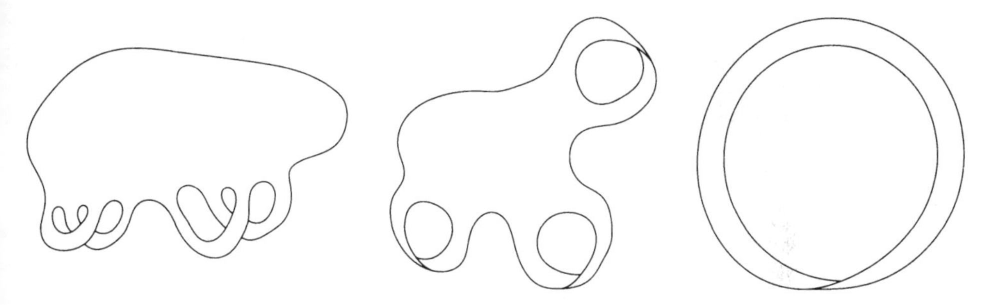

Figure 1.18. Disks with strips. The result of removing a disk from a closed surface is a disk with strips attached. On the left, we started from an oriented surface of genus two; in the middle, from a surface of non-orientable genus three; on the right, from a projective plane. The figure on the right is the Möbius strip.

In Section 1.2 we constructed a hyperbolic structure for the oriented genus-two surface. We can make an analogous construction for any surface of negative Euler number, with the help of Problem 1.3.12. In fact, such a surface is obtained by gluing the sides of a $2n$-gon, with $n \geq 3$, in such a way that all vertices are identified to a single vertex. The angle sum of the Euclidean polygon is greater than 2π, so there is a symmetric $2n$-gon of the appropriate size in $\mathbf{H}^2$, as in Figure 1.12, whose sides glue up to form a smooth hyperbolic structure on the surface.

Similarly, if a surface has Euler number zero, it can be obtained by gluing sides of a square so that all vertices are identified together. Therefore, it has a Euclidean structure. In the notation of Figure 1.14, the surface is either $a-b-$ or $a-b+$, that is, a torus or a Klein bottle.

To complete the picture, S^2 and its quotient space $\mathbf{RP}^2 = S^2/\{\pm 1\}$—the surfaces of positive Euler number—have spherical structures. Such a structure is also called *elliptic*.

Might there be some exceedingly clever construction for a hyperbolic structure on the torus or Klein bottle, or a Euclidean structure on the surface of genus two?

Problem 1.3.15 (Gauss–Bonnet signs). (a) Show that the Euler number of a closed surface with a Euclidean structure is always zero.

(b) Show that the sum of the (interior) angles of any hyperbolic triangle is always less than π, and the sum of the angles of any spherical triangle is always greater than π. (Hint: use isometries to arrange triangles, to make the comparisons easier.)

(c) Show that the Euler number of a closed surface with a hyperbolic structure is negative, and the Euler number of a closed surface with an elliptic structure is positive.

This is a good place to mention some operations that construct new surfaces from old, and whose higher-dimensional analogues will be important later. We may as well give their definitions in arbitrary dimension.

The *connected sum* of two (connected) n-manifolds M_1 and M_2 is a manifold $M_1 \# M_2$ obtained by removing copies of the n-disk D^n from M_1 and M_2 and gluing the two resulting boundary spheres together.

Exercise 1.3.16. Show that if $S_3 = S_1 \# S_2$ are surfaces, $\chi(S_3) = \chi(S_1) + \chi(S_2) - 2$. What happens for manifolds of other dimensions?

This definition can be made more precise: Choose diffeomorphic embeddings $\phi_1 : \overline{D^n} \to M_1$ and $\phi_2 : \overline{D^n} \to M_2$ of the closed n-disk, remove the two images of D^n from the union $M_1 \cup M_2$, and identify the boundaries $\phi_1(\partial D^n)$ and $\phi_2(\partial D^n)$ by the map $\phi_2 \circ \phi_1^{-1}$. To what extent does the topology of the result depend on the choice of ϕ_1 and ϕ_2? Not much, because there is essentially only one way to embed a disk in a connected manifold, up to orientation.

More precisely, if we change ϕ_1 (say) by an isotopy, the topology doesn't change, because any isotopy between embeddings of $\bar{D}^n$ into an n-manifold M can be extended to an isotopy of the whole manifold (see [Hir76, p. 185]). Now associate with an embedding $\phi : D^n \to M$ the †frame at $\phi(0)$ given by the image under $D\phi(0)$ of the canonical basis vectors of $\mathbf{R}^n$. It is easy to see (Exercise 1.3.17) that two embeddings are isotopic if and only if their associated †frames can be continuously deformed into one another (that is, they lie in the same connected component of the †frame bundle of M). This means there are two isotopy classes of diffeomorphic embeddings $\bar{D}^n \to M$ if M

is orientable, and only one if M is non-orientable. If an orientation is chosen for M, the two classes are determined by whether the embedding preserves or reverses orientation.

Exercise 1.3.17 (disk embeddings and the frame bundle). (a) Show that two embeddings of D^n that map the origin to the same point and have the same derivative there are isotopic.

(b) Show that they're isotopic if they map the origin to the same point and the frames they define at that point lie in the same connected component of $\mathrm{GL}(n, \mathbf{R})$.

(c) Show that they're still isotopic if they map the origin to different points, but the frames they define can be continuously deformed into one another.

If the two manifolds are oriented, it makes sense to form the connected sum so that the resulting manifold has an orientation which agrees with the original orientations away from the disks which are removed. This condition requires that one of ϕ_1 and ϕ_2, but not both, be orientation-preserving. With this convention, the connected sum of two oriented two-manifolds is a well-defined oriented manifold.

If one of the manifolds is non-orientable, the result of the connected sum does not depend on the choice of orientation for the gluing map, and the result of the operation is again well-defined.

However, when the two manifolds are orientable but not oriented, there is a difficulty: the two possible choices of sign for the gluing map may yield different results. (But Exercise 1.3.18 shows that this does not actually happen for surfaces.)

Exercise 1.3.18 (surface semigroup). (a) Show that every closed orientable surface admits an orientation-reversing homeomorphism.

(b) Show that the operation # is a well-defined, commutative, and associative operation on the set of homeomorphism classes of surfaces, making it into a commutative †semigroup.

(c) Show that S^2 acts as an identity element for #.

(d) Show that the torus T^2 and the projective plane $\mathbf{RP}^2$ generate the commutative semigroup of homeomorphism classes of surfaces under #, and that $T^2 \# \mathbf{RP}^2 = \mathbf{RP}^2 \# \mathbf{RP}^2 \# \mathbf{RP}^2$.

(e) Sketch a directed labeled graph whose vertices are in one-to-one correspondence with homeomorphism classes of surfaces, and whose edges show the effect of # with T^2 (if labeled A) and with $\mathbf{RP}^2$ (if labeled B).

Exercise 1.3.19 (blowing up a point). (a) Let M be a smooth n-dimensional manifold and x a point of M. The operation of replacing x by the set of tangent lines at x, resulting in the topology described in Figure 1.19, is called *blowing up* the point x. Show that the resulting topological space is a smooth manifold homeomorphic to $M \# \mathbf{RP}^n$.

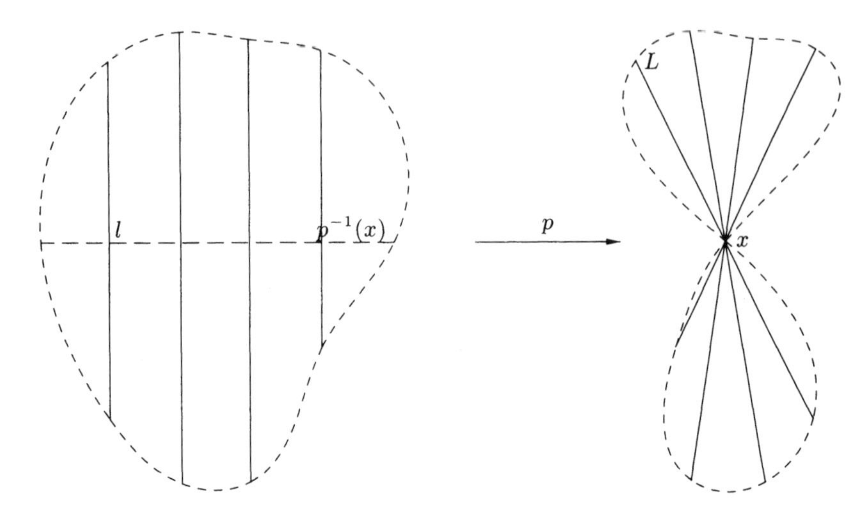

Figure 1.19. Blowing up a point. The blowup of M at x (left) has points of two types: points of $M \setminus \{x\}$, and one-dimensional subspaces of the tangent space $T_x M$. There is a natural map p from the blowup into M, taking each point of the first type to itself and each point of the second type to x. The topology of the blowup is defined by the following conditions: p is a local homeomorphism away from $p^{-1}(x)$; and a neighborhood of a point $l \in p^{-1}(x)$, corresponding to a line L, consists of points corresponding to lines near L, plus points of the first type along those lines and close to x.

(b) If $n = 2$ this amounts to cutting out a disk and gluing in a Möbius strip.

(c) What happens if we do our "blowing up" by using tangent rays instead of tangent lines?

Exercise 1.3.20 (autosum). The connected sum operation has a variation where one removes two disjoint disks on the same manifold, and connects the resulting boundary spheres together. There are two possibilities for the orientation with which the spheres are identified. Analyze what happens to surfaces under this operation of *autosum*.

Exercise 1.3.21. Show that the semigroup of (connected) closed surfaces can be generated from the identity element S^2 by the operations of blowing up and autosum.

1.4. Some Three-Manifolds

Now that we are well grounded on surfaces, we are ready for a quick flight through a few three-manifolds.

Example 1.4.1 (the three-torus). Probably the easiest three-manifold to understand is the three-torus, which can be obtained by gluing just like the two-torus: start with a cube, and glue each face of the cube to the parallel face, by parallel translation.

To visualize the three-torus, imagine the cube as a rectangular room where you are standing. Imagine what it would be like if the opposite walls were identified with each other, and the floor were identified with the ceiling. When your line of sight arrives at one wall, it continues from the corresponding point on the opposite wall, in the same direction as before. Therefore, if you look straight ahead, you see your back. If you turn to the left, you see your right side. If you look straight down, you see the top of your head. There are six images of yourself which appear to be in immediately neighboring rooms, but there are also rooms which neighbor diagonally, and, really, the lines of sight would continue indefinitely. The appearance would be identical to that of an infinite repeating array of images of yourself (and anybody or anything else in the room) in all three dimensions (Figure 1.20). The effect is reminiscent of a barber shop

Figure 1.20. A view from inside the three-torus. If you live in a three-torus, each object appears to be repeated at every point of a three-dimensional lattice.

with mirrors on facing walls, so that you see long lines of repeated images. The difference is that in a barber shop, when you look at an adjacent image, it is facing toward you, while in the torus, it is facing away. In a torus, as you turn toward any image, the image turns away. As you fly toward the image, the image flies away, and you never can meet it.

Example 1.4.2 (the three-sphere from inside). Another easily described three-manifold is the three-sphere S^3. The easiest definition of S^3 is the unit sphere $x_1^2 + x_2^2 + x_3^2 + x_4^2 = 1$ in $\mathbf{R}^4$. Unfortunately, this formula does not immediately communicate a picture of S^3 to people who are not adept at visualizing four-dimensional space. But there is another way to imagine S^3, from the point of view of an inhabitant.

To prepare the way, think first of what an inhabitant of S^2, the two-sphere, would see. By some mechanism, light rays are supposed to curve around to follow the surface. For instance, you can imagine that the "surface" is really a very thin layer of air between two large concentric glass spheres, which channel light by reflection in much the same way as fiber optics. (Unfortunately, the ecology of this model is not so clear. At best, there is just enough room for one to crawl around on one's stomach.)

Imagine creature A resting at the north pole, and another creature B creeping away. You can work out the visual images in terms of which geodesics (great circles) from the eyes of the A intersect B. As B creeps away, its image as seen by A at first grows smaller, although not quite as fast as it would in the plane. Once B reaches the equator, however, its image grows larger again with continued progress, until at the south pole, its image fills up the entire background of the field of vision of A in every direction.

The same phenomenon would take place in the three-sphere. Let's give ourselves more breathing room than we had in T^3, and imagine that we are in a three-spherical world where a great circle is about two miles in circumference. There is no gravity, and we won't fuss about food, shelter, light or other minor details just now. We have little jets on our backs for flying around wherever we please. If I fly away from you, in any direction, my visual image to you shrinks in size at first fairly rapidly, but as I approach the half-mile mark my visual size changes very slowly: it probably looks to you as though I have stopped making progress. After the half-mile mark, I gradually start to increase in visual size once more. As I approach your antipode, one mile from you, I start to grow rapidly again. When I am three feet from your antipode, the size of my visual image is exactly the same as if I were three feet from you. If I turn around and shout back, it will hurt your ears. We quickly learn that we can carry on a conversation with normal voices, for sound converges again at antipodal points just as light does.

Even though I have the same visual size to you when I hover three feet from your antipode as when I hover three feet from you, there is a difference in my visual image: you see further around to my sides. (There is also a difference in focal distance, but let's put that aside: imagine the light is very bright, so that your pupils are contracted and you don't notice this effect.) The difference becomes very dramatic if I now continue three feet further, so I cover the antipode of your eyes: you now see my image in every direction, and it is as if I were turned inside out onto the inside surface of a great hollow sphere totally surrounding you. You appear to me the same way, as the inside of a hollow sphere surrounding me.

In this description, we have left out an important part of the image. Light does not stop after traveling only a mile, it continues further. When I am a half mile from you, my image to you is as small as possible, but your lines of sight continue unimpeded completely around the three-sphere, to arrive back near where they started on yourself. In the background of everything else, you see an image of yourself, turned inside out on a great hollow sphere, with the back of your head in front of you.

There's another thing we left out: whenever I am at a distance other than one mile from you, you can actually see me in two opposite directions. For instance, when I was three feet from your antipode, had you turned around rather than me, you would have seen a perfectly normal image of me as if I were hovering three feet away and facing you, only slightly faded by the blue haze of the water vapor in the intervening air. You would also appear almost completely normal to me. But if we were to try to shake hands, they would pass through each other.

Example 1.4.3 (elliptic space). Accounts of spherical geometry are marred by exceptions arising from the existence of antipodes—for instance, two points determine a unique line, but only if they're not antipodal. Things work out more cleanly in *elliptic space*, the sphere with antipodal points identified. Topologically, n-dimensional elliptic space is just the projective n-space $\mathbf{RP}^n$, but as the quotient of the sphere by a group of isometries (a very small group, just the identity and the antipodal map), it inherits a geometry, just as the torus inherits a geometry from the Euclidean plane. Geometric assertions can easily be translated back and forth between the sphere and elliptic space, and they're often cleaner in elliptic space: for example, any two points in elliptic space determine a unique line.

We've seen that in the sphere an object moving away from you decreases in apparent size until it reaches a distance of $\pi/2$, then starts increasing again until, when it reaches a distance of π, it appears so large that it seems to surround you entirely. In elliptic space, on the other hand, the maximum distance is $\pi/2$, so that apparent size is a monotone decreasing function of distance. It would nonetheless be distressing to live in elliptic space, since the entire background of your field of view would be filled up with an image of yourself. Looking straight ahead you would see the back of your head, turned upside down and greatly magnified. Everything else would still be visible twice, in opposite directions.

Example 1.4.4 (Poincaré dodecahedral space). This famous example, discovered by Poincaré, is obtained from a dodecahedron by gluing opposite faces. The corners of the pentagons making up opposite faces of a dodecahedron are out of phase: they interleave each other, so there is no question of gluing each face to its opposite using a translation, as in the torus. Consider what happens when we glue them with as little twist as possible: a rotation by $1/10$ of a full turn, say in the clockwise direction from front to back, as in a right-handed screw (Figure 1.21, left). This prescription is symmetric, so that when we turn the dodecahedron around $180°$, the identification appears the same: right-handed screws are right-handed screws in either direction.

A dodecahedron has five edges around each face and each edge is along two faces, so it has $12 \cdot 5/2 = 30$ edges. The face pairings force various identifications of the edges which you can chase through, to check that the edges are glued together in ten groups of three.

To see what happens to the $12 \cdot 5/3 = 20$ vertices of the dodecahedron, consider the spherical triangles obtained by intersecting tiny spheres about the vertices with the dodecahedron. The way these spherical triangles are glued together is such that four triangles come together at each vertex, so the pattern is necessarily that of a tetrahedron. Therefore, the vertices of the dodecahedron are glued together in five groups of four, and the space obtained by this gluing is a manifold since it is locally homeomorphic to Euclidean

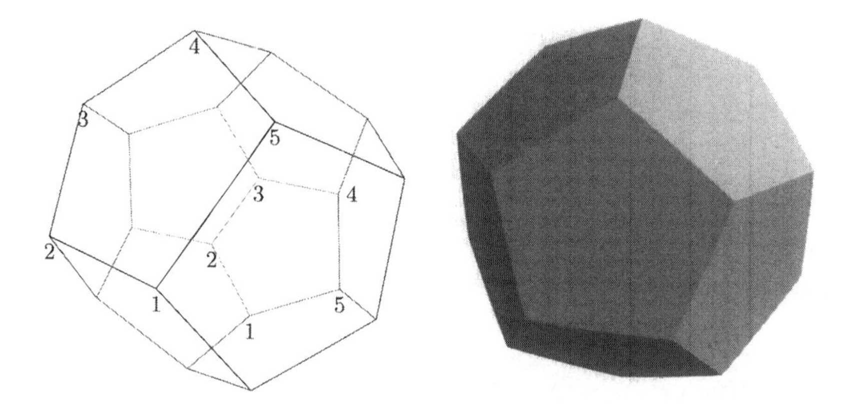

Figure 1.21. The Poincaré dodecahedral space. Each pentagonal face of a dodecahedron has an opposite face, but the corners of opposite pentagons interleave each other. If a face is glued to its opposite face by one-tenth of a clockwise revolution, the resulting space is the Poincaré dodecahedral space. The edges are glued in triples in this pattern. In order for the gluing to work geometrically, we must start with a spherical dodecahedron, with dihedral angles equal to 120°. This slightly puffy solid is shown on the right under stereographic projection (see Exercise 2.2.8)—it is almost indistinguishable from its Euclidean counterpart, which has dihedral angles 116.565°!

space, even in neighborhoods of the vertices and neighborhoods of points on the edges.

Can this gluing be done so that the geometry, and not just the topology, is locally Euclidean? If so, the angles between the faces around an edge of the glued-up manifold would add up to 2π, so (assuming everything is symmetric) they would each equal $2\pi/3$ or 120°. But the dihedral angles of a dodecahedron are slightly less—they equal $\arctan(-2)$, or 116.565°.

Here the geometry of the three-sphere comes to the rescue. Just as the angles of a geodesic triangle on the two-sphere has angles somewhat larger than the angles of a Euclidean triangle, so the dihedral angles of a polyhedron in the three-sphere are somewhat larger than those of a Euclidean polyhedron. A very small regular dodecahedron in the three-sphere has angles very close to the Euclidean angles. As the dodecahedron grows, the angles increase. In fact, when the distance from the center to the vertices is $\pi/2$ (or a half-mile, in the scale of Example 1.4.2), the dodecahedron is geometrically a two-sphere: the angles are π.

Therefore, somewhere in between there is a dodecahedron with angles exactly $2\pi/3$ radians or $120°$ (Figure 1.21, right). For this dodecahedron, the gluing will work geometrically.

We can unroll the Poincaré dodecahedral space to obtain a tiling of S^3, just as we unrolled T^2 and T^3 to obtain tilings of $\mathbf{E}^2$ and of $\mathbf{E}^3$. This is the same as saying that the Poincaré dodecahedral space, like elliptic space, is a quotient of S^3 by a group of isometries. It would be possible to mark out the tiles exactly using coordinates on $S^3 \subset \mathbf{E}^4$, and it would also be possible to work out the combinatorial structure of the tiling by logic, since we know its local structure. Either approach would involve a fair amount of work at present, but we will return to this question later, in Problem 4.4.17. It turns out that it takes 120 dodecahedra to tile the three-sphere. In $\mathbf{E}^4$, this pattern defines a 120-hedron whose faces are regular three-dimensional dodecahedra.

Example 1.4.5 (Seifert–Weber dodecahedral space). If the opposite faces of a dodecahedron are glued together using clockwise twists by three-tenths of a revolution, instead of one-tenth (Figure 1.22), a bit of chasing around the diagram shows that edges are identified in six groups of five. All twenty vertices are glued together, and small spherical triangles around the vertices (obtained by intersection of the dodecahedron with tiny spheres) are arranged in the pattern of a regular icosahedron after the gluing. The resulting space is a manifold known as the Seifert–Weber dodecahedral space.

The angles of a Euclidean dodecahedron are much larger than the $72°$ angles needed to do the gluing geometrically. In this case, we can use three-dimensional hyperbolic space $\mathbf{H}^3$, which can be mapped into the interior of a three-dimensional ball, just as in two dimensions. This description in the ball is the *Poincaré ball model* of hyperbolic space. Planes are represented as sectors of spheres orthogonal to the boundary of the ball. The angle between two hyperbolic planes is the same as the angle between the two spheres. Since the spheres intersect the boundary of the ball S^2_∞ perpendicularly, it is also the angle between their circles of intersection with this sphere.

The angles of a regular dodecahedron in $\mathbf{H}^3$ are clearly smaller than they would be in $\mathbf{E}^3$, but can they be as small as $72°$? In the limiting case, as a hyperbolic dodecahedron became very large, its vertices would appear to nearly touch S^2_∞. There is a limit, the *ideal*

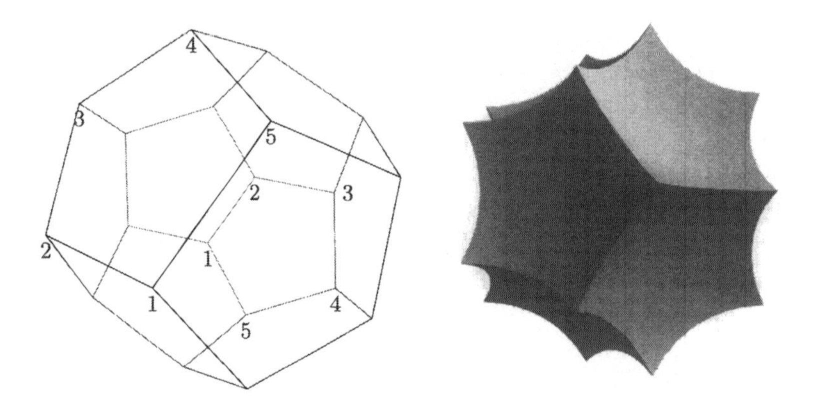

Figure 1.22. The Seifert–Weber dodecahedral space. If opposite faces of a dodecahedron are glued by three-tenths of a clockwise revolution, the edges are glued in quintuples, and the resulting space is the Seifert–Weber dodecahedral space. The gluing can be realized geometrically if we use a hyperbolic dodecahedron with dihedral angles of 72°—the solid shown on the right, in the Poincaré ball model.

dodecahedron, whose vertices are missing: it has *ideal vertices* on S^2_∞.

By looking at the symmetric placement of three circles on S^2_∞ coming from three adjacent face planes of the ideal dodecahedron which meet at an ideal vertex, it is easy to see that an ideal dodecahedron has 60° dihedral angles.

Therefore, intermediate between a very small hyperbolic dodecahedron with angles approximately 116.565° and a very large dodecahedron with angles tending toward 60° there is a dodecahedron whose dihedral angles are exactly 72°. This dodecahedron can be glued together to make a geometric form of the Seifert–Weber dodecahedral space. It corresponds to a tiling of $\mathbf{H}^3$ by dodecahedra all meeting five to an edge and twenty to a vertex.

Example 1.4.6 (lens spaces). Consider a ball with its surface divided into two hemispheres along the equator. What happens when we glue one hemispherical surface to the other?

If we glue with no twist at all, so that the identification is the identity on the equator, the resulting manifold is S^3. This is analogous to the way S^2 can be formed by dividing the boundary of a disk into two intervals, and gluing one to the other so as to match each endpoint with itself.

On the other hand, if the hemispheres are glued with a q/p clockwise revolution, where q and p are relatively prime integers, each point along the equator is identified to $p-1$ other points. A neighborhood of such a point in the resulting identification space is like p wedges of cheese stuck together to form a whole cheese, so the identification space is a manifold, called a *lens space* $L_{p,q}$.

To form a geometric model for a lens space, we need a solid something like a lens, where the angle between the upper surface and the lower surface is $2\pi/p$. This is easy to do within S^3. Any great circle in S^3 has a whole family of great two-spheres passing through it. From this family it is easy to choose two that meet at the desired angle $2\pi/p$. Now when the two faces are glued together, neighborhoods of p points on the rim of the lens which are identified fit exactly. This corresponds to a tiling of S^3 by p lenses: see Figure 1.23, left.

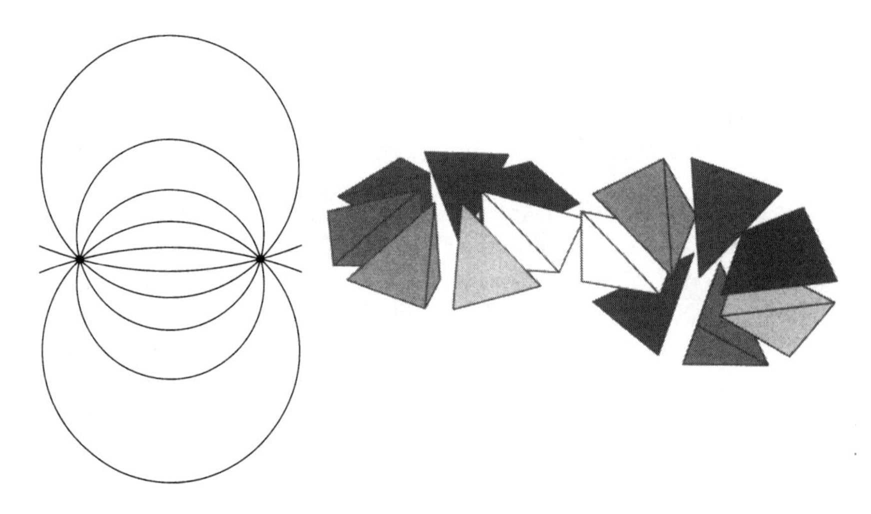

Figure 1.23. Lens spaces. On the left, S^3 is seen in cross section, tiled with twelve copies of the fundamental region for $L_{12,q}$ (we're using stereographic projection: see Exercise 2.2.8). On the right, $L_{7,2}$ is disassembled and reassembled in a different way, showing that it equals $L_{7,3}$.

Problem 1.4.7 (reworking lens spaces). (a) The lens that was glued to form $L_{p,q}$ can be cut up into p tetrahedra, meeting around one edge through its central axis. When this is done, the p tetrahedra can be assembled by gluing first the faces which came from the surface of the lens (Figure 1.23, right). What figure does this form? What identities among lens spaces can you construct?

(b) Cut out a solid cylinder around the central axis of the lens used to form $L_{p,q}$. Its upper face is glued to its lower face to form a solid torus, under the identifications. What happens to the rest of the lens when the part of its boundary on the surface of the lens is glued together? Sketch a picture for $L_{3,2}$. Describe how lens spaces can be constructed by gluing together two solid tori.

(c) Show that two lens spaces $L_{p,q}$ and $L_{p',q'}$ are homeomorphic if and only if $p = p'$ and either $q' = \pm q \mod p$ or $qq' = \pm 1 \mod p$. (This is much harder; see [Bro60] for a proof.)

Example 1.4.8 (a knottier example). Figure 1.24 shows one of the simplest possible three-dimensional gluing patterns. Start with two tetrahedra T and T' with labeled faces and directed edges divided into two types, thick and thin. Then glue faces in pairs, respecting not only face labels but also edge types and directions: for example, face A of T and face A of T' are glued together so that thick edges match thick edges and one thin edge matches the other.

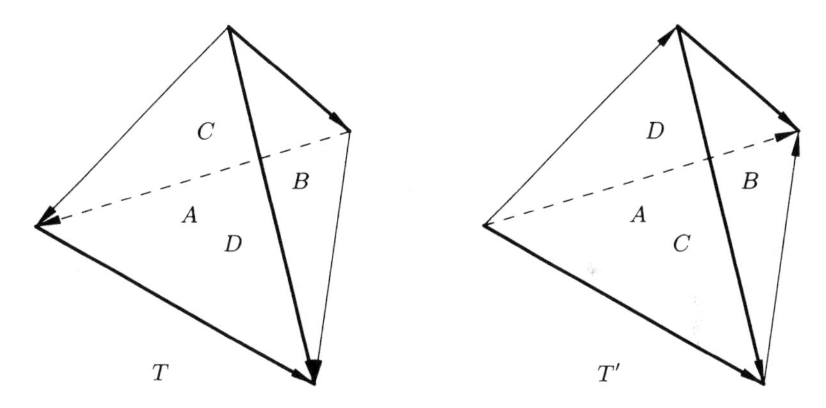

Figure 1.24. A simple three-dimensional gluing pattern. A simple pattern for gluing two tetrahedra. Each face has a label (centered on it), and faces with the same label are identified in a way that is unambiguously determined by the requirement that edge types (thick and thin) and directions match. (In fact, even the pairing of faces could be reconstructed from this requirement.)

In the resulting complex K, all thick edges end up identified, as do all thin edges. Furthermore, all vertices are identified together into one vertex V.

Is K a manifold? We need to check that each point in K has a neighborhood homeomorphic to a 3-ball. This is obvious for points coming from the interiors of the tetrahedra, and for those coming from the interior of a face, where two half-spaces meet. A point in the interior of an edge, too, has a neighborhood consisting of several wedge-neighborhoods of its preimages on the edges of T and T', glued together cyclically to make a ball.

But we run into trouble at the vertex V. Imagine a small neighborhood of V that intersects each tetrahedron in small tetrahedral neighborhoods of its vertices. The inside faces of these tetrahedra piece together to form a surface called the *link* of V. One can check explicitly that it is a torus. Alternatively, we can see by counting that the Euler number of the link is 0, and since it is oriented it must be a torus. The neighborhood of V in K is therefore a †cone on a torus, and in particular V has no neighborhood homeomorphic to a ball and K is not a manifold. In Section 3.2 we will treat in more generality and detail the issue of when complexes formed by gluing polyhedra have manifold structures.

Although K is not a manifold, by removing the recalcitrant vertex we get a non-compact manifold $M = K - \{V\}$. Since the gluing map for each pair of faces reverses orientation, M is oriented. By removing an open neighborhood of V, we obtain a compact manifold whose boundary is a torus.

Exercise 1.4.9. Construct the link of V by gluing together the eight triangles which are links of the vertices of the two tetrahedra.

It turns out that M is homeomorphic with the complement (with respect to $S^3 = \mathbf{R}^3 \cup \{\infty\}$) of a figure-eight knot, shown in Figure 1.25. To see this, we start by arranging the figure-eight knot along the one-skeleton of a tetrahedron, as in Figure 1.26(a). We see that the knot can be spanned by a two-complex, with two edges (the arrows) and four two-cells (the faces of the tetrahedron, each together with a strip and a twisted tab). This complex encloses a compact region of space, whose interior R is homeomorphic to an open ball.

Now imagine the thin edge in Figure 1.26(a) coated with red printing ink, the thick edge coated with green, and the knot coated with black. Insert a balloon into the region R and inflate it, until it fills up all the nooks and crannies of R. Then remove the balloon and examine it; it will be covered by four regions, corresponding to the four two-cells, and arranged as in Figure 1.26(b).

The boundary of each region has five parts: two, shown in the figure without arrows, are colored black and come from the knot. The other three come from the edges of the two-complex: two thin

Figure 1.25. Two views of the figure-eight knot. Two views of the figure eight knot, the second most commonly occurring knot in extension cords, dog leashes and garden hoses (after the trefoil).

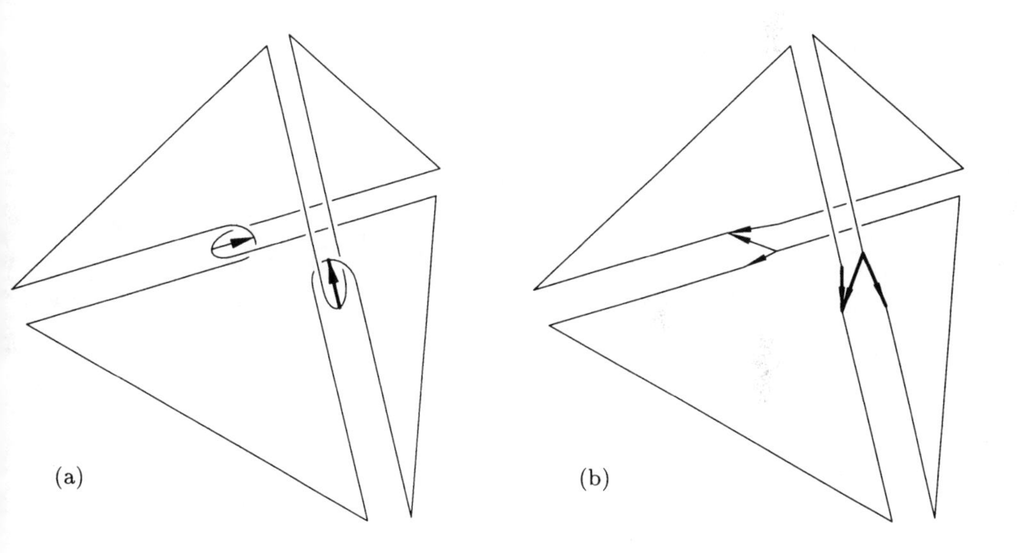

Figure 1.26. A two-complex spanning the figure-eight knot. Left: the figure-eight knot can be spanned by a complex with four two-cells—the faces of a tetrahedron, each augmented by a strip comming off of a vertex and a twisted tab comming off of the opposite edge—and two one-cells, shown as arrows. The region R inside this cell complex is homeomorphic to an open ball. Right: We extend the homeomorphism to a map from the closed ball to the closure of R. Each one-cell is represented by three edges on the boundary, and each two-cell is bounded by three edges and two pieces of the knot. Shrinking the knot pieces to points gives the tetrahedron T from Figure 1.24.

and one thick, or two thick and one thin. Now the knot is not part of the manifold M, so we may expand or contract the black curves without changing M. If we choose to contract each of them to a point, so that the four regions become three-sided, the interior of the balloon, together with the triangular faces, becomes a tetrahedron T exactly as in Figure 1.24.

T' is formed similarly, by inflating a balloon that contains the point at infinity in $S^3 = \mathbf{R}^3 \cup \{\infty\}$ (that is, shrink-wrapping the knot).

This decomposition can be hard to visualize. Two alternative and beautifully illustrated descriptions for it can be found on [Fra87, 151–155].

We will take up this space again in Example 3.3.7. For now, the striking fact that the complement of a particular knot can also be expressed as the union of two tetrahedra in a particular way serves to illustrate the need for a systematic way to compare and to recognize manifolds.

Exercise 1.4.10. Suggestive pictures can also be deceptive. Figure 1.27 shows that a trefoil knot, too, can be arranged along the one-skeleton of a tetrahedron, and spanned by a two-complex similar to the one in Figure 1.26. Blow balloons in the regions inside and outside this complex, and draw the imprints made by the knot and arrows. Do you get tetrahedra?

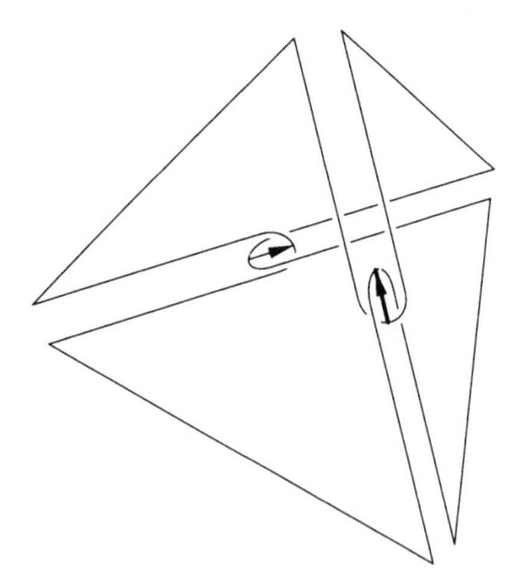

Figure 1.27. A two-complex spanning the trefoil knot. This arrangement for the trefoil knot, although very similar to Figure 1.26, does not lead to a decomposition of the complement of the knot into tetrahedra.

Chapter 2

Hyperbolic Geometry and Its Friends

"Geometry" can mean a number of different things in different contexts. In this chapter we will study geometry in the classical sense: our geometries will be analogous to standard Euclidean geometry, with concepts of straight lines or geodesics, angles and planes, a measure of distance, and a large degree of homogeneity.

One way to capture the structure is with a Riemannian metric. Any Riemannian manifold possesses a group of *isometries*, transformations of the space which preserve lines, angles and distances; however, for a typical Riemannian manifold, this group is the trivial group. Here, in contrast, we require *homogeneity* of the geometry, that is, transitivity of its group of symmetries: for any two points of the space there must be at least one isometry taking one to the other, so that each point "looks like" every other point.

In addition to homogeneity, the geometries we will consider in this chapter satisfy another property, *isotropy*. A space is isotropic if it looks the same no matter what position your head is in, or, more precisely, if for any two *frames* (ordered bases of orthonormal tangent vectors) at a point in the space, there is an isometry of the space fixing the point and taking one frame to the other.

Homogeneity and isotropy together are very strong conditions—in particular they imply that sectional curvatures are the same at every point and in every tangent two-plane. There are only three essentially distinct simply connected isotropic geometries in any dimension: with zero sectional curvature, constant positive sectional curvature K, or constant negative sectional curvature $-K$ (after scaling, K may be taken to be 1). They are called Euclidean, spherical and hyperbolic geometry, respectively.

Euclidean geometry is familiar to all of us, since it very closely approximates the geometry of the space in which we live, at least up to fairly large distances. Spherical geometry is the standard geometry of the n-sphere—geodesics are great circles, angles and distances are inherited from $(n+1)$-dimensional Euclidean space, and so on. We discussed the three-sphere briefly in example 1.4.2, and will return to it at the end of this chapter. Hyperbolic geometry is the least familiar to most people. It is also the most interesting, and by far the most important for two- and three-dimensional topology. For this reason, we will spend the next several sections giving different pictures of hyperbolic geometry.

In chapter 3 we will extend our analysis to include five other three-dimensional geometries that have significance for three-dimensional topology. These other five geometries are homogeneous but not isotropic: they are the same at every point, but not with your head at any angle. In fact, something like a notion of up and down can be geometrically defined in each of these geometries, and for some of them, a notion of north and south as well.

Two-dimensional geometry can be easily visualized "from the outside", by sketching pictures on paper. In three dimensions and higher, the best insight is often gained by imagining yourself inside the space. To formalize this intuitive concept, we need the idea of a visual sphere. Think of an observer as a point somewhere in an n-dimensional space, with light rays approaching this point along geodesics (see figure 2.1). Each of these geodesics determines a tangent vector at the point, and the $(n-1)$-sphere of tangent vectors

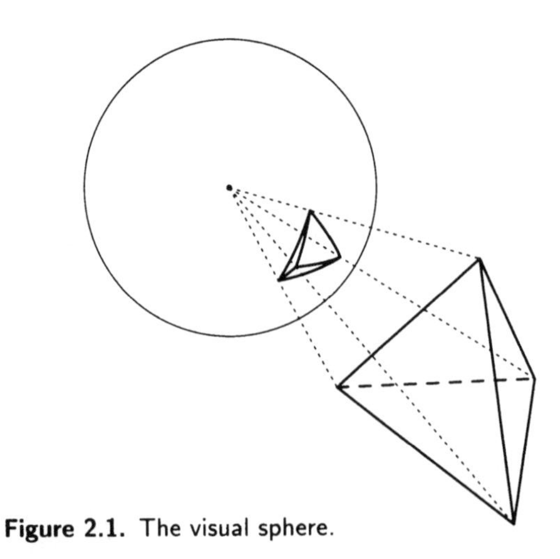

Figure 2.1. The visual sphere.

is called the *visual sphere*. Alternatively, one can think of a very small sphere centered at the observation point, and each geodesic from that point determines an intersection point with the sphere. An object is perceived as a segment of a straight line if its image on the visual sphere is an arc of a great circle. The apparent size of an object is determined by the amount of angle it takes up in the visual sphere.

2.1. Negatively Curved Surfaces in Space

Before developing abstract models for hyperbolic geometry, it pays to describe some constructions of a more physical nature. For this we need some basic concepts from the differential geometry of surfaces, which we'll introduce as required. We will not take the time here to develop this beautiful subject in detail, but you are encouraged to consult one of several readily available good sources, for example, [O'N66, Hic65, dC76].

The *Gaussian curvature*, or simply *curvature*, of a surface is a measure of its intrinsic geometry. We use the term "intrinsic" to denote those properties of a surface that are invariant when the surface is bent without being stretched: that is, they depend only on measurement of lengths of curves along the surface itself. By contrast, extrinsic properties depend on the embedding of the surface in space.

There are significant qualitative differences between surfaces with positive curvature, zero curvature, and negative curvature. Near the point of tangency, a surface of positive curvature lies to one side of any of its tangent planes. An example is a rubber ball. A surface of zero curvature has a line along which the surface agrees with its tangent plane. To illustrate this, try holding up a sheet of paper and bending it in different directions, and notice how you can find a straight line on the surface through any point. Surfaces of negative curvature cut through their tangent planes, as in a saddle.

The precise measurement of the curvature depends on the behavior of the surface to second order. One way to perform it is to arrange our coordinate system so that the point under scrutiny on our surface is at the origin, and the tangent plane at that point is horizontal. This means the surface is locally the graph of a function $f(x, y)$ such that f and its first-order partial derivatives f_x and f_y are zero at the origin. The Gaussian curvature at the origin is the determinant of the hessian matrix

2.1.1.
$$H = \begin{pmatrix} f_{xx} & f_{xy} \\ f_{yx} & f_{yy} \end{pmatrix}.$$

For example, the Gaussian curvature of the graph of the polynomial $f(x, y) = Ax^2 + 2Bxy + Cy^2$ at the origin is $K = 4(AC - B^2)$. For a paraboloid of revolution (Figure 2.2) this number is positive, whereas for a hyperbolic paraboloid (Figure 2.3) it is negative.

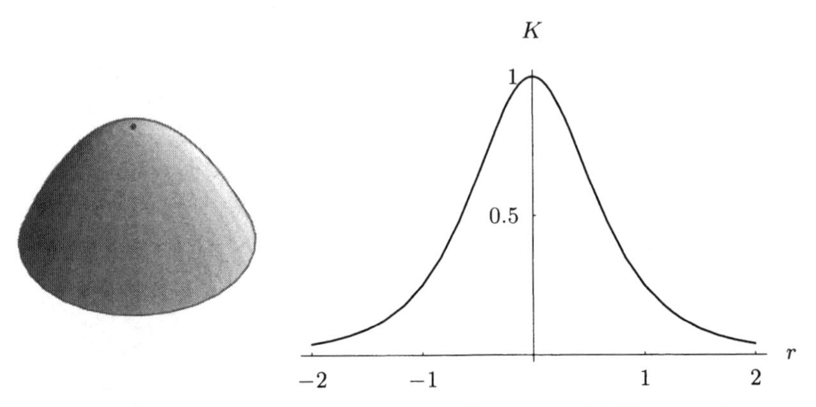

Figure 2.2. A paraboloid of revolution. This paraboloid has equation $z = -(x^2 + y^2)/2$. As a surface of revolution, its curvature depends only on the distance r from the origin. Computation shows that the curvature at distance r is $(1 + r^2)^{-2}$, whose graph is shown at right.

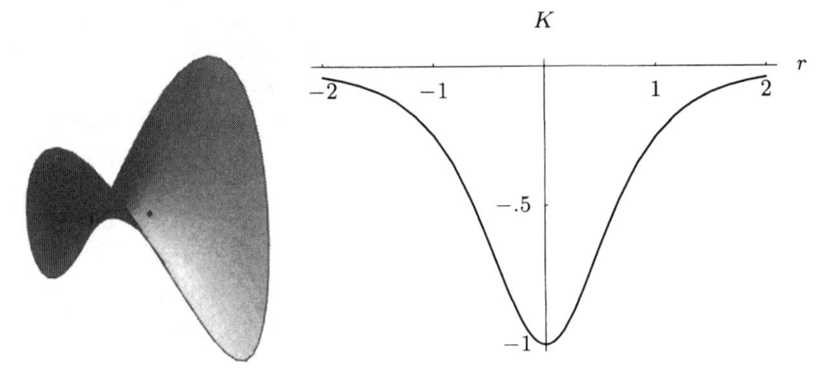

Figure 2.3. A saddle-shaped surface. This is the surface $z = (x^2 - y^2)/2$, a hyperbolic paraboloid. Perhaps surprisingly, its curvature, too, depends only on the distance r from the origin. It is -1 at the origin, and falls rapidly toward 0 as r increases—the surface appears flatter near the edges of the plot.

Notice that in these two examples $f(x, y)$ coincides with the quadratic form given by one-half its Hessian matrix, $f(x,y) = \frac{1}{2}(x,y)H(x,y)^t$. In general, $f(x, y)$ is approximated to second order by the same quadratic form.

The definition given above is based on extrinsic properties. A fundamental theorem, which Gauss called his "Theorema Egregium" or notable theorem, is that the Gaussian curvature is actually an intrinsic invariant of the surface: it does not change when the surface is bent without stretching. Here is an intrinsic way to arrive at the same number: draw a circle of radius r on the surface, centered at the point under scrutiny. If the curvature is positive there, this circle will have (for small r) a length c smaller than the length $2\pi r$ of a corresponding circle on the plane. Conversely, at a point where the surface is negatively curved, the ratio $c/2\pi r$ is greater than 1. It turns out that the second derivative of the ratio $c/2\pi r$ at $r = 0$ is $-\frac{1}{3}$ times the Gaussian curvature as defined above.

The curvature of the surfaces in figures 2.2 and 2.3 is not constant: it goes rapidly to zero away from the origin (keeping the same sign). Neither the intrinsic geometry nor the extrinsic geometry of these surfaces is homogeneous. It is easy to construct a surface in space with constant positive curvature: the sphere. It is both extrinsically and intrinsically homogeneous.

Surfaces of constant negative curvature are less obvious, but some are nevertheless well-known. The simplest of them is the pseudosphere (figure 2.4), the surface of revolution generated by a tractrix (figure 2.5). A tractrix is the track of a box of stones that starts at

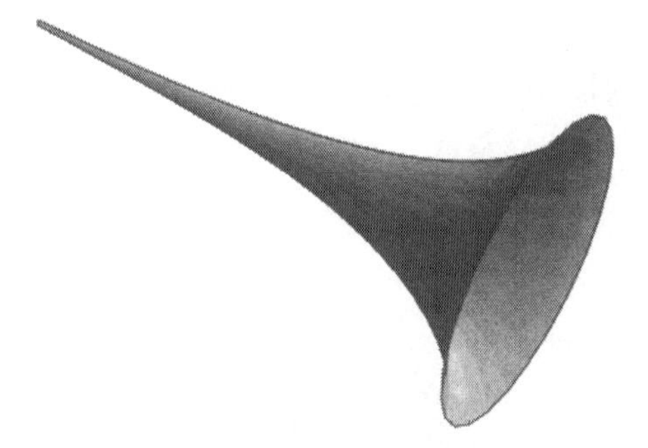

Figure 2.4. The pseudosphere. The pseudosphere is the figure obtained by rotating the tractrix about the x-axis. It has constant curvature -1. Any small patch of the surface can be placed isometrically—bending but not stretching—anywhere else on the surface.

$(0,1)$ and is dragged by means of a chain of unit length by a team of oxen walking along the x-axis. In other words, it is a curve determined (up to translation parallel to the x-axis) by the property that its tangent lines meet the x-axis a unit distance from the point of tangency.

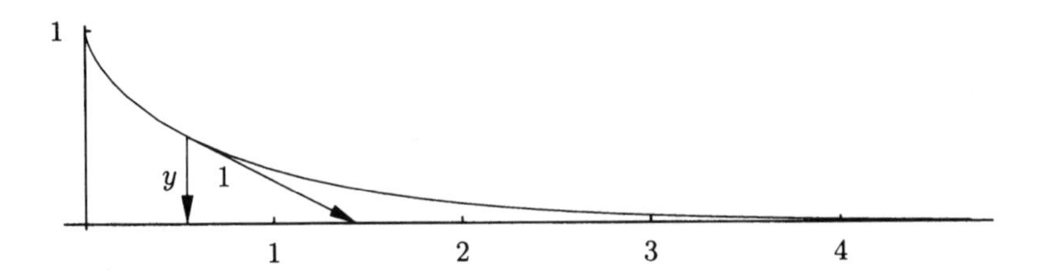

Figure 2.5. A tractrix. The tractrix is a curve whose tangent always meets the x-axis a unit distance away from the point of tangency. Therefore the derivative of the y-coordinate with respect to arc length s is $-y$, and $y(s) = e^{-s}$.

Note that the tractrix is not C^2 at the point $(0,1)$: its tangent is turning instantaneously at an infinite rate with respect to arclength. The edge of the pseudosphere is therefore an essential edge, beyond which it cannot be extended smoothly.

Problem 2.1.2 (surfaces of revolution of constant curvature). Show that the curvature K of a surface of revolution generated by rotating a plane curve $\big(x(s), y(s)\big)$, where s is arc length, around the x-axis, is given by

$$K = -\frac{1}{y}\frac{d^2y}{ds^2}.$$

Use this to verify that the curvature of the pseudosphere is -1. Then solve for $y(s)$ when K is 0, -1, and 1, and sketch a few examples of curves $\big(x(s), y(s)\big)$ in each case (see figure 2.6). Show that, for $K = -1$, the curve has a singularity at one or both ends, so the corresponding surface has an essential boundary beyond which it cannot be extended smoothly.

The intrinsic geometry of the pseudosphere, like that of the sphere, is locally homogeneous: any point has a neighborhood isometric to a neighborhood of any other point. To see this, parametrize the pseudosphere by coordinates (s, θ), where s comes from arc length along the tractrix, and θ is the angle around the x-axis. The defining property of the tractrix implies that the derivative of its y-coordinate with respect to arc length is $-y$, so that $y(s) = e^{-s}$. It follows that

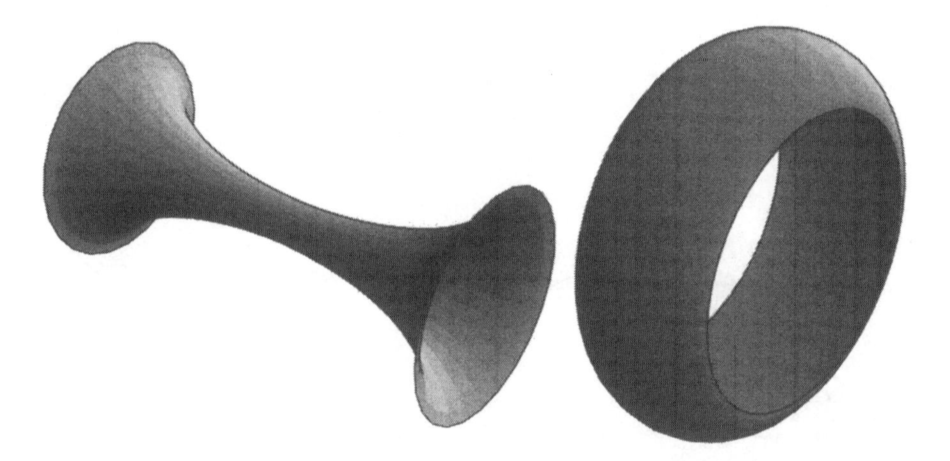

Figure 2.6. Surfaces of revolution of constant curvature. Each of the two planes containing a boundary circle is tangent to the surface along the boundary circle, and the surface has no smooth extension in space.

for any a and any θ_0, the locally defined map $(s,\theta) \mapsto (s+a,\ \theta_0+e^a\theta)$ is an isometry.

The intrinsic geometry of the pseudosphere is also locally isotropic. A small disk on the surface of the pseudosphere can be rotated about its center without stretching. This is a consequence of exercise 2.2.13, but you can also see it directly by tackling exercise 2.1.3.

Exercise 2.1.3 (making hyperbolic paper). (a) Approximate a pseudosphere by a union of truncated cones, each formed from a flat sheet of paper by cutting out a portion of an annulus along two radii (figure 2.7) and joining its radial edges. The radius of an annular segment becomes the distance from the truncated cone to the cone's vertex, and the angle subtended by the segment is proportional to the radius of the truncated cone. Thus the annular segments should all have the same radius—the length of the tangent of the tractrix extended to the x-axis—but varying angles, depending on the y-coordinate of the tractrix.

It is convenient to photocopy a drawing of several annuli on a sheet of paper. It is also worth marking an extra circle on each annulus, to indicate the extent of overlapping when the annuli are glued together.

(b) Construct a simply connected piece of surface in a similar way, starting with an annular strip that has not been made into a truncated cone. Apply this new surface to the pseudosphere; move it around and turn

Figure 2.7. Annuli for making a pseudosphere. Pieces of annuli like this can be cut out and glued together to make model surfaces of negative curvature, like the pseudosphere. The Gaussian curvature will be approximately $1/r^2$, where r is the radius of the circle that bisects the annulus.

it. Notice how much intrinsic local homogeneity the pseudosphere has, which is not visible in space.

(c) How far can you extend this piece of hyperbolic paper? Can you get it to look something like figure 2.8?

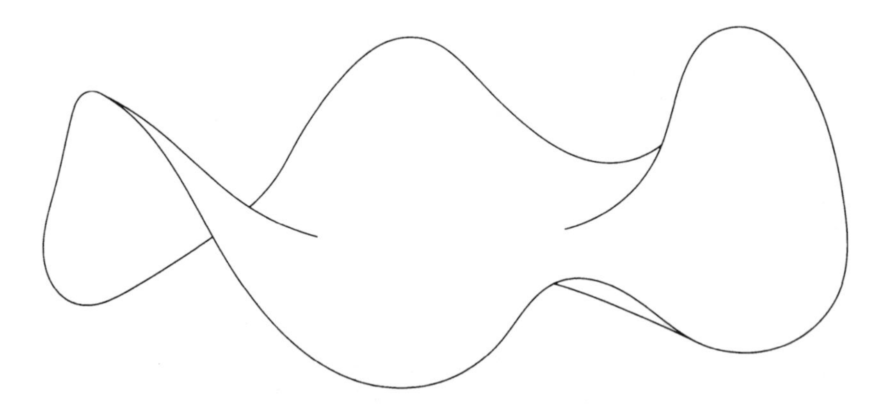

Figure 2.8. A negatively curved surface in space

An alternate medium for this construction is fabric. In fact, skirts with a nice negatively-curved flare can be (and are) made using large segments of annuli, roughly quarter-circles.

Exercise 2.1.4 (polyhedral models of negative curvature). Approximate models of a surface of constant negative curvature can also be constructed by joining triangles.

(a) Cut out a number of equilateral triangles from paper, or better, manila folders, and join them so there are seven around each vertex. Alternatively, sew pieces of cloth together.

(b) Compare these models with the paper models of exercise 2.1.3. Can you calculate or estimate what size equilateral triangle would correspond to a given size annulus?

(c) Try making similar models with eight triangles per vertex. What size equilateral triangles would be necessary to make a model comparable to that of (a)?

(d) These polyhedral models have negative curvature concentrated at their vertices. You can make smoother models by diffusing the curvature out along the edges: replace each side of the equilateral triangles by an arc of circle such that the three angles of the resulting curvilinear triangle are $2\pi/7$. (The radius of the circle should be approximately 6.69 times the sides of the triangle.)

(e) Is it possible to make models from congruent triangles that approximate the geometry of a pseudosphere? What about surfaces of revolution as in figure 2.6?

After playing with the paper models above, you may be surprised by the following result:

Theorem 2.1.5 (Hilbert). *There is no complete smooth surface in Euclidean three-space with the local intrinsic geometry of the pseudosphere.*

Actually, we've had hints of this already. We have seen that the pseudosphere and other surfaces of revolution cannot be extended beyond their edges. Other physical surfaces having negative curvature, such as leaves of many kinds of plants, brims of floppy hats, or the paper models of exercise 2.1.3, all develop wavy edges as in figure 2.8, as soon as they grow big enough. This implies that they are not mathematically smooth, but at best of class C^1.

To see this, we need a little more qualitative differential geometry. Imagine you approximate a surface at a point by a †quadratic form as we did in order to define Gaussian curvature near the beginning of this section. By a rotation of the xy-plane, this quadratic form can be diagonalized, that is, put in the form $Ax^2 + Cy^2$, with $A \geq C$, say. Then the x-axis is the direction in which the surface curves upward the most (or downward the least), and the y-axis is the direction in which it curves downward the most (or upward the least). These two directions are the *principal directions* of the surface at that point. If $A = C$, the principal directions are undefined, and the point is called an *umbilic*.

For a C^2 surface of negative curvature, the principal directions are defined everywhere and change continuously, so they can be distinguished from one another. In other words, they define two families of curves tangent to them and perpendicular to one another; these are the so-called *lines of curvature* of the surface, and they're illustrated in Figure 2.9. For a surface with many waves around the edge, the lines of curvature typically turn through one additional half-revolution for each wave, so they cannot be defined everywhere: they must have branch points in the interior, by a reasoning similar to the one used to show the Poincaré index theorem (Proposition 1.3.10). Therefore such a surface cannot be of class C^2.

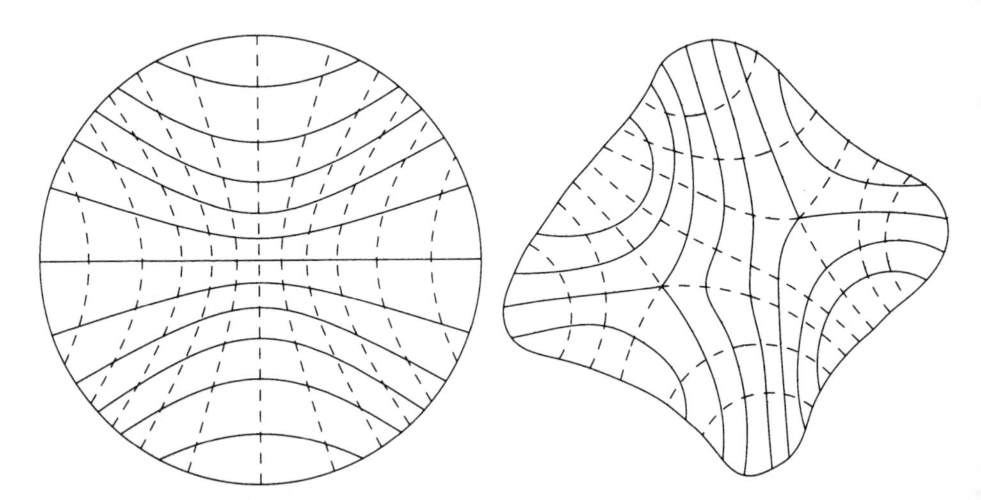

Figure 2.9. Lines of curvature. View from above of the lines of curvature of the hyperbolic paraboloid of Figure 2.3 and of the surface in Figure 2.8. The surface on the right is seen not to be smooth, because the lines of curvature of a C^2 negatively curved surface cannot branch.

Hilbert's original result in [Hil01] was for real analytic immersions, but the arguments actually work in class C^4. The extension to class C^2 requires some care [Mil72]. In class C^1 an embedding of a complete surface of negative curvature in space is no longer impossible: [Kui55] gives an explicit construction for one.

However, any such embedding would be incredibly unwieldy, and pretty much useless in the study of the surface's intrinsic geometry, as one quickly learns from trying to extend the paper models beyond a certain point. For instance, the length of a circle of radius R feet in a surface of constant curvature $-1/\mathrm{ft}^2$ is $2\pi \sinh R$. (One foot is 30.48 cm.) The circle of radius 1 foot has length 7.38 feet, which is fine: this would be a model of moderate curvature, like a sphere of radius 1 foot. But the lengths grow rapidly. The circle of radius 2 feet has length 22.78 feet, the circle of radius 10 feet has length 13.1 miles, and the circle of radius 20 feet has length 288,673 miles—more than the distance to the moon.

We must therefore resort to distorted pictures of the hyperbolic plane and of hyperbolic space. Just as it is convenient to have different maps of the earth for understanding various aspects of its geometry—for seeing shapes, for comparing areas, for plotting geodesics in navigation—so it is useful to have several maps of hyperbolic space at our disposal.

The several models of hyperbolic space that we shall look at have another important role, besides assisting our imagination. As one of the simple and basic structures of mathematics, hyperbolic geometry shows up in disguise in diverse places. The disguises it wears are usually related to one or another of these models.

2.2. The Inversive Models

One of the models, the Poincaré disk model, we're already somewhat familiar with from Section 1.2. In this section we generalize it to arbitrary dimension, and study its first cousins, the upper half-space and hemisphere models. These models share the property that they can be obtained from one another by inversions, so our first task is to make sure we understand inversions in n dimensions. Their definition is essentially the same as definition 1.2.1:

Definition 2.2.1 (inversion in a sphere). If $S \subset \mathbf{E}^n$ is an $(n-1)$-sphere in Euclidean space, the *inversion* i_S in S is the unique map from the complement of the center of S into itself that fixes every point of S, exchanges the interior and exterior of S and takes spheres orthogonal to S to themselves.

As in the two-dimensional case, the image $i_S(P)$ of a point P in a circle S with center O and radius r is the point on the ray $\overrightarrow{OP}$ such that $OP \cdot OP' = r^2$.

It is somewhat annoying that inversion in a sphere in $\mathbf{E}^n$ does not map its center anywhere. We can remedy this by considering the one-point compactification $\widehat{\mathbf{E}^n} = \mathbf{E}^n \cup \{\infty\}$ of $\mathbf{E}^n$ (Problem 1.1.1), which is homeomorphic to the sphere S^n. An inversion i_S can then be extended to map the center of S to ∞ and vice versa, so it becomes a homeomorphism of $\widehat{\mathbf{E}^n}$.

Exercise 2.2.2 (lines are circles). Show that the homeomorphism $h : \mathbf{E}^n \cup \{\infty\} \to S^n$ can be chosen in such a way that it maps circles to circles and lines to circles minus $h(\infty)$. (Hint: see stereographic projection, below.)

In view of exercise 2.2.2, it is natural to think of lines and planes as circles and spheres passing through ∞. Many properties of inversions and of the inversive models can be expressed more simply under this convention, so we will use it throughout this section. When we do want to exclude lines and planes, we'll talk about proper circles and spheres. For instance, here's the condensed version of proposition 1.2.3:

Proposition 2.2.3 (properties of inversions II). *Let S be an $(n-1)$-dimensional proper sphere in $\mathbf{E}^n$. Then the inversion i_S is conformal, and takes spheres (of any dimension) to spheres.*

Proof of 2.2.3. For conformality, notice that any two vectors based at a point are the normal vectors to two $(n-1)$-spheres orthogonal to S, so both the angle between them and the angle between their images equal the dihedral angle between the spheres.

The second statement follows from the plane case (proposition 1.2.3) for spheres of codimension one by considering the symmetries around the line joining the centers of the inverted and inverting spheres; and for lower-dimensional spheres because they are intersections of spheres of codimension one. 2.2.3

Exercise 2.2.4. (a) Since planes are special cases of spheres, what is the natural definition of inversion in a plane?

(b) What do you get when you successively invert in two concentric spheres? What if the spheres are planes?

(c) Show that composition of the inversions in two intersecting Euclidean planes is a Euclidean rotation. How would you define an inversive rotation?

The *Poincaré ball model* of hyperbolic space is what we get by taking the unit ball D^n in $\mathbf{E}^n$ and declaring to be hyperbolic geodesics all those arcs of circles orthogonal to the boundary of D^n. We also declare that inversions in $(n-1)$-spheres orthogonal to ∂D^n are hyperbolic isometries, which we will call *hyperbolic reflections*. According to problem 2.2.17 it would be enough to specify just the geodesics or just the isometries, but we won't take the minimalist approach.

Thanks to Proposition 2.2.3, we can retrace the arguments in Section 1.2 and conclude that the geodesics and isometries define the hyperbolic metric on D^n up to a constant factor, and that the Poincaré model is conformal, that is, hyperbolic angles and Euclidean angles are equal. Furthermore, it is easy to see that spheres of dimension k that meet the boundary orthogonally represent totally geodesic hyperbolic k-planes (a Riemannian submanifold is *totally geodesic* if all geodesics in the big manifold that are tangent to the submanifold are entirely contained in it).

To actually write down a formula for the metric, we look again at figure 1.11, where the hyperbolic length of a vector v is related to the angle of the banana built on it. Because of conformality we can assume that v is orthogonal to a diameter, in which case it is easy to see that the angle is $2/(1-r^2)$ times the Euclidean length of v, in the limit of small v, where r is the Euclidean distance from the basepoint of v to the origin.

Exercise 2.2.5. Draw a diagram and convince yourself of this formula.

Recall that in this construction we had the choice of a multiplicative factor. The choice we made in Section 1.2—setting the hyperbolic length of v equal to the banana angle in the limit when both go to 0—gives the following formula for the hyperbolic metric ds^2 as a function of the Euclidean metric dx^2:

2.2.6.
$$ds^2 = \frac{4}{(1-r^2)^2}dx^2.$$

Exercise 2.2.7 (curvature of the Poincaré model). (a) Find the hyperbolic distance from the origin to a point at Euclidean distance r from the origin in the Poincaré model.

(b) Find the hyperbolic length of a circle whose radius is as above.

(c) Find the Gaussian curvature of the Poincaré model at the origin (use the criterion given on page 47). The curvature at any other point is the same, since hyperbolic space is homogeneous. So it turns out that our choice of a constant factor in equation 2.2.6 was particularly fortunate.

We see that distances are greatly distorted in the Poincaré model: the Euclidean image of an object has size roughly proportional to its

Euclidean distance from the boundary ∂D^n, if this distance is small (figures 2.10 and 1.13). A person moving toward ∂D^n at constant speed would appear to be getting smaller and smaller and moving more and more slowly. She would never get there, of course; the boundary is "at infinity," not inside hyperbolic space.

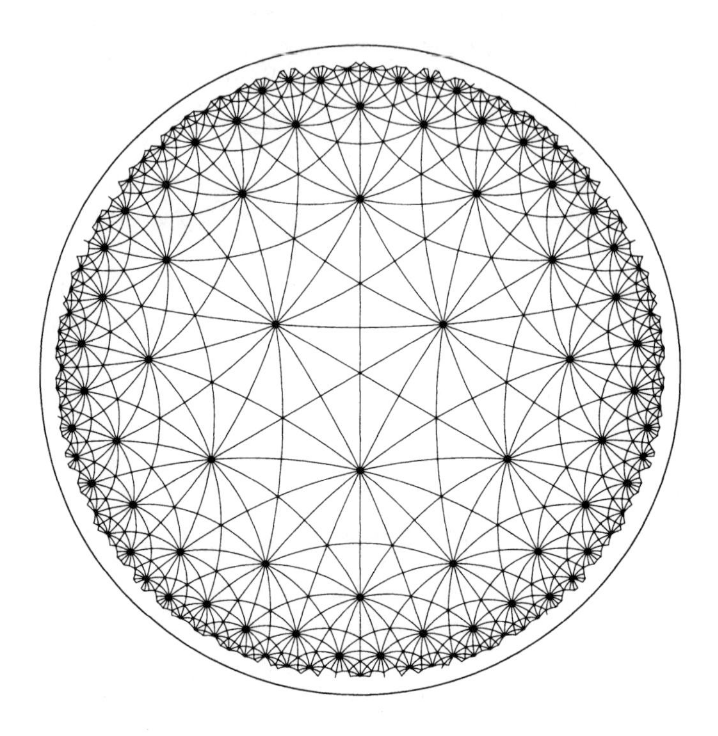

Figure 2.10. Hyperbolic tiling by 2-3-7 triangles. The hyperbolic plane laid out in congruent tracts, as seen in the Poincaré model. The tracts are triangles with angles $\pi/2$, $\pi/3$ and $\pi/7$. Courtesy HWG Homestead Bureau.

Nonetheless, ∂D^n can be interpreted purely in terms of hyperbolic geometry as the visual sphere. For a given basepoint p in D^n, each line of sight, that is, each hyperbolic ray from p, tends to a point on ∂D^n. If q is another point in D^n, each line of sight from q appears, as seen from p, to trace out a segment of a great circle in the visual sphere at p, since p and the ray determine a hyperbolic two-plane. This visual segment converges to a point in the visual sphere of p; in this way, the visual sphere of q is mapped to the visual sphere at p. The endpoint of a line of sight from p, as seen by q, gives the inverse map. In this way the visual spheres of all observers in hyperbolic space can be identified. This construction is independent of the model, and so associates to hyperbolic space $\mathbf{H}^n$ the *sphere at infinity* S^{n-1}_∞.

We now turn to a very useful construction, closely related to inversions. The *stereographic projection* from an n-dimensional proper sphere $S \subset \mathbf{E}^{n+1}$ onto a plane tangent to S at x is the map taking each point $p \in S$ to the intersection q of the line $\overline{px'}$ with the plane, where x' is the point opposite x on S.

Exercise 2.2.8 (stereographic projection). Show that stereographic projection can be extended to an inversion. (Hint: see figure 2.11.) Consequently, it is conformal, and takes spheres to spheres.

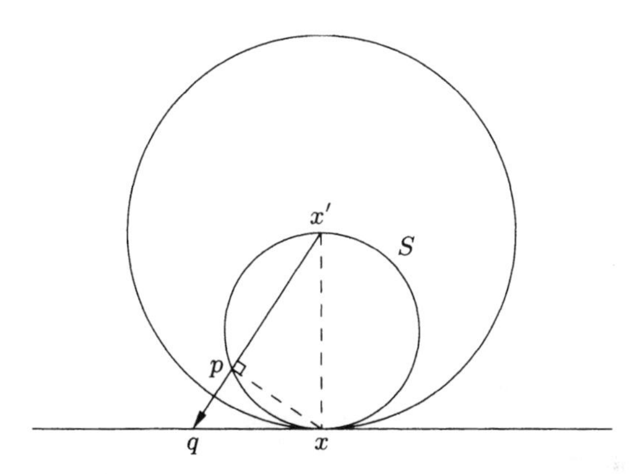

Figure 2.11. Stereographic projection and inversion. Stereographic projection from a sphere to a plane is identical to inversion in a sphere of twice the radius.

Our next model of $\mathbf{H}^n$ is derived from the Poincaré ball model by stereographic projection. We place the Poincaré ball D^n on the plane $\{x_0 = 0\}$ of $\mathbf{E}^{n+1}$, surrounded by the unit sphere $S^n \subset \mathbf{E}^{n+1}$, and we project from D^n to the northern hemisphere of S^n with center at the south pole $(-1, 0, \ldots, 0)$, as shown in figure 2.12(a). This is an inverse stereographic projection, at least up to a dilatation (since the projection plane is equatorial rather than tangent). In this way we transfer the geometry from the equatorial disk to the northern hemisphere to get the *hemisphere model*. Since stereographic projection is conformal and takes circles to circles, the hemisphere model is conformal and its geodesics are semicircles orthogonal to the equator $S^{n-1} = \partial D^n$.

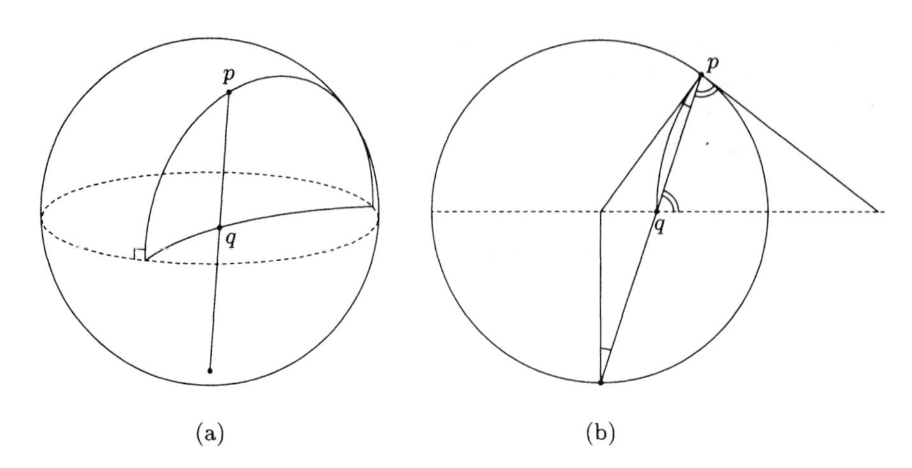

Figure 2.12. The hemisphere model. (a) By stereographic projection from the south pole of a sphere we map the equatorial disk to the northern hemisphere. Transfering the Poincaré disk metric by this map we get a metric on the northern hemisphere whose geodesics are semicircles perpendicular to the equator. (b) The circle going through p and q and orthogonal to the equatorial disk is also orthogonal to the sphere. This shows that the projection of part (a) can also be obtained by following hyperbolic geodesics orthogonal to the equatorial disk.

It is easy to see from figure 2.12(b) that for each point $q \in D^n$, the circle orthogonal to the equatorial disk $D^n \subset D^{n+1}$ and to $S^n = \partial D^{n+1}$ meets the northern hemisphere at the same point p as the image of q under the projection above. This means we can interpret this projection purely in terms of hyperbolic geometry: fix a totally geodesic n-space $\mathbf{H}^n$ inside hyperbolic $(n+1)$-space $\mathbf{H}^{n+1}$, and choose one of the half-spaces it determines. For each $q \in \mathbf{H}^n$, the hyperbolic ray from q perpendicular to $\mathbf{H}^n$ and pointing into the half-space we chose converges to a point in the corresponding visual hemisphere, so we get a map $\mathbf{H}^n \to S^n_\infty$. By making $\mathbf{H}^n = D^n$ be the equatorial disk in the Poincaré ball model of $\mathbf{H}^{n+1} = D^{n+1}$, we see that this map $\mathbf{H}^n \to S^n_\infty$ coincides with the projection above.

From the hemisphere model we get the third important inversive model of hyperbolic space, also by stereographic projection. This time we project from a point on the equator, say $(0, \ldots, 0, 1)$, onto a vertical subspace, say $\{x_n = 0\}$, which we identify with $\mathbf{E}^n$. The open northern hemisphere maps onto the open upper half-space $\{x_0 > 0\}$, and the equator—the sphere at infinity of the hemisphere model—maps onto the bounding plane $\mathbf{E}^{n-1} = \{x_0 = 0\}$, except for the center of projection, which is mapped to the point at infinity. In other words, the sphere at infinity here is given by the one-point compactification of the bounding plane, $S^{n-1}_\infty = \mathbf{E}^{n-1} \cup \{\infty\}$. Geodesics are given by semicircles orthogonal to the bounding plane $\mathbf{E}^{n-1}$ (figure 2.13), and hyperbolic reflections are inversions in spheres orthogonal to the bounding plane. Clearly, this model, too, is conformal.

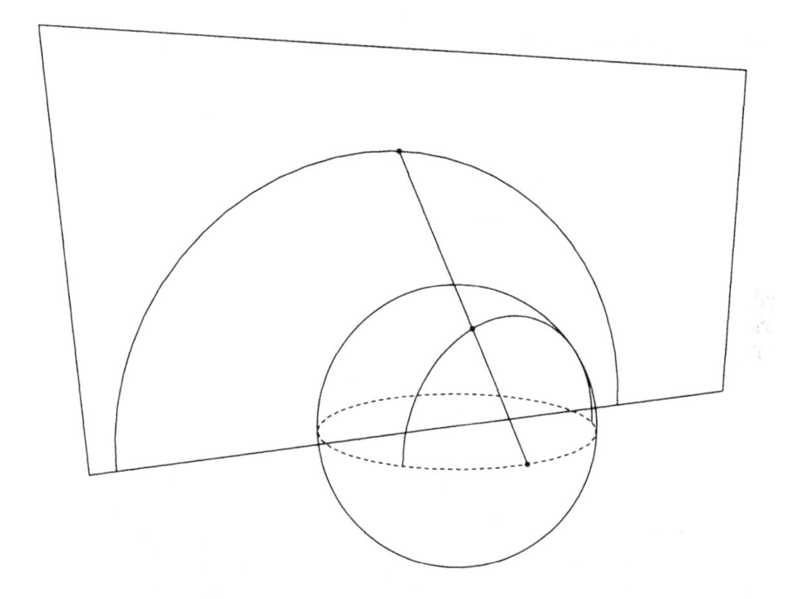

Figure 2.13. Geodesics in the upper half-space model. Geodesics in the upper half-space model of hyperbolic space appear as semicircles orthogonal to the bounding plane, or half-lines perpendicular to it.

The hyperbolic metric in the upper half-space model has an especially simple form, which is well suited to many computations. We could compute it by writing down an explicit formula for the composition of projections that goes from the Poincaré disk model to the upper half-space model, and then using equation 2.2.6 above; but we can save a lot of energy by observing that this transition map is conformal, so our argument about banana angles applies here too. If we position the banana so that one of its vertices is at infinity, it becomes a solid cone, and then it's easy to compare the angle at the vertex with the Euclidean length of the vector the cone is built on. In the limit when both are small, the Euclidean length is just x_0 times the angle, where x_0 is the Euclidean distance from the vector's basepoint to the bounding plane. In other words, the relation between the hyperbolic metric ds^2 and the Euclidean metric dx^2 is

2.2.9. $$ds^2 = \frac{1}{x_0^2} dx^2.$$

Thus the Euclidean image of an object has size exactly proportional to its Euclidean distance from the bounding plane $\mathbf{E}^{n-1}$. Figure 2.14 shows the same congrent tracts as figure 2.10, but seen in the upper half-space model.

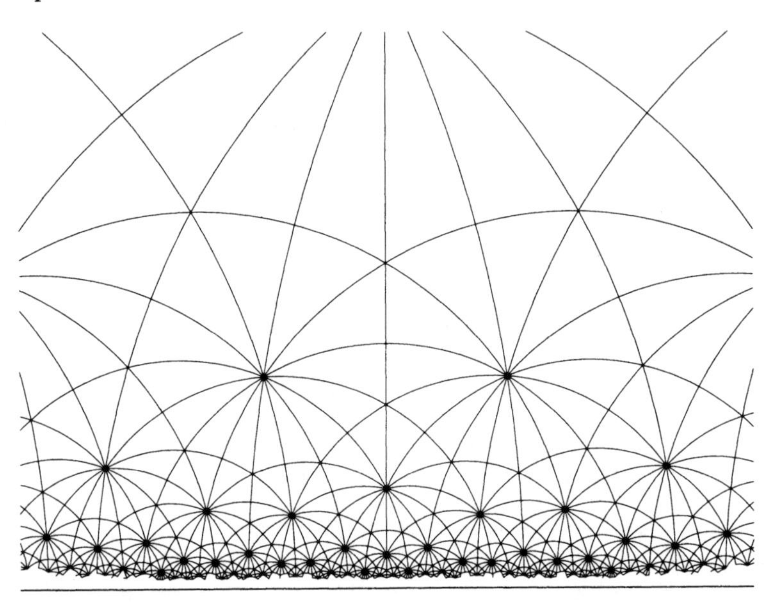

Figure 2.14. Hyperbolic tiling by 2-3-7 triangles. Another view of the hyperbolic world divided into congruent tracts. Upper half-plane projection.

Exercise 2.2.10 (Euclidean similarities are hyperbolic isometries). A *similarity* of $\mathbf{E}^n$ is a transformation that multiplies all distancees by the same (non-zero) factor. Any similarity of $\mathbf{E}^n$ can be composed from an element of $O(n)$ (where we fix an origin for $\mathbf{E}^n$ arbitrarily), an expansion or contraction by a scalar factor, and a translation.

A similarity of $\mathbf{E}^{n-1}$ extends in a unique way to a similarity preserving upper half-space. Show that such a similarity is a hyperbolic isometry, by expressing it as a composition of reflections.

The easy visibility of this significant subgroup of isometries of $\mathbf{H}^n$ is a frequently useful aspect of the upper half-space model.

Exercise 2.2.11. We already know that hyperbolic space is homogeneous and isotropic, but explicit formulas are especially easy to write down in the upper half-space model.

(a) Given two points in upper half-space, write a composition of hyperbolic reflections that will map one to the other.

(b) Let $O(n)$ be the group of isometries of the tangent space to $\mathbf{H}^n$ at a point p. Show that any element of $O(n)$ can be realized by a composition of hyperbolic reflections. (Hint: Show by induction that $O(n)$ is generated by reflections, and that any reflection in $O(n)$ can be realized by a reflection in a sphere orthogonal to $\mathbf{E}^{n-1}$ in the upper half-space model.)

(c) Show that reflections generate the whole group of isometries of $\mathbf{H}^n$. (Hint: an isometry is determined by its derivative at a point.)

A horizontal Euclidean plane $\{x_0 = c\}$, for $c > 0$, is not a plane in hyperbolic geometry: it lies entirely on one side of a true hyperbolic plane tangent to it, which is a Euclidean sphere. (One can also notice that pushing a horizontal surface up along orthogonal geodesics shrinks the hyperbolic metric, so the surface must curve upwards.) These horizontal surfaces are examples of *horospheres* (or *horocycles* if $n = 2$). Horospheres are characterized by the property that their parallel surfaces are all congruent—in our case, by arbitrary dilatations centered at points in the plane at infinity. Another characteristic property is that a horosphere is orthogonal to all planes passing through a certain point on the sphere at infinity; we say the horosphere is *tangent* to S^{n-1}_∞ at that point.

Problem 2.2.12. Show that these two characterizations are equivalent, and that according to them a horosphere in upper half-space appears as either a horizontal Euclidean plane or a Euclidean sphere tangent to the bounding plane. How does a horosphere appear in the Poincaré disk model?

The intrinsic geometry of a horosphere is Euclidean. This is easiest to see when the horosphere appears as $\{x_0 = c\}$, by examining the form of the hyperbolic metric given by equation 2.2.9. In fact, it follows from the same equation that any Euclidean isometry acting on a horosphere extends to an isometry of $\mathbf{H}^n$ which preserves it and all of its parallel horospheres.

The region $\{x_0 \geq c\}$ above a horosphere is called a *horoball*. If a horosphere appears as a Euclidean sphere tangent to the bounding plane, its corresponding horoball is just the Euclidean ball bounded by that sphere.

Using horocycles we can describe the relation of the hyperbolic plane to the pseudosphere (figure 2.4).

Exercise 2.2.13 (pseudosphere is locally hyperbolic). Consider the map that wraps the region $y \geq 1$ of the upper half-plane around the pseudosphere, taking horocycles $y = C$ to meridians and vertical geodesics $x = C$ to generating curves (tractrixes). Show that, if the map is periodic with period 2π in the x-direction, it is a local isometry. (Hint: arc length along horocycles, measured between fixed vertical geodesics, decreases exponentially with hyperbolic distance from the line $y = 1$, and arc length along meridians on the pseudosphere similarly decreases exponentially with distance from the edge.)

The paper models of exercise 2.1.3 are based on horocycles. In the limit, when the annuli are infinitely thin, the metric becomes the hyperbolic metric, and the circles become horocycles. This comparison demonstrates something that was hard to explain before: the isotropy of the pseudosphere's metric.

We can use the upper half-space model to study the group of isometries of hyperbolic space. Consider a reflection of $\mathbf{H}^n$ given by inversion in an $(n-1)$-sphere S orthogonal to $\mathbf{E}^{n-1}$. The restriction of this inversion to the sphere at infinity $S^{n-1}_\infty = \mathbf{E}^{n-1} \cup \infty$ is just the inversion in the $(n-2)$-sphere $S \cap S^{n-1}_\infty$, and every inversion of S^{n-1}_∞ can be so expressed. A transformation of S^{n-1}_∞ that can be expressed as a composition of inversions is known as a *Möbius transformation*, and the group of all such transformations is the *Möbius group*, denoted Möb_{n-1}. Since, by exercise 2.2.11(c), all hyperbolic isometries can be generated by reflections, it follows that the group of isometries of $\mathbf{H}^n$ is isomorphic to Möb_{n-1}.

Problem 2.2.14 (the Möbius group). Analyze and become familiar with the Möbius group. Show that:

(a) The subgroup of the Möbius group that fixes ∞ is isomorphic to the group of Euclidean similarities.

(b) The subgroup of the Möbius group Möb_n that takes an $(n-1)$-sphere to itself and fixes a point not on that sphere is isomorphic to the group $O(n)$.

(c) For $n > 1$, the Möbius group consists exactly of those homeomorphisms of S^n_∞ that take $(n-1)$-spheres to $(n-1)$-spheres.

(d) Any Möbius transformation that takes a sphere S to a sphere R conjugates i_S to i_R.

(e) The subgroup of the Möbius group that takes a k-sphere to itself is isomorphic to $\text{Möb}_k \times O(n-k)$.

(f) There is a subgroup of the Möbius group isomorphic to $O(n+1)$.

What is the dimension of the Möbius group?

Some geometric problems involving spheres can be greatly simplified by artful application of a Möbius transformation.

Exercise 2.2.15 (Steiner's porism). Suppose you are given an arrangement of circles in the plane consisting of two non-intersecting circles A and B and a chain of circles $X_0, X_2, \dots, X_{n-1}$, where each X_i is tangent to A, B, and $X_{i+1 \bmod n}$. The circles are disjoint except for tangencies.

Show that if the X_i are erased, and any circle Y_0 tangent to A and to B is constructed, the analogous chain of circles it determines closes up after exactly n circles.

Exercise 2.2.16 (tangent spheres). Let A, B and C be mutually tangent two-spheres. Let X_0 be a fourth sphere tangent to A, B and C. Construct a chain of spheres, beginning with X_0, each sphere being tangent to A, B and C and to its neighbors in the chain.

Show that the chain closes up on the sixth sphere.

Problem 2.2.17 (minimal hyperbolic properties). In the discussion of hyperbolic geometry above, there was no attempt to characterize hyperbolic geometry using a minimal amount of structure. Here are some steps in this direction:

(a) Show that hyperbolic lines can be characterized in terms of the metric as curves that minimize distance between any two points. (Hint: in the upper half-space model, reduce to the case that p and q are on a vertical line.)

(b) Show how to characterize hyperbolic lines directly in terms of the group of isometries, as fixed-point sets.

(c) Show that the only diffeomorphisms of upper half-space to itself that take all hyperbolic lines to hyperbolic lines are the hyperbolic isometries. (This contrasts with the Euclidean case, where affine transformations take lines to lines.)

(d) Conclude that any of these three structures is sufficient as a base structure to define hyperbolic geometry: the set of lines, the group of isometries, or the metric up to a constant multiple.

(e) (Harder.) Show that the measure of angle also is sufficient to define hyperbolic geometry.

2.3. The Hyperboloid Model and the Klein Model

A sphere in Euclidean space with radius r has constant curvature $1/r^2$. By analogy, since hyperbolic space has constant curvature -1, hyperbolic space should be a sphere of radius $i = \sqrt{-1}$.

This seemingly impossible condition can actually be given a reasonable interpretation. Let's see how far analogy can take us. To get an n-sphere, we start with the †positive definite quadratic form Q^+ on $\mathbf{R}^{n+1}$ given by $Q^+(x) = x_0^2 + x_1^2 + \cdots + x_n^2$, where $x = (x_0, \ldots, x_n)$. This gives $\mathbf{R}^{n+1}$ a Euclidean metric $dx^2 = dx_0^2 + dx_1^2 + \cdots + dx_n^2$, making it into $\mathbf{E}^{n+1}$. Restricting to the unit sphere $S = \{Q^+ = 1\}$, we get a Riemannian metric of constant positive curvature 1. The isometries of S come from linear transformations of $\mathbf{E}^{n+1}$ preserving Q^+; the group of these *orthogonal transformations* is denoted $O(n+1)$.

Now let's start instead with the indefinite metric

2.3.1.
$$ds^2 = -dx_0^2 + dx_1^2 + \cdots + dx_n^2$$

in $\mathbf{R}^{n+1}$ associated to the quadratic form $Q^-(x) = -x_0^2 + x_1^2 + \cdots + x_n^2$. With this metric, $\mathbf{R}^{n+1}$ is often referred to as *Lorentz space*, and denoted $\mathbf{E}^{n,1}$. When $n = 3$, this is the universe of special relativity, although physicists usually reverse the sign of Q^-. In this interpretation, the vertical direction x_0 represents time, and the horizontal directions represent space. A vector x is *space-like*, *time-like* or *light-like* depending on whether $Q^-(x)$ is positive, negative, or zero. By analogy with the Euclidean case, the length of a vector x is $\sqrt{Q^-(x)}$, so light-like vectors have zero length, and time-like vectors have imaginary length (which we take to be a positive multiple of i).

The sphere of radius i about the origin in $\mathbf{E}^{n,1}$ is the hyperboloid $H = \{Q^- = -1\}$. When restricted to this hyperboloid, the indefinite metric ds^2 of equation 2.3.1 becomes a *bona fide*, positive definite Riemannian metric: it is easy to check that any tangent vector to the hyperboloid has real length. Topologically, H is the union of two open disks.

As in the case of the sphere, which in example 1.4.3 is turned into elliptic space, we really want to identify antipodal points of H, to get a subset of projective space $\mathbf{RP}^n$. Unlike the case of the sphere, here antipodal points lie in disjoint components of H, so this subset of $\mathbf{RP}^n$ can be modeled by one component of the hyperboloid, say, the upper sheet H^+, where $x_0 > 0$. This is the *hyperboloid model* of hyperbolic space.

Another way to model the same subset of $\mathbf{RP}^n$ is by inhomogeneous coordinates, where a point $(x_1, \ldots, x_n) \in \mathbf{R}^n$ stands for the line with parametric equation $(t, x_1 t, \ldots, x_n t)$. Then H maps onto the unit disk, giving the *projective model*, or *Klein model* of hyperbolic space (although it was first introduced by Beltrami: cf. page 8). Notice that, although we can look at the Klein disk as embedded in $\mathbf{E}^{n,1}$ (figure 2.15), its metric is not the induced metric, but rather the push-forward of the hyperboloid's metric by central projection.

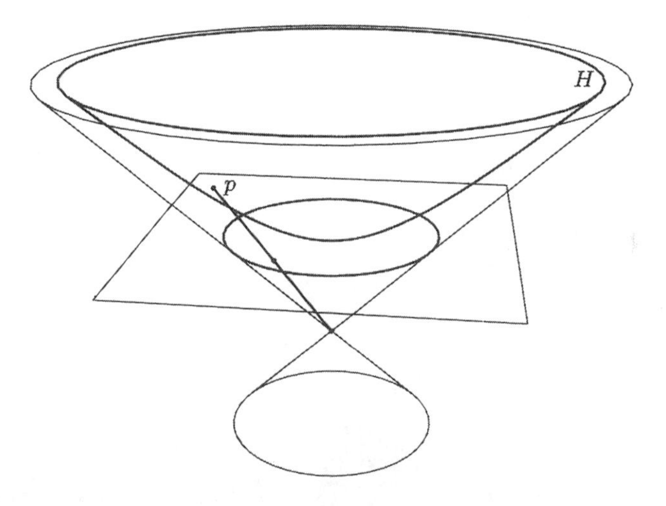

Figure 2.15. The hyperboloid model and the Klein model. A point $p=(x_0, x_1, \ldots, x_n)$ on the hyperboloid maps to a point $(x_1/x_0, \ldots, x_n/x_0)$ in $\mathbf{R}^n$, shown here as the horizontal hyperplane $\{x_0 = 1\}$. This transfers the metric from the hyperboloid to the unit disk in $\mathbf{R}^n$, giving the projective model, or Klein model, for hyperbolic space.

How can we see that these are indeed models for n-dimensional hyperbolic space? The best way is not by direct calculation but, as for the inversive models, through the study of lines and isometries.

In $\mathbf{E}^{n,1}$ we still have a notion of orthogonality, given by the inner product $-x_0 y_0 + x_1 y_1 + \cdots + x_n y_n$. The *orthogonal complement* of any non-zero vector x is an n-dimensional subspace, denoted by $x^\perp$; it contains x if and only if $Q^-(x) = 0$. The orthogonal complement of a subspace is the intersection of the orthogonal complements of its points.

Exercise 2.3.2 (characterization of tangent vectors). If $x \in H$ is a point on the hyperboloid, the tangent space of H at x coincides with $x^\perp$.

We also have the notion of orthogonal transformations, that is, linear transformations of $\mathbf{R}^{n+1}$ preserving Q^-. Just as the group of isometries of S is identical with $O(n+1)$, the group of orthogonal transformations of $\mathbf{E}^{n+1}$, so also the group of isometries of H is $O(n,1)$, the group of orthogonal transformations of $\mathbf{E}^{n,1}$.

Exercise 2.3.3. The proof is basically the same as for $\mathbf{E}^{n+1}$. First an orthogonal transformation of $\mathbf{E}^{n,1}$ clearly induces an isometry on H, non-trivial if the transformation is non-trivial. Next, all isometries of H come from such transformations, that is, $O(n,1)$ acts transitively and isotropically on H:

(a) If p and q are points on H, find an element of $O(n,1)$ interchanging p and q. (Hint: consider the orthogonal complement of $p-q$.)

(b) If $p \in H$, show that any isometry of the tangent space of p is induced by an element of $O(n,1)$. (Hint: by part (a), it is enough to consider the case where p is the "north pole" $(1,0,\ldots,0)$.)

Exercise 2.3.4 (Lorentz transformations). An element of $O(n,1)$ that takes each component of H to itself is called a *Lorentz transformation*. Show that the group of Lorentz transformations, or *Lorentz group*, has index two in $O(n,1)$, and coincides with the group of isometries of H^+.

This gives a way to describe geodesics in the hyperboloid and Klein models: if $p \in H^+$ is a point and v is a non-zero tangent vector to H^+ at p, the geodesic through p in the direction of v is the intersection L of H^+ with the two-plane P determined by p, v and the origin (figure 2.16). The same is true with the Klein disk K instead of H^+.

To see this, consider the Lorentz transformation that fixes P and equals -1 on the orthogonal complement of P. The corresponding isometry of H^+ or K fixes exactly those points that lie on L. By uniqueness, the geodesic through p in the direction of v is fixed by this isometry, and so must be contained in L. But L is a connected curve, so it must be the geodesic.

For the hyperboloid model, L is a branch of a hyperbola, whose asymptotes are rays in the *light cone* $\{Q^- = 0, x_0 > 0\}$—so called because, in the relativistic interpretation, it is the union of the trajectories in space-time of light rays emanating at time 0 from a point source at the origin. Rays in the light cone, then, are the points at infinity for this model.

Exercise 2.3.5 (parametrization of geodesics). Show that if v has unit length, L is parametrized with velocity 1 by $p\cosh t + v\sinh t$. What is the analogous formula for the sphere?

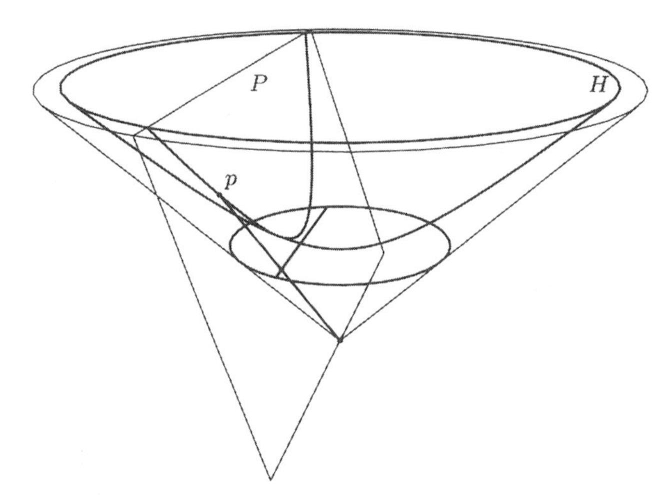

Figure 2.16. Geodesics in the hyperboloid and Klein models. In the hyperboloid model geodesics are the intersections of two-planes through the origin with the hyperboloid. In the projective model they're straight line segements.

For the Klein model, L is a segment of a straight line, meaning that this model is projectively correct: geodesics look straight. This makes the Klein model particularly useful for understanding incidence in a configuration of lines and planes. The sphere at infinity is just the unit sphere S_∞^{n-1}, the image in $\mathbf{RP}^n$ of the light cone. Angles are distorted in the Klein model, but they can be accurately and conveniently computed in the hyperboloid model if one remembers to use the Lorentz metric of equation 2.3.1, rather than the Euclidean metric.

We now exhibit a correspondence between the Klein model and the hemisphere model of section 2.2 that takes geodesics into geodesics. Since the set of geodesics is sufficient to characterize hyperbolic geometry (problem 2.2.17), we conclude that K, and consequently also H^+, are indeed legitimate models for $\mathbf{H}^n$. If K is placed as the equatorial disk of the unit sphere in $\mathbf{R}^{n+1}$, the correspondence is given by Euclidean orthogonal projection onto K (figure 2.17): geodesics in the hemisphere model are half-circles orthogonal to the equator, so they indeed project to segments of straight lines.

Note that this arrangement is similar to the one we made to go from the Poincaré to the hemisphere model, but the projection is different. The composed map, from the Poincaré disk model to the hemisphere back to the Klein model, is a surprising transformation

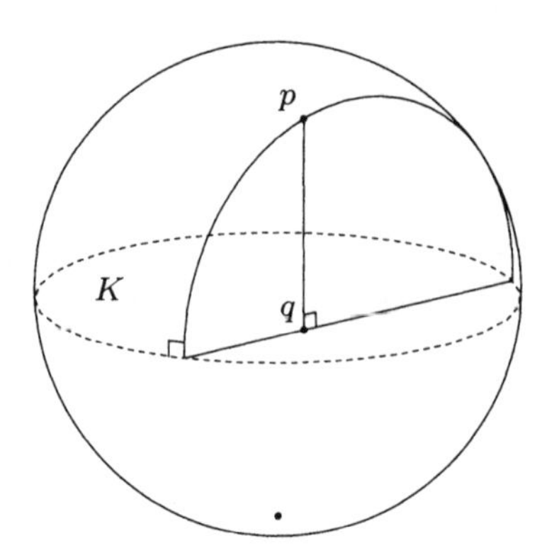

Figure 2.17. Going from the hemisphere to the Klein model. We get the Klein model of hyperbolic space from the hemisphere model by (Euclidean) orthogonal projection onto the equatorial disk. Compare figure 2.12(a).

of the unit disk to itself that maps each radius to itself, while simultaneously straightening out every circle orthogonal to the unit sphere into a straight line segment.

Exercise 2.3.6. Find the formula for this Poincaré-to-projective transformation, in polar coordinates (r, θ), where $0 \leq r < 1$ and $\theta \in S^{n-1}$. Find the formula for the inverse transformation.

Exercise 2.3.7. The Poincaré model maps to the hemisphere model by stereographic projection (figure 2.12). Show that the same projection, if extended to H^+, gives a direct correspondence between these two models and the hyperboloid model.

As a subset of projective space $\mathbf{RP}^n$, the Klein model has a natural interpretation in terms of hyperbolic perspective. In fact, it embeds in *visual projective space*, the visual sphere with antipodal points identified. Imagine you are in $\mathbf{H}^{n+1}$, hovering above a copy of $\mathbf{H}^n$. Since light rays coming to you from a geodesic in $\mathbf{H}^n$ lie in a hyperbolic two-plane, their tangent vectors lie in a great circle of your visual sphere, so you see the geodesic as a straight line. The hyperplane $\mathbf{H}^n$ looks like the Klein model! The sphere at infinity of $\mathbf{H}^n$ looks like a sphere to you. In contrast to the situation in Euclidean space, the visual radius of a hyperbolic plane in hyperbolic space is always strictly less than π, provided the plane does not contain the eye. Your image of $\mathbf{H}^n$ shrinks as you move away from it and expands as you move closer. See figure 2.18.

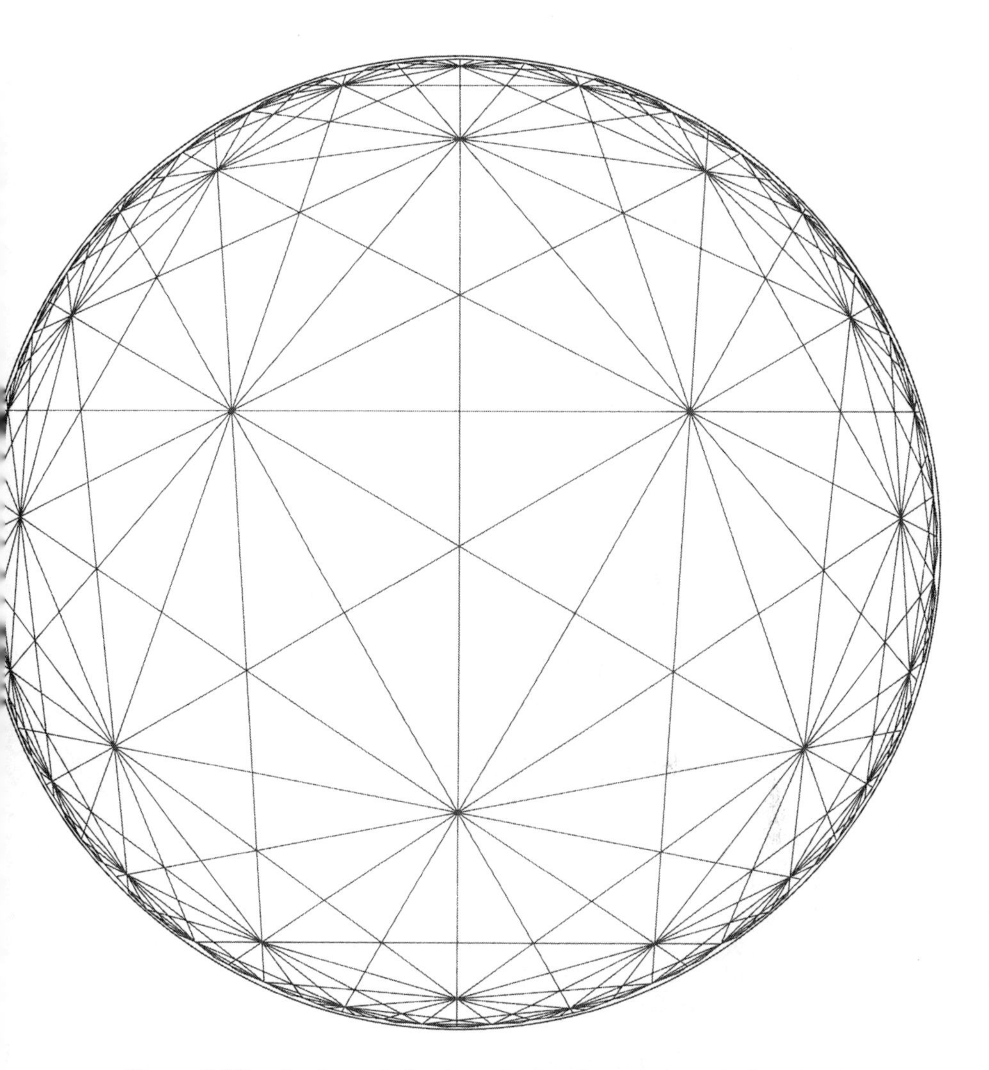

Figure 2.18. A view of the hyperbolic plane from a helicopter in hyperbolic space. The same tracts in the hyperbolic plane shown in figures 2.10 and 2.14, this time in the projective disk projection. This picture is in true perspective in hyperbolic three-space. Stare at the plane near its horizon and try to sense the way it slopes away from you and the way the area of the plane grows very rapidly (exponentially) with distance.

You can pursue this line of thought, and really develop a sense of what it would look like to live in $\mathbf{H}^3$. To get the correct visual geometry, position yourself at the center of either the Poincaré ball model or the Klein model. From this point, hyperbolic objects appear at the correct position in the visual sphere. Our Euclidean stereoscopic vision would mislead us: it would lead us to judge distances just as they are in the Klein model, with everything inside a finite sphere. The change in appearance of distant objects as you move towards or away from them is quite different, though. They diminish or increase in size exponentially fast with distance.

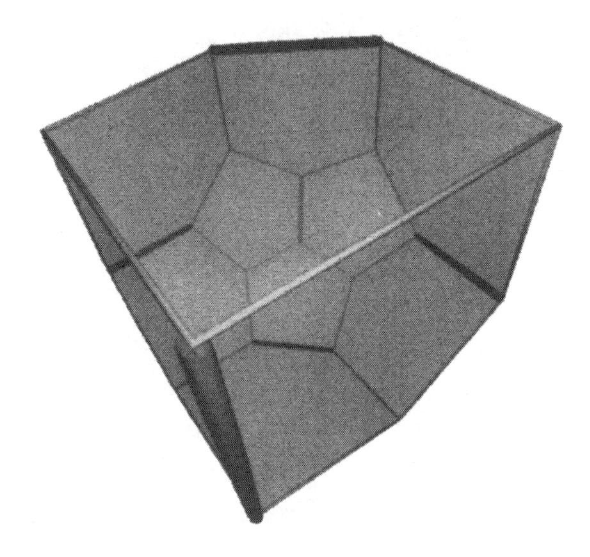

Figure 2.19. A right-angled hyperbolic dodecahedron in true hyperbolic perspective. Note that the edges appear straight. From a scene in *Not Knot* [GM91].

This can be seen in figure 2.19 and on the cover figure of this book, both taken from the final scenes of the video *Not Knot* [GM91], and programmed by Charlie Gunn. *Not Knot*, produced by the Geometry Supercomputer Project (now Geometry Center), was the first place to really create the sense of living and moving in hyperbolic space. Notice on the cover image (which shows a tiling of $\mathbf{H}^3$ by congruent right-angled dodecahedra) how, for many of the dodecahedra, one face is in front of all the others. This is a consequence of the rapid diminution of the angular size of an image with distance. Other dodecahedra are seen with an edge or a vertex closest to us. Each pentagon is part of a hyperbolic plane tiled by right-angled pentagons. Notice how these planes slope away toward the horizon. The correct hyperbolic lighting helps show this effect. The reflection of light also helps distinguish the appearance of hyperbolic space from the appearance of a Euclidean ball from its interior.

Exercise 2.3.8. How can you tell from a distance at what angle two planes meet in $\mathbf{H}^3$?

We now have a geometric interpretation for points in $\mathbf{RP}^n$ that lie in the unit ball and on the sphere. How about points outside the closed ball? If $x \in \mathbf{RP}^n$ is such a point, Q^- is positive on the associated line $X \subset \mathbf{E}^{n,1}$. This means that Q^- is indefinite on the orthogonal complement $X^\perp$ of X, and that the corresponding hyperspace $x^\perp \subset \mathbf{RP}^n$ intersects hyperbolic space. We call $x^\perp$ the *dual hyperspace* of x. The hyperbolic significance of projective duality is that any line from x to $x^\perp$ is perpendicular to $x^\perp$. This is best seen in the hyperboloid model, as shown in figure 2.20(a): if $p \in X^\perp \cap H^+$

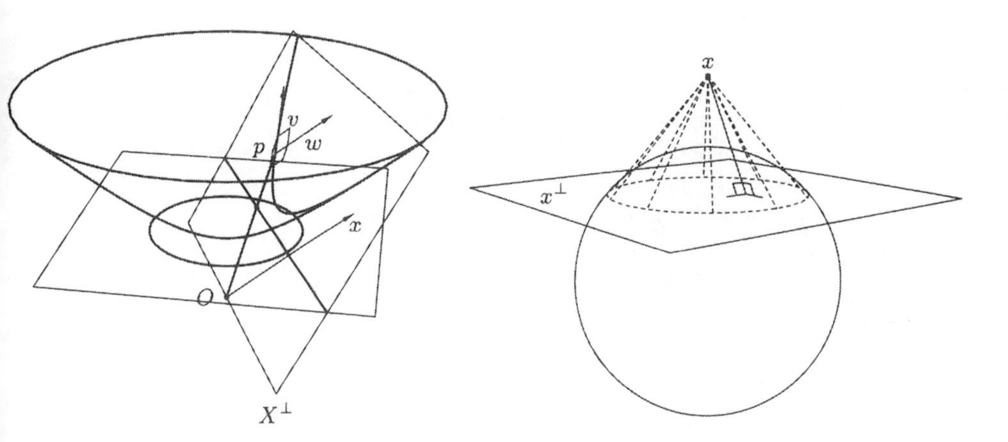

Figure 2.20. Duality between a hyperplane and a point. The dual of a point x outside $\mathbf{H}^n$ is a hyperplane $x^\perp$ intersecting $\mathbf{H}^n$. (a) Lines from $x^\perp$ to x are perpendicular to $x^\perp$, and lines perpendicular to $x^\perp$ go through x (see text for detailed argument). (b) In $\mathbf{RP}^n$, the point x is the vertex of the cone tangent to S^{n-1}_∞ at the $(n-2)$-sphere where $x^\perp$ meets S^{n-1}_∞.

represents a point in $x^\perp$ and $v \in X^\perp$ is any tangent vector at p that represents a direction in $x^\perp$, we want to show that v is perpendicular to the tangent vector w that represents the direction from p to X.

But w lies in the plane determined by p and X, and, by exercise 2.3.2, is orthogonal to p; since p is orthogonal to X, this implies that w is in fact parallel to X, and consequently orthogonal to v.

Exercise 2.3.9. (a) Prove the assertion in the caption of figure 2.20(b).

(b) What is the dual of a point in $\mathbf{H}^n$? Of a point on S^{n-1}_∞?

(c) What is the dual of a k-plane?

(d) Show that the dual of a k-plane P is the intersection of the duals of points in P. Write down a dual statement.

In the two-dimensional case the picture is especially simple. Any two lines intersect somewhere in $\mathbf{RP}^2$. If the intersection is inside S^1_∞, the lines meet in the conventional sense, from the point of view of a hyperbolic observer. If the intersection is on S^1_∞, the lines converge together on the visual circle of the observer, and they are called *parallels*. Otherwise, they are called *ultraparallels*, and have a unique common perpendicular in $\mathbf{H}^2$, dual to their intersection point outside S^1_∞.

Exercise 2.3.10 (parallelism in hyperbolic space). Extend parallelism and ultraparallelism to k-planes in $\mathbf{H}^n$. Show that ultraparallel $(n-1)$-hyperplanes in $\mathbf{H}^n$ have a unique common perpendicular line. For a related statement about lines in $\mathbf{H}^3$, see proposition 2.5.3.

Problem 2.3.11 (projective transformations of hyperbolic space). A *projective transformation* is a self-map of $\mathbf{RP}^n$ obtained from an invertible linear map of $\mathbf{R}^{n+1}$ by passing to the quotient. An orthogonal transformation of $\mathbf{E}^{n,1}$ clearly gives rise to a projective transformation taking S^{n-1}_∞ to itself; show that the converse is also true.

This implies that any projective transformation of $\mathbf{RP}^n$ that leaves $\mathbf{H}^n$ invariant is an isometry, in contrast with the Euclidean situation, where there are many projective transformations that are not isometries: the affine transformations (compare problem 2.2.17).

Problem 2.3.12 (shapes of Euclidean polygons). The angles of a regular pentagon in plane Euclidean geometry are all 108°, but not all pentagons with 108° angles are regular. Consider the space of (not necessarily simple) pentagons having 108° angles and sides parallel to the corresponding sides of a model regular pentagon, and parametrize this space by the (signed) side lengths $s_1, \ldots, s_5$.

(a) Show that the the s_i are subject to a linear relation that confines them to a three-dimensional linear subspace V of $\mathbf{R}^5$.

(b) Show that the area enclosed by a pentagon is a quadratic form on V which is isometric to $\mathbf{E}^{2,1}$. How do you measure area for a non-simple pentagon?

(c) Describe a model for the hyperbolic plane in terms of the subset of V consisting of pentagons of unit area. Does your model have a single component?

(d) Show that the space of simple pentagons of unit area is a right-angled pentagon in the hyperbolic plane.

(e) There is an area-preserving "butterfly operation" on pentagons that replaces a side of length s by a side of length $-s$, changing the lengths of the two neighboring sides to make it fit. Interpret this operation in hyperbolic geometry.

(f) Given a non-simple pentagon in V, when is it possible to modify it by a sequence of butterfly moves until it is simple?

(g) Generalize to higher dimensions. Show that the space of simple normalized Euclidean polygons of $n+3$ sides, having unit area and edges parallel to and in the same direction as some convex model polygon, is parametrized by a convex polyhedron in hyperbolic n-space. Can you describe the three-dimensional hyperbolic polyhedron when the model polygon is a regular hexagon?

Problem 2.3.13 (the paraboloid model). To obtain other projectively correct models of hyperbolic space, one can transform the Klein model by any projective transformation.

(a) Write down a projective transformation that maps the unit sphere of $\mathbf{R}^n$ to the paraboloid $x_n = x_1^2 + \cdots + x_{n-1}^2$. Applying this projective transformation to the Klein model K, we obtain the *paraboloid model* of hyperbolic space.

(b) The paraboloid model is to the Klein model as the upper half-space is to the Poincaré model. More precisely, the upper half-space model singles out a point on the sphere at infinity of hyperbolic space, and the group of hyperbolic isometries preserving this point appears as the group of similarities preserving upper half-space. How does it appear in the paraboloid model?

(c) Show that orthogonal projection from the paraboloid $x_n = x_1^2 + \cdots + x_{n-1}^2$ to the hyperplane $x_n = 0$ in $\mathbf{E}^n$ induces an isomorphism between the group of affine maps preserving the paraboloid and the group of similarities preserving upper half-space.

Problem 2.3.14. We have seen that Lorentz transformations correspond to isometries of $\mathbf{H}^3$, but we did not give a physical interpretation to $\mathbf{H}^3$. Is there any way that people might actually see or experience $\mathbf{H}^3$ in the relativistic universe?

2.4. Some Computations in Hyperbolic Space

Ultimately, what we seek when we study mathematics is a qualitative understanding. But precise, quantitative manipulations—the nitty-gritty of mathematics—are also important as a way to reach this end, and as a test that our qualitative understanding is correct. The models of hyperbolic space that we developed over the last two sections provide precise representations of hyperbolic objects, but they're intrinsically limited by their lack of symmetry. Informal pictures, on the other hand, can be invaluable as a help to intuition, and with time and experience they become pretty clear and undistorted. However, they are by nature imprecise.

To fill in the gap between these two ways of understanding, it is convenient to know the formulas for measurement in hyperbolic space, so that one can deal with it from the point of view of a surveyor or a builder. Working with formulas tends to be slow and pedestrian, but at least they are precise and the pictures they evoke are undistorted. In this section, then, we develop trigonometric formulas and formulas for the area in the hyperbolic plane.

We work along the lines of Section 2.3, investigating the similarities and differences between Lorentz space $\mathbf{E}^{2,1}$ and Euclidean space $\mathbf{E}^3$, and between hyperbolic and spherical geometry. Spherical trigonometry is sometimes presented as an array of easily confused formulas, but these formulas are, in fact, equivalent to statements about dot products of unit vectors in three-space, and can be neatly derived from the formula for inversion of a 3×3 matrix.

We start with any triple (v_1, v_2, v_3) of unit vectors lying in $S^2 \subset \mathbf{E}^3$. If they are linearly independent, so that no great circle, or spherical line, contains all three, they determine a spherical triangle, formed by joining each pair v_i, v_j by a spherical line segment of length $d(v_i, v_j) = \theta_{ij} < \pi$. The *dual basis* to (v_1, v_2, v_3) is another triple (w_1, w_2, w_3) of vectors—but not necessarily unit vectors—in $\mathbf{E}^3$, defined by the conditions $v_i \cdot w_i = 1$ and $v_i \cdot w_j = 0$ if $i \neq j$, for $i, j = 1, 2, 3$. If we let V and W be the matrices with columns v_i and w_i, this can be expressed as $W^t V = I$.

Geometrically, w_i points in the direction of the normal vector of the plane spanned by v_j and v_k, where i, j and k are distinct; it follows that the angle ϕ_i of the spherical triangle $v_1 v_2 v_3$ at v_i is $\pi - \angle(w_j, w_k)$, since the angle between two planes is the supplement of the angle between their outward normal vectors (Figure 2.21).

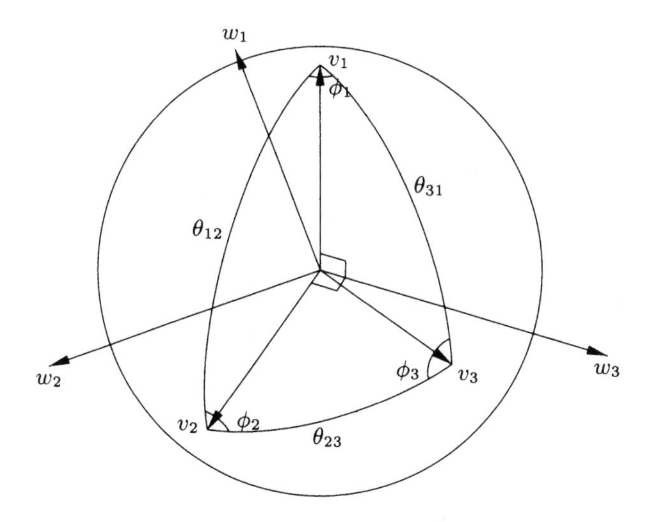

Figure 2.21. Proving the spherical law of cosines.

To relate all these angles, we consider the matrices V^tV and W^tW of inner products of the two bases, and notice that they are inverse to one another and that

$$V^tV = \begin{pmatrix} 1 & c_{12} & c_{13} \\ c_{12} & 1 & c_{23} \\ c_{13} & c_{23} & 1 \end{pmatrix},$$

where $c_{ij} = v_i \cdot v_j = \cos\theta_{ij}$. Thus $W^tW = (V^tV)^{-1}$ is a multiple of the matrix of cofactors of V^tV,

2.4.1.

$$W^tW = \frac{1}{\det(V^tV)} \begin{pmatrix} 1-c_{23}^2 & c_{13}c_{23}-c_{12} & c_{12}c_{23}-c_{13} \\ c_{13}c_{23}-c_{12} & 1-c_{13}^2 & c_{12}c_{13}-c_{23} \\ c_{12}c_{23}-c_{13} & c_{12}c_{13}-c_{23} & 1-c_{12}^2 \end{pmatrix}.$$

From this we can easily compute, say,

2.4.2.

$$\cos\phi_3 = -\cos\angle(w_1, w_2) = -\frac{w_1 \cdot w_2}{|w_1|\,|w_2|} = \frac{\cos\theta_{12} - \cos\theta_{13}\cos\theta_{23}}{\sin\theta_{13}\sin\theta_{23}},$$

or, in the more familiar notation where A, B, C stand for the angles at v_1, v_2, v_3 and a, b, c stand for the opposite sides,

2.4.3. $$\cos c = \cos a \cos b + \sin a \sin b \cos C.$$

This is called the *spherical law of cosines*. The *dual spherical law of cosines* is obtained by reversing the roles of (v_1, v_2, v_3) and

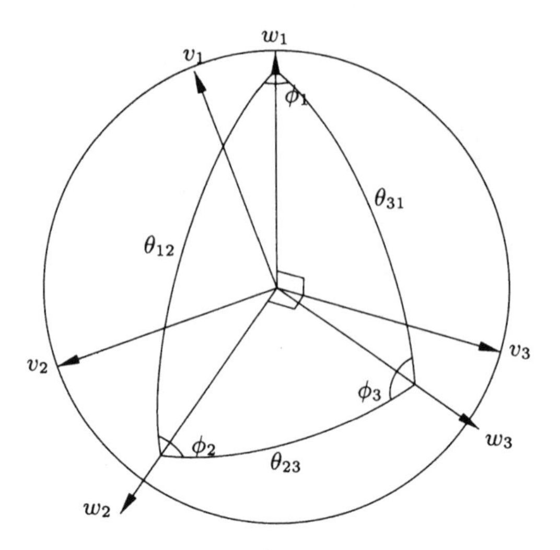

Figure 2.22. The dual spherical law of cosines.

(w_1, w_2, w_3) with respect to the triangle (Figure 2.22): we set the v_i not to the vertices of the triangle, but to unit vectors orthogonal to the planes containing the sides. Then $\phi_k = \pi - \angle(v_i, v_j)$ for i, j, k distinct, and $\theta_{ij} = \angle(w_i, w_j)$. We obtain, for example,

$$\cos\theta_{12} = \cos\angle(w_1, w_2) = \frac{w_1 \cdot w_2}{|w_1|\,|w_2|} = \frac{\cos\phi_2\cos\phi_1 + \cos\phi_3}{\sin\phi_2\sin\phi_1},$$

or

$$\cos C = -\cos A\cos B + \sin A\sin B\cos c.$$

Exercise 2.4.4. In the limiting case of a very small triangle, we should be able to recover formulas of Euclidean trigonometry. What do you get from the series expansion of the spherical law of cosines when a, b, c are very small? What do you get from the dual spherical law of cosines?

We now turn to Lorentz space $\mathbf{E}^{2,1}$. The situation is similar, but somewhat complicated by the fact that non-zero vectors can have real, zero or imaginary length. To cut down the number of cases, we consider only vectors of non-zero length. We may as well assume that they are *normalized* in the sense that they have length 1 or i and their x_0-coordinate is positive if they have length i. We recall from Section 2.3 that if a normalized vector $x \in \mathbf{E}^{2,1}$ has length i, it stands for a point on the hyperboloid model H^+ of the hyperbolic plane, just as a unit vector in $\mathbf{E}^3$ gives a point in S^2. If x has length

1, it lies outside the hyperbolic plane, and we denote by $x^\perp$ the trace in the hyperbolic plane of its dual line.

Suppose, then, that x and y are normalized and distinct. The quadratic form Q^-, restricted to the plane spanned by x and y, can have signature $(2,0)$, $(1,0)$, or $(1,1)$, corresponding to the cases where the plane intersects the hyperboloid, is tangent to the cone at infinity, or avoids both (Figure 2.23). In each case, we need to

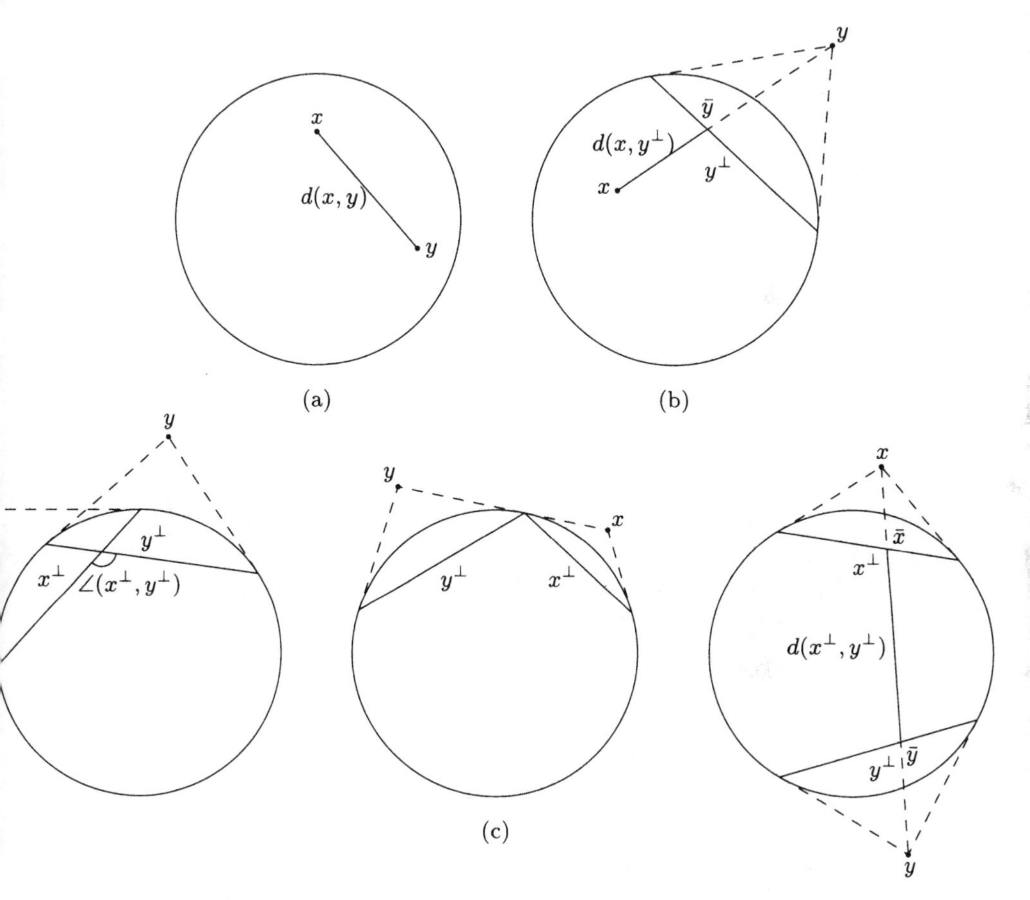

Figure 2.23. Interpretation of the inner product for various relative positions of two points. The labels correspond to the cases in Proposition 2.4.5; the figures are drawn in the projective model.

interpret the quantity $x \cdot y$, which previously gave the cosine of the angle between two vectors (here the inner product is the one associated with the form Q^-, of course). We may as well do it in arbitrary dimension:

Proposition 2.4.5 (interpretation of the inner product). *If x and y are normalized vectors of non-zero length in $\mathbf{E}^{n,1}$, either*

(a) *$x, y \in H^+$ have length i, and $x \cdot y = -\cosh d(x, y)$; or*

(b) *$x \in H^+$ has length i, y has length 1 and $x \cdot y = \pm \sinh d(x, y^\perp)$; or*

(c) *x and y have length 1, and the hyperplanes $x^\perp, y^\perp \subset \mathbf{H}^n$ are secant, parallel or ultraparallel depending on whether Q^- has signature $(2,0)$, $(1,0)$, or $(1,1)$ on the plane spanned by x and y. In the first case, $x \cdot y = \pm \cos \angle(x^\perp, y^\perp)$; in the second, $x \cdot y = \pm 1$; and in the third, $x \cdot y = \pm \cosh d(x^\perp, y^\perp)$.*

Proof of 2.4.5. Let P be the plane spanned by x and y. In cases (a) and (b), P intersects H^+ in a hyperbolic line $L \ni x$, and by exercise 2.3.5 this line is parametrized with velocity 1 by $x \cosh t + v \sinh t$, where v is a unit tangent vector to H^+ at x.

If $y \in H^+$, this implies that $y = x \cosh t + v \sinh t$ for $t = \pm d(x, y)$, depending on the way we chose v. Since x and v are orthogonal (exercise 2.3.2), we get

$$x \cdot y = x \cdot (x \cosh t + v \sinh t) = -\cosh t = -\cosh d(x, y).$$

If, on the other hand, $y \notin H^+$, exercise 2.4.6 shows that the distance $d(x, y^\perp)$ is achieved for the point $\bar{y} = L \cap y^\perp$, because L is the unique perpendicular from x to $y^\perp$. Thus $\bar{y} = x \cosh t + v \sinh t$ for $t = d(x, y^\perp)$, and y, being a linear combination of x and v orthogonal to $\bar{y}$, must be of the form $\pm(x \sinh t + v \cosh t)$ (recall that x and y are normalized). We conclude that

$$x \cdot y = \pm x \cdot (x \sinh t + v \cosh t) = \pm \sinh t = \pm \sinh d(x, y^\perp).$$

The third possibility in (c) is a variation on (a) and (b). Here $L = P \cap H^+$ contains neither x nor y, but we can parametrize it starting at $\bar{x} = L \cap x^\perp$. Then $x = \pm v$, $\bar{y} = L \cap y^\perp = \bar{x} \cosh t + v \sinh t$ for $t = d(x^\perp, y^\perp)$ (again by exercise 2.4.6), and $y = \pm(\bar{x} \sinh t + v \cosh t)$, so $x \cdot y = \pm \cosh d(x^\perp, y^\perp)$.

We're left with the first two possibilities in (c). If Q^- is positive definite on P, it is indefinite on the orthogonal complement $P^\perp$, so $P^\perp \cap H^+ = x^\perp \cap y^\perp$ is non-empty. Let p be a point in this intersection; to measure $\cos \angle(x^\perp, y^\perp)$ (which is only defined up to sign) it is enough to find tangent vectors to H^+ at p that are normal to $x^\perp$ and $y^\perp$, and take the cosine of their angle. But x and y themselves can serve as such tangent vectors, so $\cos \angle(x^\perp, y^\perp) = \pm x \cdot y$.

If Q^- is †positive semidefinite on P, it is also positive semidefinite on $P^\perp$, so $P^\perp \cap H^+ = x^\perp \cap y^\perp$ is empty, but $P^\perp \cap S^{n-1}_\infty$ consists of a single line through the origin. Thus $x^\perp$ and $y^\perp$ meet at infinity—they are parallel. The value of $x \cdot y$ follows from the fact that this case is a limit between the previous two. $\boxed{2.4.5}$

Exercise 2.4.6 (minimum distance implies perpendicularity). (a) If $L \subset \mathbf{H}^n$ is a line, $y \in \mathbf{H}^n$ is a point outside L and x is a point on L such that the distance $d(x, y)$ is minimal, the line xy is perpendicular to L.

(b) If $L, M \subset \mathbf{H}^n$ are non-intersecting lines and $x \in L$ and $y \in M$ are points on L and M such that the distance $d(x, y)$ is minimal, xy is perpendicular to L and M.

Exercise 2.4.7. (a) To resolve the ambiguities in signs in Proposition 2.4.5, we must assign an orientation to the dual hyperplane of a vector v of real length. This can be done by distinguishing between the two half-spaces determined by $v^\perp$ in $\mathbf{E}^{n,1}$, on the basis of which one contains v; this also distinguishes between the two half-spaces determined by $v^\perp$ in $\mathbf{H}^n$. This done, we can define $d(x, y^\perp)$ in part (b) as a signed quantity; how does the formula read then? What about the various cases in part (c)?

(b) Is there a sensible normalization for vectors of zero length? How would you interpret $v \cdot w$ if either or both vectors have zero length?

Now we can calculate the trigonometric formulas for a triangle in $\mathbf{H}^2$, or, more generally, the intersection with $\mathbf{H}^2$ of a triangle in $\mathbf{RP}^2$. As before, we let (v_1, v_2, v_3) be a basis of normalized vectors in $\mathbf{E}^{2,1}$, forming a matrix V, and we look at its dual basis (w_1, w_2, w_3), whose vectors form a matrix W. Here, V and W are no longer inverse to each other; instead, we can write $W^tSV = I$, where S is a symmetric matrix expressing the inner product associated with Q^- in the canonical basis—here the diagonal matrix with diagonal entries $(-1, 1, 1)$. However, the matrices of inner products, V^tSV and W^tSW, are still inverse to each other:

$$(V^tSV)(W^tSW) = (V^tSV)(V^{-1}W) = V^tSW = (W^tSV)^t = I.$$

Since some of the v_i may have imaginary length, V^tSV no longer has all ones in the diagonal; instead, it looks like this:

$$V^tSV = \begin{pmatrix} \varepsilon_1 & c_{12} & c_{13} \\ c_{12} & \varepsilon_2 & c_{23} \\ c_{13} & c_{23} & \varepsilon_3 \end{pmatrix},$$

where $\varepsilon_i = v_i \cdot v_i = \pm 1$. It follows, as before, that the matrix of inner products of the w_i is

2.4.8.

$$W^t S W = \frac{1}{\det V^t S V} \begin{pmatrix} \varepsilon_2\varepsilon_3 - c_{23}^2 & c_{13}c_{23} - \varepsilon_3 c_{12} & c_{12}c_{23} - \varepsilon_2 c_{13} \\ c_{13}c_{13} - \varepsilon_3 c_{12} & \varepsilon_1\varepsilon_3 - c_{13}^2 & c_{12}c_{13} - \varepsilon_1 c_{23} \\ c_{12}c_{23} - \varepsilon_2 c_{13} & c_{12}c_{13} - \varepsilon_1 c_{23} & \varepsilon_1\varepsilon_2 - c_{12}^2 \end{pmatrix}.$$

This can be used in the same way as Equation 2.4.1, but we have to be careful about signs, since normalizing a vector of imaginary length can require multiplication by a negative scalar.

Take the case when v_1, v_2, v_3 all have imaginary length, so they form a triangle with all three vertices in $\mathbf{H}^2$, as in Figure 2.24(a). By Proposition 2.4.5(a), the interpretation of $c_{ij} = v_i \cdot v_j$ is in terms of the distances d_{ij}, namely, $c_{ij} = -\cosh d_{ij}$. The vectors w_i have real length, and their duals $w_i^\perp$ represent the sides of the triangle; by Proposition 2.4.5(c) and exercise 2.4.7(a), $w_i \cdot w_j / (|w_i|\,|w_j|) = -\cos\phi_k$, where i, j, k are distinct and ϕ_k is the interior angle at v_k.

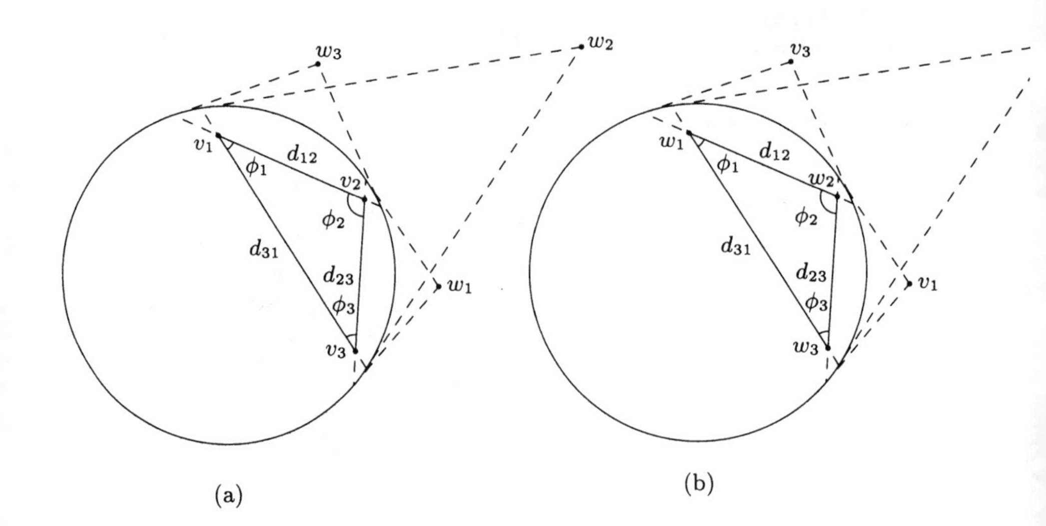

Figure 2.24. Two ways to see a triangle in the hyperbolic plane. We first arrange the basis of unit vectors (v_1, v_2, v_3) to match the triangle's vertices (a), and derive a formula relating side lengths. We then make (v_1, v_2, v_3) correspond to the duals of the edges (b), and obtain a formula relating angles.

Setting all the ε_i to -1 in Equation 2.4.8, we see that we must switch the sign of the matrix before normalizing, so the diagonal entries are positive. We then get

$$\cos\phi_3 = -\frac{w_1 \cdot w_2}{|w_1|\,|w_2|} = \frac{c_{13}c_{23} + c_{12}}{\sqrt{c_{13}^2 - 1}\sqrt{c_{23}^2 - 1}}$$
$$= \frac{\cosh d_{13}\cosh d_{23} - \cosh d_{12}}{\sinh d_{13}\sinh d_{23}}.$$

Letting A, B, C stand for the angles at v_1, v_2, v_3 and a, b, c for the opposite sides, we obtain the *hyperbolic law of cosines*

2.4.9. $$\cosh c = \cosh a \cosh b - \sinh a \sinh b \cos C.$$

To obtain the dual law, we start with the v_i outside hyperbolic space, dual to the sides of the triangle under consideration, as shown in Figure 2.24(b). Then $c_{ij} = -\cos\phi_k$ for i, j, k distinct, and $|(w_i \cdot w_j)/(|w_i|\,|w_j|)| = \cosh d_{ij}$. Thus we have

$$\cosh d_{12} = \left|\frac{w_1 \cdot w_2}{|w_1|\,|w_2|}\right| = \frac{c_{13}c_{23} + c_{12}}{\sqrt{1 - c_{13}^2}\sqrt{1 - c_{23}^2}} = \frac{\cos\phi_1\cos\phi_2 + \cos\phi_3}{\sin\phi_1\sin\phi_2},$$

or

$$\cos C = -\cos A \cos B + \sin A \sin B \cosh c.$$

The formulas for a right triangle are worth mentioning separately, since they are particularly simple. Switching A and C (as well as a and c) in the previous formula and setting $C = \pi/2$ we get

$$\cosh a = \frac{\cos A}{\sin B} \qquad \text{if } C = \pi/2$$

(note that $\cos A = \sin B$ in a Euclidean right triangle). From Equation 2.4.9 we obtain the *hyperbolic Pythagorean theorem*:

$$\cosh c = \cosh a \cosh b \qquad \text{if } C = \pi/2.$$

From the formula for $\cosh b$ analogous to Equation 2.4.9, by making the substitutions $\cosh c = \cosh a \cosh b$ and $\cos B = \cosh b \sin A$ and using the identity $\cosh^2 a = 1 + \sinh^2 a$, we get

$$\sin A = \frac{\sinh a}{\sinh c} \qquad \text{if } C = \pi/2.$$

Thus given any triangle, the altitude h corresponding to side c satisfies $\sin B = \sinh h/\sinh a$ and also $\sin A = \sinh h/\sinh b$. This proves the *hyperbolic law of sines*, valid for any hyperbolic triangle:

$$\frac{\sinh a}{\sin A} = \frac{\sinh b}{\sin B} = \frac{\sinh c}{\sin C}.$$

So far we've applied Equation 2.4.8 to triangles entirely inside hyperbolic space and to their duals. Mixed cases can also be interesting; for example, Figure 2.25(a) shows how a pentagon having five right angles can be thought of as a right triangle with two vertices outside the circle at infinity; they are represented in the pentagon by their duals, which form two non-adjacent sides. By using Proposition 2.4.5 and keeping track of signs you can convince yourself that the hyperbolic Pythagorean theorem acquires the form

$$\sinh a \sinh b = \cosh d.$$

There follows a *hexagonal law of sines* for all-right hexagons, since

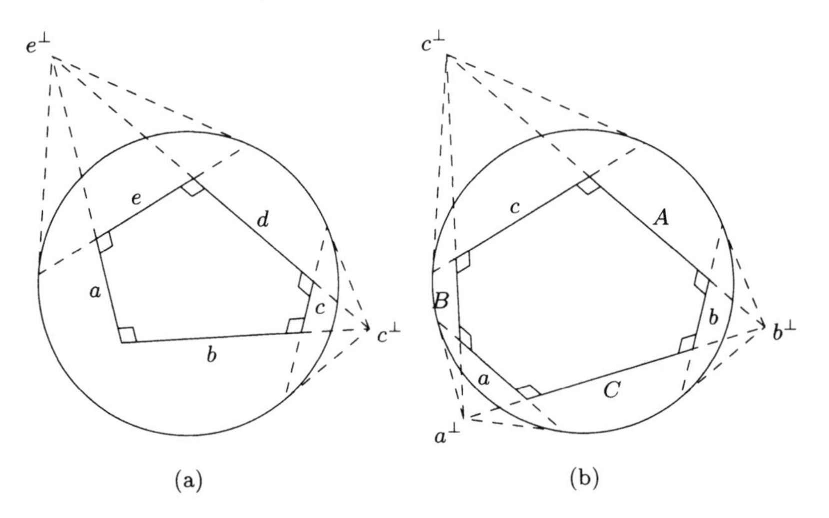

Figure 2.25. Trigonometry in an all-right pentagon and in an all-right hexagon.

we can draw a line perpendicular to two opposite sides to obtain two all-right pentagons:

2.4.10.
$$\frac{\sinh a}{\sinh A} = \frac{\sinh b}{\sinh B} = \frac{\sinh c}{\sinh C},$$

in the notation of Figure 2.25(b). Right-angled hexagons can also be seen as triangles with all three vertices outside infinity, as shown in the same figure.

Exercise 2.4.11 (existence and uniqueness of right hexagons). Given three numbers $A, B, C > 0$, we show that there exists an essentially unique right-angled hexagon H in $\mathbf{H}^2$ with alternate side lengths A, B, C. This will be useful in Section 4.6, when we study the space of hyperbolic structures on surfaces.

(a) Suppose we have H. As in Figure 2.25(b), let $a^\perp \subset \mathbf{E}^{2,1}$ be the dual of the side of H opposite the side of length A, and likewise for $b^\perp$ and $c^\perp$. Show that we can choose $x_a \in a^\perp$, $x_b \in b^\perp$ and $x_c \in c^\perp$ so that the matrix of inner products of x_a, x_b, x_c has 1's along the diagonal and negative entries off the diagonal. What is the absolute value of the off-diagonal entries?

(b) Any other right-angled hexagon with alternate side lenghts A, B, C is isometric to H. (Hint: Show that the linear transformation taking x_a, x_b, x_c to the corresponding points for the new hexagon is an element of $O(2, 1)$.)

(c) Conversely, given only A, B, C, we can find x_a, x_b, x_c with the desired inner products, and therefore H. (Hint: Diagonalize the matrix of inner products.)

For other proofs, see Problem 2.6.9 and Figure 4.15.

We are accustomed to the notion that choice of scale in Euclidean space is arbitrary: it does not essentially matter whether we measure in feet or meters. Figures can be scaled up or down arbitrarily.

The same is not true in hyperbolic and spherical geometry. If you double the sides of a hyperbolic or spherical triangle, its angles, given by equations 2.4.9 and 2.4.3, are no longer the same. There is no hyperbolic or spherical analogue for similarity transformations (see problem 2.3.11).

This strong dependency between size and angles can be seen even more clearly in terms of area. For a nice open region in the Euclidean plane—say one with a connected, piecewise smooth boundary—the total amount of curvature of the boundary is always 2π, and in particular the sum of the (signed) exterior angles of a polygon is 2π. For a region of the hyperbolic plane, the total curvature increases with the area; this is a special case of the Gauss–Bonnet theorem, a very general and profound result of differential geometry. We won't state the Gauss–Bonnet theorem in any more generality here; instead we present an elegant method, also due to Gauss, to calculate the area of a hyperbolic triangle using only elementary means. We start with an *ideal triangle*, one whose three "vertices" are at infinity. Although unbounded, ideal triangles have finite area:

Proposition 2.4.12 (ideal triangles). *All ideal triangles are congruent, and have area* π.

Proof of 2.4.12. Using the upper half-plane model of $\mathbf{H}^2$, it is easy to see that any ideal triangle can be transformed by isometries so as to match a model triangle with vertices ∞, $(-1,0)$ and $(1,0)$ (Figure 2.26).

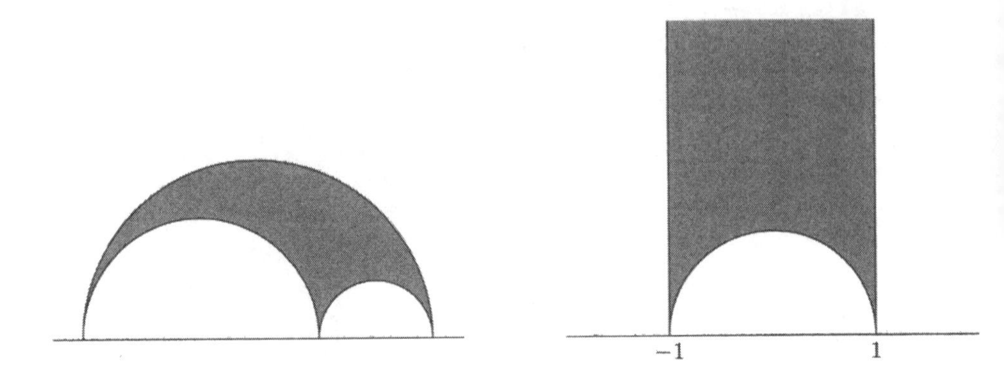

Figure 2.26. All ideal triangles are congruent. Given any ideal triangle, we can send one of its vertices to ∞ by inversion, then apply a Euclidean similarity to send the remaining two vertices to $(-1,0)$ and $(1,0)$.

Now let the coordinates of the upper half-plane be x and y, with the x-axis as the boundary. The model triangle is the region given by $-1 \le x \le 1$ and $y \ge \sqrt{1-x^2}$, with hyperbolic area element $(1/y^2)\,dx\,dy$ (Equation 2.2.9). Thus the area is

$$\int_{-1}^{1}\int_{\sqrt{1-x^2}}^{\infty}\frac{1}{y^2}\,dy\,dx = \int_{-1}^{1}\frac{1}{\sqrt{1-x^2}}\,dx = \int_{-\pi/2}^{\pi/2}\frac{1}{\cos\theta}\cos\theta\,d\theta = \pi.$$

2.4.12

Proposition 2.4.13 (area of hyperbolic triangles). *The area of a hyperbolic triangle is π minus the sum of the interior angles (the angle being zero for a vertex at infinity).*

Proof of 2.4.13. When all angles are zero we have an ideal triangle. We next look at $\frac{2}{3}$-ideal triangles, those with two vertices at infinity. Let $A(\theta)$ denote the area of such a triangle with angle $\pi-\theta$ at the finite vertex. This is well-defined because all $\frac{2}{3}$-ideal triangles with the same angle at the finite vertex are congruent—the reasoning is similar to that for ideal triangles.

Gauss's key observation is that A is an additive function, that is, $A(\theta_1+\theta_2) = A(\theta_1)+A(\theta_2)$, for $\theta_1, \theta_2, \theta_1+\theta_2 \in (0,\pi)$. The

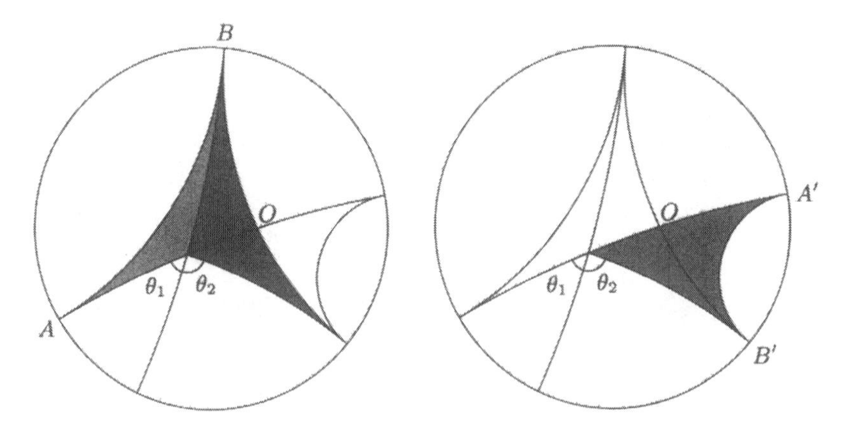

Figure 2.27. Area of $\frac{2}{3}$-ideal triangles. By definition, the areas of the shaded triangles on the left are $A(\theta_1)$ and $A(\theta_2)$. Likewise, the area of the shaded triangle on the right is $A(\theta_1+\theta_2)$. But the shaded areas in the two figures coincide, because the triangles OAB and $OA'B'$ are congruent by a reflection through O. Therefore $A(\theta)$ is an additive function of θ; this is used to compute the area of $\frac{2}{3}$-ideal triangles.

proof of this follows from Figure 2.27. It follows that A is a **Q**-linear function from $(0,\pi)$ to **R**. It is also continuous, so it must be **R**-linear. But $A(\pi)$ is the area of an ideal triangle, which is π by Proposition 2.4.12; it follows that $A(\theta)=\theta$, and the area of a $\frac{2}{3}$-ideal triangle is the complement of the angle at the finite vertex.

A triangle with two or three finite vertices can be expressed as the difference between an ideal triangle and two or three $\frac{2}{3}$-ideal ones, as shown in Figure 2.28. You should check the details. 2.4.13

Exercise 2.4.14 (spherical area). Derive the formula for the area of a spherical triangle by an analogous procedure, starting with 4π as the area of the sphere.

Corollary 2.4.15 (area of hyperbolic polygons). *The sum S of the interior angles of a planar hyperbolic polygon is always less than the sum of the angles of a Euclidean polygon with the same number n of sides, and the deficiency $(n-2)\pi - S$ is the area of the polygon.*

Proof of 2.4.15. Subdivide the polygon into triangles, as in the Euclidean case. 2.4.15

Exercise 2.4.16. (a) What is the area of a surface of genus two made from the regular octagon of Figure 1.12(b)?

(b) (Harder.) Show that *any* surface of genus two of constant curvature -1 has the same area.

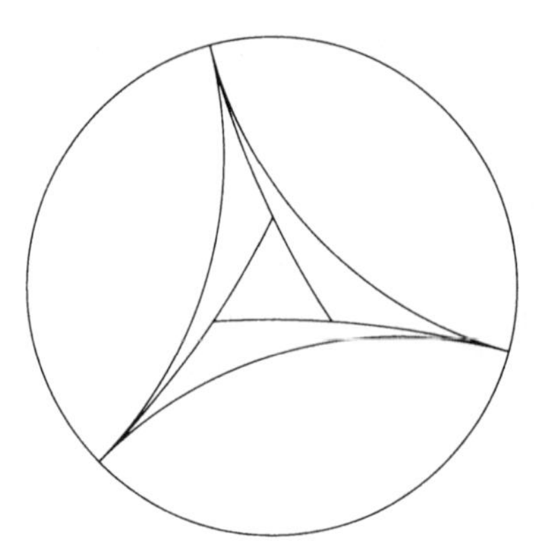

Figure 2.28. Area of general hyperbolic triangles. If you subtract a finite hyperbolic triangle from a suitable ideal triangle, you get three $\frac{2}{3}$-ideal triangles. Adding up angles and areas gives Proposition 2.4.13.

Exercise 2.4.17. What is a good definition for the area of a non-simple polygon? What is the formula for the area of a non-simple hyperbolic polygon? Compare Problem 2.3.12.

2.5. Hyperbolic Isometries

We now turn to the qualitative study of the isometries of hyperbolic space, which, as we saw in Problem 2.2.14, form a large group. We could use linear algebra for this purpose, but we'll instead use direct geometric constructions, the better to develop our hyperbolic intuition. The algebraic approach is taken in Problem 2.5.24.

We start in three dimensions. Let $g : \mathbf{H}^3 \to \mathbf{H}^3$ be an orientation-preserving isometry other than the identity. An *axis* of g is any line L that is invariant under g and on which g acts as a (possibly trivial) translation.

Proposition 2.5.1 (axis is unique). *A non-trivial orientation-preserving isometry of* $\mathbf{H}^3$ *can have at most one axis.*

Proof of 2.5.1. Suppose that L and M are distinct axes for an orientation-preserving isometry g. If g fixes both L and M pointwise, take a point x on M but not on L. Then g fixes the plane containing L and x, because it fixes three non-collinear points on it. Since g preserves orientation, it is the identity.

If, on the other hand, L is translated by g, we have $d(x, M) = d(g(x), M)$ for $x \in L$, so the function $d(x, M)$ is periodic and therefore bounded. But two distinct lines cannot remain a bounded distance from each other in both directions, since that would imply they have the same two endpoints on the sphere at infinity. 2.5.1

Exercise 2.5.2. Find a non-trivial orientation-preserving isometry of $\mathbf{H}^3$ that leaves invariant more than one line.

Any orientation-preserving isometry of $\mathbf{E}^3$ is either a translation, a rotation about some axis, or a *screw motion*, that is, a rotation followed by a translation along the axis of rotation. (Exercise 2.5.6 asks you to prove this.) The situation in $\mathbf{H}^3$ is somewhat richer, and has its own special terminology.

If a non-trivial orientation-preserving isometry g of $\mathbf{H}^3$ has an axis that is fixed pointwise, it is called an *elliptic* isometry, or a *rotation* about its axis. In this case the orbit of a point p off the axis—the set of points $g^k(p)$, for $k \in \mathbf{Z}$—lies on a circle around the axis.

If g has an axis that is translated by a non-trivial amount, it is called *hyperbolic*. There are two possibilities here: the orbit of a point off the axis may lie on a plane, always on the same side of the axis; it is in fact contained in an equidistant curve, like the one shown in Figure 1.10. In this case we say that g is a *translation*. Alternatively, the orbit can be the vertices of a polygonal helix centered around the axis; in this case g is a screw motion, as can be seen by applying a compensatory translation. (An alternate usage of the word "hyperbolic" specializes it to what we're calling translations, in which case "loxodromic" designates a screw motion. This distinction is not very useful, and we'll not adopt it.)

By the proposition above, no transformation can be at the same time elliptic and hyperbolic. But there are isometries that are neither elliptic nor hyperbolic: they are called *parabolic*. For instance, any isometry of $\mathbf{H}^3$ that appears as a Euclidean translation parallel to the bounding plane in the upper half-space model is parabolic.

The proof of the next proposition describes a geometric construction to locate the axis of a non-trivial orientation-preserving isometry of $\mathbf{H}^3$, if it has one. We will need an elementary fact about pairs of lines; recall (cf. Exercise 2.3.10) that two lines in $\mathbf{H}^3$ are called *parallel* if they have a common endpoint on S^2_∞, or, equivalently, if the distance between them approaches zero at one end or the other.

Lemma 2.5.3 (common perpendicular for lines in $\mathbf{H}^3$). *Two distinct lines in $\mathbf{H}^3$ are either parallel, or they have a unique common perpendicular.*

Proof of 2.5.3. Let the lines be X and Y, and consider the distance function $d(x, y)$ between points $x \in X$ and $y \in Y$. If the lines are not parallel, this function goes to ∞ as either x or y or both go to ∞; therefore it has a minimum, attained at points x_0 and y_0.

If the minimum is zero, that is, if the lines cross, any common perpendicular must go through the intersection point, otherwise we'd have a triangle with two right angles, which is impossible by Proposition 2.4.13. As there is a unique line orthogonal to the plane spanned by X and Y and passing through their intersection point, the lemma is proved in this case. (Notice that this part is false in dimension greater than three.)

If the minimum distance is not zero, the line between x_0 and y_0 is a perpendicular by Exercise 2.4.6. If there were another common perpendicular, we'd obtain a quadrilateral in space with all right angles (although conceivably its sides could cross). Subdividing the quadrilateral by a diagonal, we'd get two plane triangles, whose angles add to at least 2π; this is again impossible. $\boxed{2.5.3}$

If L is a line in $\mathbf{H}^3$, we denote by r_L the reflection in L, which is the rotation of π about L.

Proposition 2.5.4 (finding the axis). *Any non-trivial orientation-preserving isometry g of $\mathbf{H}^3$ can be written in the form $g = r_L \circ r_M$, where the lines L and M are parallel, secant or neither depending on whether g is parabolic, elliptic or hyperbolic. The axis of g is the common perpendicular of L and M, if it exists.*

Proof of 2.5.4. Take any point p such that $g(p) \neq p$. If $g^2(p) = p$, the midpoint q of the line segment $\overline{p\,g(p)}$ is fixed by g, and the plane through q perpendicular to the line $p\,g(p)$ is invariant. Since g reverses the orientation of this plane, it must act on it as a reflection, fixing a line K. Therefore, $g = r_K$ is elliptic of order two. In this case, one can take L and M to be two orthogonal lines, both orthogonal to K at a point $x \in K$.

If p, $g(p)$ and $g^2(p) \neq p$ are collinear, p is fortuitously on the axis of g, and g is hyperbolic. We can replace p by some other point not on this axis, and reduce to the next case.

In the remaining case, we let M be the bisector of the angle $p\,g(p)g^2(p)$, so that r_M fixes $g(p)$ and interchanges p with $g^2(p)$. To define L, we look at the dihedron with edge $g(p)g^2(p)$ whose sides contain p and $g^3(p)$, respectively, and take L as the line that bisects this dihedron and is also a perpendicular bisector of the segment $\overline{p\,g(p)}$, as shown in Figure 2.29. (The dihedral angle along $p\,g(p)$

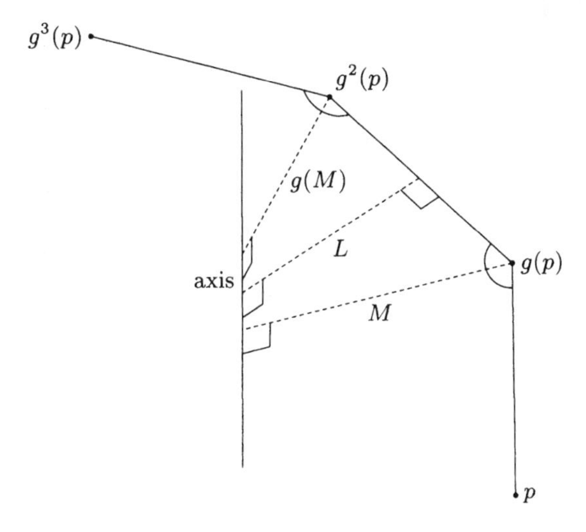

Figure 2.29. The axis of a three-dimensional isometry. The axis of an isometry g of $\mathbf{H}^3$ may be constructed, in the generic case, by connecting the orbit of a point p in a polygonal path.

may be 0 or π, but this doesn't cause problems.) By symmetry, r_L interchanges $g(p)$ with $g^2(p)$, and also p with $g^3(p)$. Therefore, $r_L \circ r_M$ sends p to $g(p)$, $g(p)$ to $g^2(p)$ and $g^2(p)$ to $g^3(p)$. Since $r_L \circ r_M$ and g agree at three non-collinear points, they agree on the whole plane containing these three points. Therefore they agree everywhere, since they both preserve orientation.

The common perpendicular of L and M, if it exists, is the axis of g, because it is invariant under $r_L \circ r_M$, which acts on it as a translation. The sorting into cases now follows from Lemma 2.5.3 and from the definitions of parabolic, elliptic and hyperbolic transformations. 2.5.4

Exercise 2.5.5. Using the decomposition of Proposition 2.5.4, show that any parabolic orientation-preserving isometry of $\mathbf{H}^3$ is conjugate to a Euclidean translation of the upper half-space model.

Exercise 2.5.6 (isometries of $\mathbf{E}^3$). Prove that any orientation-preserving isometry of $\mathbf{E}^3$ is either a translation, a rotation or a screw motion. Give a geometric construction for the axis of the transformation, in the latter two cases.

The proof of Proposition 2.5.4 also applies to any two-dimensional isometry, whether it preserves or reverses orientation, since such an isometry can be extended to a three-dimensional orientation-preserving isometry. But three-dimensional orientation-reversing isometries, or arbitrary isometries in higher dimension, can have fixed-point sets other than lines: for example, a reflection in a plane of any dimension, or a compound rotation about a plane of any even codimension. Any transformation with a fixed point in hyperbolic space is still called *elliptic*. An isometry that translates an axis is *hyperbolic*, and as before the axis is unique. Any other non-trivial transformation is *parabolic*.

Problem 2.5.7 (orientation-reversing isometries of $\mathbf{H}^3$). Classify orientation-reversing isometries of $\mathbf{H}^3$. One approach is to modify the construction above, and express an orientation-reversing isometry as the composition of reflection in a plane with reflection in a line. Another approach is to exploit the fact that the square of an orientation-reversing isometry is orientation-preserving. Which isometries of $\mathbf{H}^3$ have orientation-reversing square roots? What are the square roots of the identity? What are the square roots of other isometries that have them?

To study isometries in arbitrary dimension, we continue with the geometric point of view and develop a method, based on the convexity of the distance function, that will be important later, because it can be adapted to other negatively curved metrics on $\mathbf{R}^n$. For another, algebraic, method, see Problem 2.5.24.

A *convex function* on a Riemannian manifold is a real-valued function f such that, for every geodesic γ, parametrized at a constant speed, the induced function $f \circ \gamma$ is convex. In other words, for every $t \in (0,1)$,

$$f \circ \gamma(t) \leq t f \circ \gamma(0) + (1-t) f \circ \gamma(1).$$

If the inequality is strict for all non-constant γ, we say that f is *strictly convex*.

The *product* of two Riemannian manifolds is the product manifold, with the Riemannian metric that is the sum of the metrics on the factors. It is easy to show that a curve in the product manifold is a geodesic (parametrized at constant speed) if and only if each projection in a factor is also a geodesic (parametrized at constant speed). In particular, a geodesic in $\mathbf{H}^n \times \mathbf{H}^n$ is a curve each of whose projections is a hyperbolic line or a point.

Theorem 2.5.8 (distance function is convex). *The distance function $d(x,y)$, considered as a map $d : \mathbf{H}^n \times \mathbf{H}^n \to \mathbf{R}$, is convex.*

The composition $d \circ \gamma$ is strictly convex for any geodesic γ in $\mathbf{H}^n \times \mathbf{H}^n$ whose projections to the two factors are distinct lines.

Proof of 2.5.8. We can assume $n \geq 3$, since $\mathbf{H}^n$ is isometrically embedded in $\mathbf{H}^{n+1}$. We also assume, for now, that the projections X and Y of γ are lines that don't lie on the same plane—in particular, they don't meet, even at infinity.

We parametrize X and Y by arc length, and use x and y to refer to points on X or Y as well as their parameters. Given $x \in X$ and $y \in Y$, we let $\xi(x, y)$ be the angle between the segment $\overline{xy}$ and the positive ray determined on X by x; similarly, $\eta(x, y)$ will be the angle between $\overline{yx}$ and the positive ray determined on Y by y (Figure 2.30).

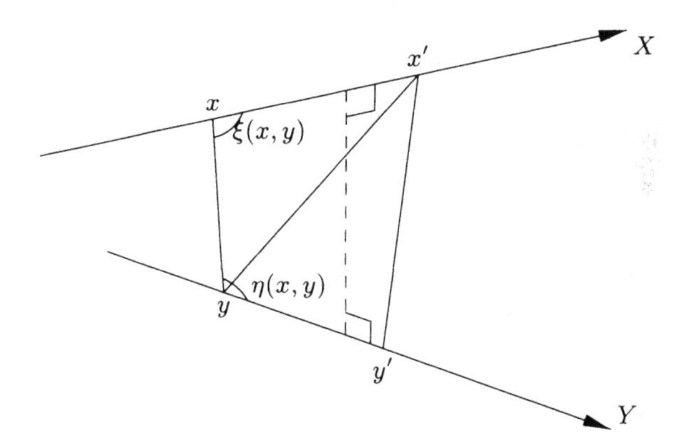

Figure 2.30. Derivative of distance is monotone. Positions along the lines X and Y uniquely determine the angles between the connecting segment and the lines. This can be used to show that the distance function between two lines is convex.

Applying the area formula (Proposition 2.4.13) to a triangle with two vertices on X and one on Y, we see that ξ increases monotonically with x when y is fixed. The map $(\xi, \eta) : \mathbf{R}^2 \to (0, \pi) \times (0, \pi)$ is clearly differentiable; we show that it is one-to-one. We do this by looking at $|\xi(x', y') - \xi(x, y)| + |\eta(x', y') - \eta(x, y)|$, where (x, y) and (x', y') are arbitrary pairs of points on X and Y. We can assume that $x' \geq x$. If we also have $y' \geq y$, as in the figure, we can write

$$\begin{aligned}
&|\xi(x', y') - \xi(x, y)| + |\eta(x', y') - \eta(x, y)| \\
&\qquad \geq \xi(x', y') - \xi(x, y) + \eta(x', y') - \eta(x, y) \\
&\qquad = (\xi(x', y') - \xi(x', y)) + (\xi(x', y) - \xi(x, y)) \\
&\qquad\qquad + (\eta(x', y') - \eta(x', y)) + (\eta(x', y) - \eta(x, y)).
\end{aligned}$$

Now

2.5.9. $$\xi(x', y') - \xi(x', y) > -\angle yx'y'$$

unless $y = y'$. This is just the triangle inequality for spherical triangles, and the inequality is strict because X and Y are not coplanar. We also have $\eta(x', y') - \eta(x', y) = \angle yx'y' + \text{area}\,\Delta yx'y'$, by the area formula (Proposition 2.4.13). Using similar relations for the other differences, we get

2.5.10.
$$|\xi(x', y') - \xi(x, y)| + |\eta(x', y') - \eta(x, y)| > \text{area}\,\Delta xyx' + \text{area}\,\Delta yx'y'$$

unless $x = x'$ and $y = y'$.

The case $y' \leq y$ is handled the same way, starting with the inequality $|\xi(x', y') - \xi(x, y)| + |\eta(x', y') - \eta(x, y)| \geq \xi(x', y') - \xi(x, y) - (\eta(x', y') - \eta(x, y))$. We conclude that $|\xi(x', y') - \xi(x, y)| + |\eta(x', y') - \eta(x, y)|$ is always positive if $(x, y) \neq (x', y')$, so (ξ, η) is a one-to-one map.

On the other hand, Exercise 2.5.13 shows that the function d has gradient $\nabla d = (d_x, d_y) = (-\cos\xi, -\cos\eta)$; it follows that ∇d is also a one-to-one, differentiable map from $\mathbf{R}^2$ to $(-1, 1) \times (-1, 1)$.

Now the derivative of ∇d is

$$H = \begin{pmatrix} d_{xx} & d_{xy} \\ d_{yx} & d_{yy} \end{pmatrix},$$

the hessian of the distance function. A smooth function is convex if and only if its second derivative along any line of the domain is everywhere non-negative, and is strictly convex if the second derivative is everywhere positive. So we have almost arrived at our desired destination. The only thing we still need to do is to show that H is positive definite.

For this we need an infinitesimal version of the argument above. It is tempting to carry this out using the terms in Equation 2.5.10 involving the area. However, we avoid doing this, in order to have a valid argument when we are not working in hyperbolic space, but in any space of non-positive curvature. For example, in Euclidean space the area terms would not be present.

Let y_v, y'_v and X_v be the points on the visual sphere at x' that correspond to y, y' and the positive endpoint of X. If d_v denotes distance on the visual sphere, we have $d_v(X_v, y_v) = \xi(x', y)$, $d_v(X_v, y'_v) = \xi(x', y')$ and $d_v(y_v, y'_v) = \angle yx'y'$, and Equation 2.5.9

follows from the triangle inequality, as already observed. But, in fact, we can write a stronger inequality, using Exercise 2.5.13(c) and the assumption that y_v, y'_v and X_v are not collinear on the sphere. Namely, there exists a positive number ε, depending only on x and y, such that

2.5.11. $$\xi(x', y') - \xi(x', y) + \angle yx'y' \geq \varepsilon |y - y'|.$$

Combining this with a similar equation for η, we get the following counterpart for Equation 2.5.10:

2.5.12. $$|\xi(x', y') - \xi(x, y)| + |\eta(x', y') - \eta(x, y)| > \varepsilon(|x - x'| + |y - y'|)$$

unless $x = x'$ and $y = y'$.

It follows that the ratio of the image area to the domain area under the differentiable homeomorphism (ξ, η) is at least ε^2. Therefore the hessian of the distance function has a determinant that is always strictly positive or always strictly negative. The distance function is a †proper map from $\mathbf{R}^2$ to $[0, \infty)$, and therefore attains its minimum. At the minimum the hessian cannot have negative determinant, so the determinant is everywhere positive. Since the diagonal entries of H are positive, H is positive definite. This completes the proof that if X and Y are not coplanar the distance function is strictly convex.

Now convexity is a closed condition, so it still holds even when the projections X and Y of γ are coplanar, or even when one or both reduce to a point, because all these cases are limits of the case above. But for strict convexity the limiting argument does not work, and indeed the distance is not strictly convex if X or Y is a point, or if they are the same line. So we must find other means to prove strict convexity when X and Y are distinct lines.

At this point we must give up any hope of a proof that works for any space of non-positive curvature—think of parallel lines in $\mathbf{E}^n$. Instead, we're free to use Equation 2.5.10, which still holds with $\geq$ instead of $>$. This inequality implies that Equation 2.5.12 holds, so long as $x \notin Y$ and $y \notin X$. Using the gradient ∇d as above we conclude that $d : X \times Y \to \mathbf{R}$ is strictly convex except possibly along two lines of the form $x = x_0$ and $y = y_0$. In particular, $d \circ \gamma : \mathbf{R} \to \mathbf{R}$ is strictly convex except at a maximum of two points; but a convex function that is strictly convex in the complement of a finite set of points is strictly convex everywhere. [2.5.8]

Exercise 2.5.13 (derivative of distance). (a) Let x_0 be a fixed point in hyperbolic space. Show that the derivative at x of the distance function

$x \mapsto d(x, x_0)$ is the unit vector at x pointing along the geodesic from x_0 to x away from x_0.

Let $\alpha(t)$ be a differentiable curve in hyperbolic space, parametrized by arc length, and suppose that $x_0 \neq \alpha(t_0)$. Prove that the derivative of the distance $d(x_0, \alpha(t))$ with respect to t at t_0 is $\cos\theta$, where θ is the angle between the geodesic from x_0 to $\alpha(t_0)$ and $\alpha'(t_0)$.

(b) Show that this is also true in Euclidean space.

(c) Show that it is true in the sphere, provided that x and x_0 are not antipodal.

(d) (For those who know some Riemannian geometry.) Show that the same result holds in any Riemannian manifold, provided that x does not lie in the cut locus of x_0.

One standard starting point for a study of spaces of negative curvature is to assume the following property as an alternative definition of negative curvature [Bus55, Chapter 5]:

Definition 2.5.14 (Busemann's definition of negative curvature). A metric space is said to have *negative curvature* if, for any triangle ABC, the distance between the midpoints of AB and AC is less than half the distance between B and C.

This definition applies in many situations, including the study of Finsler manifolds, where other definitions from Riemannian manifolds do not apply. Theorem 2.5.8 implies that hyperbolic space satisfies this property, as does any simply connected complete Riemannian manifolds of strictly negative sectional curvature.

Exercise 2.5.15. Assume that Busemann's definition of negative curvature is satisfied, and do not assume that the distance function is convex. Let AB and CD be geodesics, with midpoints X and Y. Prove that $d(X,Y) < \frac{1}{2}(d(A,C) + d(B,D))$, unless all the points lie on the same geodesic. (Hint: draw a diagonal.) What happens when all the points lie on the same geodesic?

Prove that Theorem 2.5.8 holds for any space satisfying Busemann's definition of negative curvature.

The *translation distance* of an isometry $g : \mathbf{H}^n \to \mathbf{H}^n$ is the function $d_g(x) = d(x, g(x))$. By applying Theorem 2.5.8 to the graph of g, which is a geodesic-preserving embedding of $\mathbf{H}^n$ in $\mathbf{H}^n \times \mathbf{H}^n$, we get:

Corollary 2.5.16 (translation distance is convex). *For any isometry g of $\mathbf{H}^n$, the translation distance d_g is a convex function on $\mathbf{H}^n$. It is strictly convex except along lines that map to themselves.*

Proposition 2.5.17 (classification of isometries of $\mathbf{H}^n$). *Let g be an isometry of $\mathbf{H}^n$.*

(a) *g is hyperbolic if and only if the infimum of d_g is positive. This infimum is attained along a line, which is the unique axis for g. The endpoints of the axis are the fixed points of g on S_∞^{n-1}.*

(b) *g is parabolic if and only if the infimum of d_g is not attained. This infimum is then zero; g fixes a unique point p on S_∞^{n-1}, and acts as a Euclidean isometry in the upper half-space model with p at ∞.*

(c) *g is elliptic if and only if d_g takes the value zero. The set $d_g^{-1}(0)$ is a hyperbolic subspace of dimension k, for $0 \le k \le n$.*

Proof of 2.5.17. If d_g attains a positive infimum at some point x, it also attains the infimum at $g(x)$. By convexity, d_g has the same value on the line segment joining x and $g(x)$, so by Corollary 2.5.16 the line through x and $g(x)$ is invariant. This line is translated along itself, so g is hyperbolic. The uniqueness of its axis follows just as in the second half of the proof of Proposition 2.5.1. To show that the endpoints of the axis are the only fixed points on S_∞^{n-1}, we assume that one endpoint is at infinity in the upper half-space model. Then g acts as a Euclidean similarity on the bounding hyperplane $S_\infty^{n-1} \setminus \{\infty\}$. As such it can have at most one fixed point, unless it is the identity.

If d_g does not attain an infimum, there is a sequence $\{x_i\}$ such that $d_g(x_i)$ tends toward the infimum. By compactness, we can assume that $\{x_i\}$ converges to a point $x \in S_\infty^{n-1}$, which must be fixed by g. We can take $x = \infty$ in the upper half-space projection, so g acts as a Euclidean similarity. If this similarity has no fixed point on $S_\infty^{n-1} \setminus \{\infty\}$, it is an isometry; therefore d_g goes to zero on any vertical ray, and $\inf d_g = 0$. Also, since g has no axis and no fixed point in $\mathbf{H}^n$, it is parabolic.

If, instead, g does fix a point on $S_\infty^{n-1} \setminus \{\infty\}$, it leaves invariant the vertical line L through that point. If P is a (hyperbolic) hyperplane orthogonal to L, the closed region F between P and $g(P)$ is a fundamental domain for g, that is, for any point $x \in \mathbf{H}^n$, there is some $k \in \mathbf{Z}$ such that $g^k(x) \in F$. In particular, any value of d_g is achieved inside F. Because d_g does not attain its infimum, the compactness argument of the preceding paragraph shows that g fixes a point in $\bar{F} \cap S_\infty^{n-1}$. But if g fixes three points on S_∞^{n-1}, it fixes a whole plane in $\mathbf{H}^n$, contradicting the assumption that $\inf d_g$ is not attained.

Finally, if d_g takes the value zero, g is by definition elliptic, and its zero-set is a k-dimensional subspace because the entire line joining any two fixed points is fixed. 2.5.17

Exercise 2.5.18 (conjugating fixed points). (a) If α and β are isometries of $\mathbf{H}^n$ such that α and $\beta^{-1}\alpha\beta$ have the same fixed points on S^{n-1}_∞, these points are also fixed by β.

(b) If two non-elliptic isometries of $\mathbf{H}^n$ commute, either they are both hyperbolic and have the same axis, or they are both parabolic and fix the same point at infinity.

The convexity of $d(x, y)$ as a function of one variable alone is enough to define a *hyperbolic mean*: a rule that gives, for any finite collection of points in $\mathbf{H}^n$ with associated weights, or masses, its *center of mass*. The center of mass should be preserved under hyperbolic isometries, it should depend continuously on the points and their weights, and if there is only one point it should be the point itself.

One of the characterizations of the mean in Euclidean space works here too: it associates with the collection $\{(p_i, m_i)\}$ of points p_i with mass m_i the point x that minimizes the function $\sum_i m_i d^2(p_i, x)$. Although $d(p_i, x)$ is not a strictly convex function of x, its square is. The sum is therefore strictly convex, and since it is unbounded on any line, there is a unique point where it attains its minimum.

Corollary 2.5.19 (finite hyperbolic group has a fixed point). *A finite group of isometries of* $\mathbf{H}^n$ *has at least one fixed point.*

Proof of 2.5.19. Let $x \in \mathbf{H}^n$ be any point. If F is a finite group of isometries of $\mathbf{H}^n$, the center of mass of the orbit of x, with each point equally weighted, is fixed by F. 2.5.19

Exercise 2.5.20 (compact group has fixed point). Any compact †topological group can be given a finite invariant *Haar measure*, that is, a finite measure that is invariant under the action of the group on itself by left or right translations [MZ55]. Use this fact to show that the action of any compact group G on $\mathbf{H}^n$ has a fixed point. (Hint: average the images of a point $x \in \mathbf{H}^n$, using Haar measure on G.)

Exercise 2.5.21 (the hyperbolic median). A *median* for a weighted collection of points in $\mathbf{E}^1$ is any point that minimizes the weighted sum of the distances to the points. The set of medians is either a single point, or an interval.

This definition generalizes word for word to $\mathbf{E}^n$ and to $\mathbf{H}^n$. Analyze the existence and uniqueness of the Euclidean and hyperbolic median when $n = 2$ (a typical case), and describe its qualitative properties. Give a geometric characterization of the median of three points.

Problem 2.5.22 (other hyperbolic means). There is a very simple definition of a hyperbolic mean using the hyperboloid model: Given a collection $\{(p_i, m_i)\}$, treat the p_i as vectors in $H^+ \subset \mathbf{E}^{n,1}$, take their mean, and multiply by a scalar to put it back on the hyperboloid.

What is the relation between this mean and the square-of-distance mean? Is it the same? Is it expressible as the minimum of a linear combination of convex functions of distance?

Problem 2.5.23. The Brouwer fixed-point theorem asserts that any continuous map $D^n \to D^n$ has a fixed point. Use this theorem, together with an understanding of isometries of S^{n-1} and $\mathbf{E}^{n-1}$, to classify the isometries of $\mathbf{H}^n$, at least in the cases $n = 2$ and $n = 3$.

Problem 2.5.24 (the algebraic study of isometries of $\mathbf{H}^n$). By Exercise 2.3.4, isometries of $\mathbf{H}^n$ are in one-to-one correspondence with linear maps of $\mathbf{R}^{n+1}$ that leave invariant each component of the set $\{Q^- = -1\}$, where Q^- is a quadratic form of type $(n, 1)$. We can use this correspondence to obtain information on the properties of hyperbolic isometries.

(a) As a warm-up exercise, let V be a two-dimensional real vector space, Q a (possibly degenerate) quadratic form on V, and $A : V \to V$ a linear map preserving Q. Describe the relationship between Q and the eigenvectors and eigenvalues of A.

(b) Let V be an $(n+1)$-dimensional real vector space and Q a quadratic form of type $(n, 1)$ on V. Show that there is an isomorphism from V into $\mathbf{R}^{n+1}$ that transforms Q into the form $Q^- = -x_0^2 + x_1^2 + \cdots + x_n^2$. This means that we can just study $\mathbf{E}^{n,1}$.

(c) From now on, assume that A is a linear transformation of $\mathbf{E}^{n,1}$ preserving Q^-. If W is a minimal invariant subspace of A, show that W has dimension one or two. (This is true for any transformation of any non-trivial vector space.) If the dimension is two, Q^- is positive definite on W.

(d) Factor the characteristic polynomial p of A into irreducible quadratic and linear factors over $\mathbf{R}$. For any irreducible quadratic factor q of p, the subspace annihilated by $q(A)$ must be positive definite, so the roots of q are on the unit circle.

(e) If p has any roots that are not on the unit circle, there are precisely two such roots, λ and λ^{-1}, and the isometry of hyperbolic space induced by A is hyperbolic. (We sometimes say that A itself is hyperbolic.)

(f) If A fixes some vector v with $Q^-(v) < 0$, its induced isometry is elliptic; it fixes some totally geodesic subspace $P \subset \mathbf{H}^n$ isometric to $\mathbf{H}^k$, for some $0 \le k \le n$, and "rotates" the normal space of P.

(g) If all characteristic roots are on the unit circle and A fixes no vectors v with $Q^-(v) < 0$, A fixes a non-zero v with $Q^-(v) = 0$. All such fixed vectors are multiples of one another. The isometry induced by A is parabolic. The hyperplane $P = \{w : v \cdot w = 1\}$, where $\cdot$ is the inner product associated with Q^-, is invariant by A. It intersects the region $Q^- < 0$ in the projective paraboloid model of Problem 2.3.13. The quadratic form Q^- is degenerate on the tangent space of P, and it induces a Euclidean metric on the quotient space $P/\mathbf{R}v$. The transformation induced by A on $P/\mathbf{R}v$ is a Euclidean isometry without fixed points.

Exercise 2.5.25 (parabolic adjustment). The Möbius transformation induced on the sphere at infinity S^n by an elliptic, parabolic or hyperbolic transformation of $\mathbf{H}^{n+1}$ is also called elliptic, parabolic or hyperbolic. Show that, given three distinct points a, b and c on S^n, there is a unique parabolic transformation g fixing a and taking b to c.

2.6. Complex Coordinates for Hyperbolic Three-Space

Three-dimensional hyperbolic geometry is intimately associated with complex numbers, and many geometric relations in $\mathbf{H}^3$ can be elegantly described through this association.

The complex plane $\mathbf{C}$ embeds naturally in the complex projective line $\mathbf{CP}^1$, the set of complex lines (one-dimensional complex subspaces) of $\mathbf{C}^2$. The embedding maps a point $z \in \mathbf{C}$ to the complex line spanned by $(z, 1)$, seen as a point in $\mathbf{CP}^1$; we call z the *inhomogeneous coordinate* for this point, while any pair $(tz, t) \in \mathbf{C}^2$, with $t \in \mathbf{C}^* = \mathbf{C} \setminus \{0\}$, is called a set of *homogeneous coordinates* for it. The remaining point in $\mathbf{CP}^1$, namely the subspace spanned by $(1, 0)$, is the *point at infinity*; we can make ∞ its "inhomogeneous coordinate".

Topologically, $\mathbf{CP}^1$ is the one-point compactification $\widehat{\mathbf{C}}$ of $\mathbf{C}$ (cf. Problem 1.1.1), so we can extend the usual identification of $\mathbf{E}^2$ with $\mathbf{C}$ to ∞. This shows that $\mathbf{CP}^1$ is a topological sphere, called the *Riemann sphere*.

As in the real case (Problem 2.3.11), a *projective transformation* of $\mathbf{CP}^1$ is what you get from an invertible linear map of $\mathbf{C}^2$ by passing to the quotient. Projective transformations are homeomorphisms of $\mathbf{CP}^1$. If a projective transformation A comes from a linear map with matrix $\left(\begin{smallmatrix} a & b \\ c & d \end{smallmatrix}\right)$, its expression in inhomogeneous coordinates is

2.6.1.
$$A(z) = \frac{az + b}{cz + d}$$

(naturally, this should be interpreted as giving a/c for $z = \infty$ and ∞ for $z = -d/c$). A map $A : \mathbf{CP}^1 \to \mathbf{CP}^1$ of the form 2.6.1 (with $ad - bc \neq 0$) is called a *linear fractional transformation* (or *fractional linear transformation*). Linear fractional transformations behave in a familiar way:

Exercise 2.6.2 (linear fractional transformations are Möbius transformations). (a) Show that, under the usual identification $\mathbf{CP}^1 = S^2$, any linear fractional transformation is a Möbius transformation, that is, a composition of inversions. (Hint: look at $z \mapsto 1/z$ first.)

(b) Show that any *orientation-preserving* Möbius transformation of S^2 is a linear fractional transformation.

Sometimes Möbius transformations of $\widehat{\mathbf{C}}$ are considered to be just the linear fractional transformations. To avoid confusion, we won't use this convention.

Problem 2.6.3. It follows from Proposition 1.2.3 and Exercise 2.6.2 that linear fractional transformations map circles (including lines) into circles. This can also be proved directly, by applying such a transformation to the general equation of a circle.

Is there a conceptual way to explain why circles go to circles? We'll take up this question again in Problem 2.6.5.

Two non-singular linear maps of $\mathbf{C}^2$ give the same projective transformation of $\mathbf{CP}^1$ if and only if one is a scalar multiple of the other. Thus, identifying together scalar multiples in the linear group $\mathrm{GL}(2, \mathbf{C})$ gives the group of projective transformations of $\mathbf{CP}^1$, which we naturally denote by $\mathrm{PGL}(2, \mathbf{C}) = \mathrm{GL}(2, \mathbf{C})/\mathbf{C}^*$. This group is also known as $\mathrm{PSL}(2, \mathbf{C})$, because it can be obtained by identifying together scalar multiples in the special linear group $\mathrm{SL}(2, \mathbf{C})$, consisting of linear transformations of $\mathbf{C}^2$ with unit determinant.

As we saw on page 62, a Möbius transformation of S^{n-1}_∞ can be extended to a unique isometry of $\mathbf{H}^n$. Since $\mathrm{PGL}(2, \mathbf{C})$ acts on S^2_∞ by Möbius transformations (Exercise 2.6.2), this action can be extended to all of $\mathbf{H}^3$, providing the first link between hyperbolic geometry and the complex numbers:

Theorem 2.6.4. *The group of orientation-preserving isometries of* $\mathbf{H}^3$ *is* $\mathrm{PGL}(2, \mathbf{C})$, *identified via the action on* $S^2_\infty = \mathbf{CP}^1$.

This was first proven by Poincaré, who followed essentially the reasoning above.

There is another, more intrinsic way to obtain the action of $\mathrm{PGL}(2,\mathbf{C})$ on $\mathbf{H}^3$. Consider the real vector space V of Hermitian forms on $\mathbf{C}^2$, that is, of maps $H : \mathbf{C}^2 \times \mathbf{C}^2 \to \mathbf{C}$ linear in the second variable and satisfying

$$H(w,v) = \overline{H(v,w)}$$

for $v, w \in \mathbf{C}^2$. In the canonical basis, a Hermitian form has matrix $\left(\begin{smallmatrix} r & z \\ \bar z & s \end{smallmatrix}\right)$ with $r, s \in \mathbf{R}$ and $z \in \mathbf{C}$, so V has dimension four.

The determinant of a form represented by $\left(\begin{smallmatrix} r & z \\ \bar z & s \end{smallmatrix}\right)$ is $rs - z\bar z$, so the function det is a quadratic form on V. The signature of det is $(1,3)$, because $\left(\begin{smallmatrix} 1 & 0 \\ 0 & 1 \end{smallmatrix}\right)$, $\left(\begin{smallmatrix} 1 & 0 \\ 0 & -1 \end{smallmatrix}\right)$, $\left(\begin{smallmatrix} 0 & 1 \\ 1 & 0 \end{smallmatrix}\right)$ and $\left(\begin{smallmatrix} 0 & -i \\ i & 0 \end{smallmatrix}\right)$, for example, form an orthonormal basis. Thus V, with the quadratic form $-\det$, is isomorphic to $\mathbf{E}^{3,1}$; by Section 2.3, this means that the set of definite Hermitian forms in $\mathbf{C}^2$, up to multiplication by (non-zero real) scalars, forms a model for $\mathbf{H}^3$!

The sphere at infinity S^2_∞ in this description consists of Hermitian forms of rank one, up to scalars. To each such form we associate its nullspace, which is a point in $\mathbf{CP}^1$; the nullspace determines the form up to multiplication by a scalar (check this), so we get a canonical identification of S^2_∞ with $\mathbf{CP}^1$.

Now define an action of $\mathrm{GL}(2,\mathbf{C})$ on V as follows: for any $A \in \mathrm{GL}(2,\mathbf{C})$ and $H \in V$, let $A(H)$ be such that $A(H)(v,w) = H(Av, Aw)$. The action of A preserves the form $-\det$ up to a scalar—the determinants of all elements of V get multiplied by $|\det A|^2$—so the induced projective transformation in $V/\mathbf{R}^*$ gives an isometry of $\mathbf{H}^3$, as discussed on page 66. This isometry remains the same when we multiply A by a scalar, so in effect we have an action of $\mathrm{PGL}(2,\mathbf{C})$ on $\mathbf{H}^3$ by isometries. We also have an action of $\mathrm{PGL}(2,\mathbf{C})$ on $S^2_\infty = \mathbf{CP}^1$.

Problem 2.6.5 (the action of $\mathrm{GL}(2,\mathbf{C})$ **at infinity).** (a) Show that the action of $A \in \mathrm{GL}(2,\mathbf{C})$ on $\mathbf{CP}^1$, in terms of the identification described above, is given (in homogeneous coordinates) by multiplication by A^{-1}. Are all linear fractional transformations obtained in this way?

(b) Use this to show that $\mathrm{PGL}(2,\mathbf{C})$ coincides with the group of orientation-preserving isometries of $\mathbf{H}^3$. (Hint: Orientation-preserving isometries of $\mathbf{H}^3$ act simply transitively on the set of triples of distinct points at infinity. The same is true for the action of projective transformations of $\mathbf{CP}^1$, as you should check by writing an explicit formula).

(c) If $H \in V$ represents a point of $V/\mathbf{R}^*$ outside the sphere at infinity, the signature of H is $(1,1)$, that is, H is indefinite. What is the zero-set of H in $\mathbf{C}^2$? Show that, passing to the quotient, the zero-set becomes a circle in $\mathbf{CP}^1$. Can every circle in $\mathbf{CP}^1$ be obtained in this way? Show that, under the identification of $\mathbf{CP}^1$ with the set of forms of rank one in V, all forms in the zero-set of H are orthogonal to H (for the inner product associated with $-\det$). Thus the zero-set of H is the intersection of the polar plane of H with the sphere at infinity.

(d) How do zero-sets transform under the action of $\mathrm{PGL}(2, \mathbf{C})$? (Compare part (a)). Can you answer Problem 2.6.3 now?

Exercise 2.6.6. Show in two ways that the group of orientation-preserving isometries of $\mathbf{H}^2$ is $\mathrm{PGL}(2, \mathbf{R}) = \mathrm{PSL}(2, \mathbf{R})$. (Hints: (1) Use the fact that $\mathrm{PGL}(2, \mathbf{R})$ is the subset of $\mathrm{PGL}(2, \mathbf{C})$ that preserves the real axis. (2) In the discussion above, replace Hermitian forms on $\mathbf{C}^2$ by quadratic forms on $\mathbf{R}^2$.)

Exercise 2.6.7 (classification by the trace). Classify orientation-preserving isometries of $\mathbf{H}^3$ into hyperbolic, elliptic and parabolic according to the quantity $\mathrm{tr}^2 A / \det A$, where $A \in \mathrm{GL}(2, C)$ represents the isometry. (Hint: you can assume that $\det A = 1$. By conjugation, you can also assume that A is in Jordan normal form, so it is either a diagonal matrix or $\left(\begin{smallmatrix} 1 & 1 \\ 0 & 1 \end{smallmatrix}\right)$.)

Show how to find the axis of a hyperbolic or elliptic isometry, and the angle of rotation of an elliptic isometry, in terms of a representative $A \in \mathrm{GL}(2, C)$. What is the trace of a rotation of order two?

Problem 2.6.8 (complex trigonometry). The set of oriented lines in $\mathbf{H}^3$ has the structure of a complex two-manifold, namely $(\mathbf{CP}^1)^2 \setminus \Delta$, where Δ is the diagonal $\{(z, z) : z \in \mathbf{CP}^1\}$. Derive formulas for the geometry of lines in $\mathbf{H}^3$, as follows:

(a) Let W be the vector space of linear transformations of $\mathbf{C}^2$ with trace 0. There is a one-to-one correspondence between unoriented lines in $\mathbf{H}^3$ and one-dimensional subspaces of W whose non-zero representatives are in $\mathrm{GL}(2, \mathbf{C})$. (Hint: see Exercise 2.6.7.)

(b) Consider the inner product associated with the quadratic form $-\det$ on W: it is given by $A \cdot B = \frac{1}{2} \mathrm{tr}\, AB$. Using Proposition 2.5.4 and Exercise 2.6.7, interpret the inner product $A \cdot B$ of two vectors of unit length in terms of the geometry of the pair of lines given by A and B (compare Proposition 2.4.5).

(c) Derive a formula for the relationship between a triple of lines in $\mathbf{H}^3$ and the dual triple of common orthogonals to pairs of these lines. This formula should generalize the squared form of many formulas from Section 2.4—for instance, the spherical formulas come from the situation that three lines intersect in a point in $\mathbf{H}^3$.

(d) Interpret $W \cap \mathrm{SL}(2, \mathbf{C})$ as the set of *directed* lines in $\mathbf{H}^3$. Refine parts (b) and (c) to take into account the extra information.

Problem 2.6.9 (trace relations). (a) Show that any two elements A and B of $\mathrm{SL}(2, \mathbf{C})$ satisfy

$$\operatorname{tr} AB + \operatorname{tr} AB^{-1} = \operatorname{tr} A \operatorname{tr} B.$$

(Hint: use the fact that A satisfies its characteristic polynomial.)

(b) Show that the trace of every element of the group generated by A and B is expressible as a polynomial in $\operatorname{tr} A$, $\operatorname{tr} B$, and $\operatorname{tr} AB$.

(c) Prove that for any complex numbers x, y, and z, such that the symmetric matrix

$$\begin{pmatrix} 1 & x & z \\ x & 1 & y \\ z & y & 1 \end{pmatrix}$$

is nonsingular, there exist A and B in $\mathrm{SL}(2, \mathbf{C})$ such that $\operatorname{tr} A = 2x$, $\operatorname{tr} B = 2y$ and $\operatorname{tr} AB = 2z$, and the subgroup of $\mathrm{SL}(2, \mathbf{C})$ generated by A and B is unique up to conjugacy.

(Hint: Interpret this matrix as the matrix of inner products for a triple of unit vectors in the space W of Problem 2.6.8, representing order-two rotations a, b and c. If the matrix is non-singular, the three vectors form a basis; use this to reconstruct the group G generated by a, b and c. Then set $A = bc$ and $B = ca$, and show that the group generated by A and B has index two in G, and can also be determined up to conjugacy.)

(d) Compare the results when this analysis is applied to two subgroups of $\mathrm{SL}(2, \mathbf{C})$ that have the same image in $\mathrm{PSL}(2, \mathbf{C})$.

(e) What interpretation can be given when the matrix of (c) is singular?

(f) Show that the trace of every element of the group generated by A, B and C is expressible as a polynomial in the traces of A, B, C, AB, BC, CA and ABC. In fact, the first six traces are almost enough to express everything: show, by expanding the trace of $(ABC)^2$, that the trace of ABC satisfies a quadratic equation in terms of the other six traces, the other root of which is the trace of ACB.

(g) Show how (c) can be applied to give another proof of the existence and uniqueness of a right hexagon with given alternating sides (Exercise 2.4.11).

2.7. The Geometry of the Three-Sphere

Just like the circle and the two-sphere, the three-sphere is very round. But there are some beautiful, classical aspects to its roundness that are not easy to guess from its lower-dimensional sisters.

The most direct way to define S^3 is as the unit sphere $\{x_1^2 + x_2^2 + x_3^2 + x_4^2 = 1\}$, but visualizing this set directly is hard for most people. Stereographic projection (page 57) gives a picture of S^3, as $\widehat{\mathbf{R}^3} = \mathbf{R}^3 \cup \{\infty\}$, that makes the sphere much more tangible, and preserves some aspects of its geometry: for example, the roundness of circles and spheres. But this picture suffers from a loss of symmetry: $\widehat{\mathbf{R}^3}$ is not as round as it should be, and objects the same size in S^3 are not the same size in $\widehat{\mathbf{R}^3}$. To overcome this drawback, it helps to practice imagining the rigid motions of S^3 as they appear in $\widehat{\mathbf{R}^3}$.

Identifying $\mathbf{R}^4$ with $\mathbf{C}^2$, we obtain new pictures. If the coordinates in $\mathbf{C}^2$ are z_1 and z_2, the equation of a unit sphere becomes $|z_1|^2 + |z_2|^2 = 1$. Each complex line (one-dimensional subspace) in $\mathbf{C}^2$ intersects S^3 in a great circle, called a *Hopf circle*. Since exactly one Hopf circle passes through each point of S^3, the family of Hopf circles fills up S^3, and the circles are in one-to-one correspondence with the complex lines of $\mathbf{C}^2$, that is, with the Riemann sphere $\mathbf{CP}^1 = \widehat{\mathbf{C}}$. Informally, the three-sphere is a two-sphere's worth of circles. Formally, we get a *fiber bundle* $p : S^3 \to S^2$, with fiber S^1. (Fiber bundles will be defined and discussed extensively in Section 3.6, but the basic idea is that for small neighborhoods in S^2, the part of S^3 that projects to that neighborhood has the structure of a product with S^1.) We call this structure the *Hopf fibration*—this name is traditional, although "fibration" has a specific meaning in homotopy theory that is much more general than a fiber bundle.

Figure 2.31 shows what the Hopf fibration looks like under stereographic projection. In this figure the vertical axis is the intersection of S^3 with the complex line $z_1 = 0$, and the horizontal circle is the intersection with $z_2 = 0$. The locus $\{|z_2| \leq a\}$, for any $0 < a < 1$, is a solid torus neighborhood of the $z_2 = 0$ circle. Its boundary $\{|z_2| = a\}$, a torus of revolution, is filled up by Hopf circles, each winding once around the z_1-circle and once around the z_2-circle. Any torus of revolution in $\mathbf{R}^3$ can be transformed into a torus of this form by a similarity. The fact that any torus of revolution has curves winding around in both directions that are geometric circles

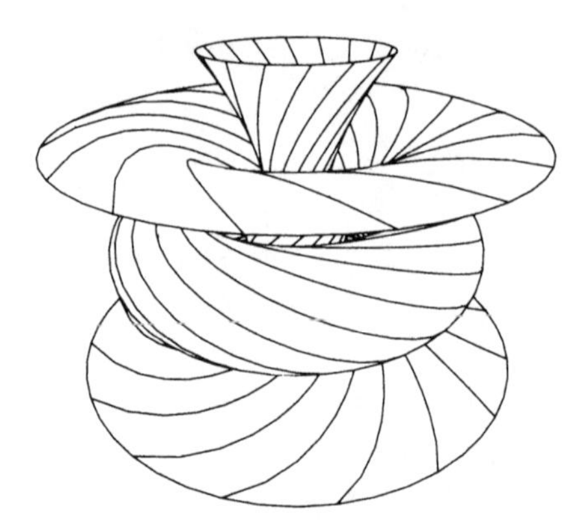

Figure 2.31. The Hopf fibration in S^3. The three-sphere $|z_1|^2+|z_2|^2 = 1$ fibers over the Riemann sphere: each fiber is a great circle, the locus of $z_1/z_2 =$ constant. Each torus $\{|z_2| = a\}$, for $0 < a < 1$—which can also be expressed as $\{|z_1| = a\}$—divides S^3 into two solid tori, one of them containing the point at infinity in the picture.

is quite surprising, so much so that such circles on the torus have a special name: *Villarceau circles*.

Exercise 2.7.1 (the Hopf fibration). (a) Show that the maps $g_t : S^3 \to S^3$ given by multiplication by e^{it}, for $t \in \mathbf{R}$, are isometries, and that they leave the Hopf fibration invariant. Thus S^3 has isometries that don't have an axis: the motion near any point is like the motion near any other point. This is one way in which S^3 seems "rounder" than S^2. The one-parameter family $\{g_t\}$ is called the Hopf †flow.

(b) Show that two Hopf circles C and C' are parallel in the sense that any two points in C' are at the same distance from C (in the metric of S^3).

(c) Show that the metric in $\mathbf{CP}^1 = S^2$ induced by the Hopf map is the standard, round metric. What are the images of great circles?

The three-sphere, like the circle, is a topological group. This is easiest to see using *quaternions*, which are, so to speak, an extension of complex numbers. We spend some time here investigating their properties, because they turn out to be useful in several ways.

The space $\mathcal{H}$ of quaternions is simply $\mathbf{R}^4$, together with a certain non-commutative multiplication $\mathcal{H} \times \mathcal{H} \to \mathcal{H}$. This multiplication is bilinear over $\mathbf{R}$, so in order to define it we can just specify its effect on a basis of $\mathcal{H}$. The basis is traditionally denoted $\{1, i, j, k\}$, and the action is as follows: 1 is the identity, and

$$i^2 = j^2 = k^2 = -1,$$

$$ij = k = -ji, \quad jk = i = -kj, \quad ki = j = -ik.$$

The subspace spanned by 1 is identified with $\mathbf{R}$, and its elements are called *real*; the subspace $\mathbf{R}i + \mathbf{R}j + \mathbf{R}k$ is the space of *pure quaternions*. It is easy to see that the quaternion product is associative, but not commutative; that $\mathbf{R}$ is the center of $\mathcal{H}$ (so the structure of $\mathcal{H}$ as a ring determines its structure as a real vector space), and that a quaternion is pure if and only if its square is a non-positive real number (so the ring structure also determines the decomposition into real and pure subspaces).

Exercise 2.7.2. Show that any ring automorphism of $\mathbf{H}$ is a linear map. This differs from the situation in $\mathbf{C}$, which has many automorphisms.

The *conjugate* of a quaternion $q = a+bi+cj+dk$ is $\bar{q} = a-bi-cj-dk$. The product $q\bar{q}$ is positive real, and its square root is the *absolute value*, or *norm*, of q, denoted by $|q|$. This coincides with the norm of q as a vector in $\mathbf{E}^4$. A quaternion of norm 1 is a *unit quaternion*. Any non-zero quaternion q has an inverse $q^{-1} = |q|^{-2}\bar{q}$; thus $\mathcal{H}$ is a *skew field*. If p and q are quaternions, we have $\overline{pq} = \bar{q}\bar{p}$ and $|pq| = |p|\,|q|$. Since the set of unit quaternions is the unit three-sphere $S^3 \subset \mathbf{E}^4$, we have proved the first statement in the following theorem:

Theorem 2.7.3 (the group structure of S^3). *The three-sphere has the structure of a non-commutative group, with center $\{\pm 1\}$. Left or right multiplication gives a self-action of S^3 by orientation-preserving isometries. Conjugation gives a self-action by isometries which, in addition, takes any two-sphere with center* 1 *onto itself. The quotient $S^3/\{\pm 1\}$ is isomorphic to the group* SO(3) *of orientation-preserving isometries of S^2.*

Proof of 2.7.3. The distance along S^3 between two points depends solely on the norm of the difference between them; since the norm is preserved by left or right multiplication by a unit quaternion, these map are isometries. They're orientation-preserving by continuity, because S^3 is connected.

Conjugation is also an isometry; since it fixes 1, it leaves invariant the sets of points at constant distance from 1, which are two-spheres.

The action of S^3 on these two-spheres by conjugation defines a homomorphism $\rho : S^3 \to \mathrm{SO}(3)$, whose kernel is the center of S^3. Surjectivity is shown in exercise 2.7.4. $\boxed{2.7.3}$

Exercise 2.7.4 (rotating S^2 with quaternions). Let S^2 be the sphere of pure quaternions of unit norm. Show that conjugation by a unit quaternion $r+p$, where r is real and p is pure, rotates S^2 around the axis in the direction of p by an angle $2\arctan(|p|/|r|)$. (Hint: assume first that $r=\cos\theta$, $p=i\sin\theta$. Extend to arbitrary p by showing that i can be taken to any point on S^2 by conjugation.)

The descriptions of S^3 via quaternions and via complex numbers can be combined. If we look at $\mathcal{H}$ as a complex vector space, multiplication on the left by the quaternion i is the same as multiplication by the complex number i, so the vector field $X_i(p)=ip$, for $p\in S^3$, induces the Hopf flow on S^3. There is added symmetry in this description, because i plays no special role among the quaternions: any pure quaternion of unit norm can be used in lieu of i to impart a complex vector space structure to $\mathbf{R}^4$ (compare exercise 2.7.4). Thus there are many Hopf flows and many Hopf foliations on S^3. In particular, we can take three mutually orthogonal vector fields X_i, X_j and X_k to get three mutually orthogonal families of Hopf circles.

Exercise 2.7.5. Using stereographic projection as in figure 2.31, with the identity element $1\in\mathcal{H}$ at the origin and i, j and k on the three coordinate axis, try to get an idea of what the three orthogonal Hopf foliations look like.

There is another way to describe the group structure on S^3, via *unitary transformations* of $\mathbf{C}^2$, that is, complex linear transformations that preserve the standard Hermitian form $(z_1,z_2)\cdot(w_1,w_2)=z_1\bar{w}_1+z_2\bar{w}_2$ on $\mathbf{C}^2$. Here again the case of the circle is analogous: the group $\mathrm{SO}(2)$ of orthogonal linear transformations of the Euclidean plane having determinant 1 acts simply transitively on S^1, so fixing a point $x\in S^1$ provides an identification $\mathrm{SO}(2)\cong S^1$ that takes each $g\in\mathrm{SO}(2)$ to $g(x)\in S^1$.

Similarly, the group $\mathrm{SU}(2)$ of unitary transformations of $\mathbf{C}^2$ of determinant 1 acts simply transitively on S^3, so by fixing a point $x\in S^3$ we get an identification $\mathrm{SU}(2)\cong S^3$. If $x=(1,0)\in S^3\subset\mathbf{C}^2$, any point $(z_1,z_2)\in S^3$ gets identified to

$$\begin{pmatrix} z_1 & -\bar{z}_2 \\ z_2 & \bar{z}_1 \end{pmatrix}.$$

Exercise 2.7.6. Consider $\mathcal{H}$ with the complex structure defined by multiplication on the left by a given unit quaternion. Show that multiplication on the right by any quaternion is a complex linear map. Show that multiplication on the right by a unit quaternion is a unitary map. This again provides an identification $S^3 \cong \mathrm{SU}(2)$, which makes S^3 the group that preserves, simultaneously, *all* the complex structures defined by multiplication on the left by unit quaternions.

The connection between the group structure on S^3 and the geometry of S^2, too, can be recovered from this point of view. Any element of $\mathrm{SU}(2)$ acting as an isometry of S^3 takes fibers to fibers, so its action passes to the quotient, giving a projective transformation of $\mathbf{CP}^1$ that is also an isometry of S^2. As a map of S^2, a projective transformation of $\mathbf{CP}^1$ is orientation-preserving, so we get a homomorphism $\mathrm{SU}(2) \to \mathrm{SO}(3)$. An element of $\mathrm{SU}(2)$ induces the identity map on $\mathbf{CP}^1$ if and only if it is a scalar multiple of the identity map on $\mathbf{C}^2$; since the only two such elements are $\pm\mathrm{Id}$, we get an injective homomorphism $\mathrm{PSU}(2) \to \mathrm{SO}(3)$, where $\mathrm{PSU}(2) = \mathrm{SU}(2)/\{\pm\mathrm{Id}\}$. You should check that this map is also surjective, and consequently an isomorphism.

Exercise 2.7.7. Topologically, $\mathrm{PSU}(2)$ is just $S^3/\{\pm 1\} = \mathbf{RP}^3$. Describe a direct correspondence between $\mathbf{RP}^3$ and $\mathrm{SO}(3)$, by realizing $\mathbf{RP}^3$ as the unit ball in $\mathbf{R}^3$ with antipodal points on its boundary identified. (Hint: for a point $r, \theta \in \mathbf{R}^3$, with $\theta \in S^2$ and $r \in [0,1]$, consider the rotation through an angle πr around the axis that contains θ.)

Another intriguing aspect of S^3 is the structure of its own group of rigid motions. The two actions of S^3 on itself, by left and right multiplication, commute, so we can define a homomorphism $\tau : S^3 \times S^3 \to \mathrm{SO}(4)$ by

$$\tau(g,h)(x) = gxh^{-1}.$$

By Theorem 2.7.3 some transformation $\tau(h,h)$ will get us from any orthonormal frame at 1 to any other such frame, so further composition with a transformation of the form $\tau(gh^{-1},1)$ will get us to any orthonormal frame at any point. This shows that τ is surjective, and also that the kernel of τ is $\{\pm(1,1)\}$, so we get an isomorphism

2.7.8. $$\mathrm{SO}(4) = (S^3 \times S^3)/C_2,$$

where C_2 is the cyclic group of order 2.

How can the group of isometries of a space so round as S^3 be almost a product? To understand this better, consider more carefully the nature of right and left multiplication.

If $g \in S^3$ is not ± 1, the transformation $x \mapsto gx$ fits into the unique Hopf flow generated by the Hopf field $x \mapsto px$, where p is the unit quaternion in the direction of the g purely imaginary component of g. Similarly, when $h \in S^3$ is not ± 1, the transformation $x \mapsto xh$ fits into a unique Hopf flow, generated by a vector field $x \mapsto xp$. The two kinds of Hopf flows are distinguished as left-handed or right-handed: the circles near a given circle wind around it in a left-handed sense or in a right-handed sense (as the threads of a common screw or jar lid). As we have seen algebraically, any right-handed Hopf flow commutes with any left-handed Hopf flow.

Problem 2.7.9. The commutation of right-randed and left-handed Hopf flows can also be seen geometrically.

(a) Describe the sensation (i.e., the motion and rate of turning) of a person in S^3 when being left-multiplied by a one-parameter subgroup of S^3.

(b) Let X and Y be right- and left-handed Hopf fields. Show that, by starting at any point in S^3 and moving in the direction orthogonal to both X and Y, you reach a point where $X = Y$. Consequently, there is an entire circle C^+ where $X = Y$. Similarly, there is a circle C^- where $X = -Y$, and that $d(C^+, C^-) = \pi$.

(c) Show that the locus of points that lie at a distance a from C^+ is a torus invariant under X and Y. This is also the locus of points at a distance $\pi - a$ from C^-. Show that X and Y act as rigid motions of these tori, and consequently commute. Compare figure 2.31.

(d) Interpret C^+ and C^- algebraically, in terms of the element of $\mathrm{SO}(4)$ associated with multiplication on the left by an element that generates X and on the right by an element that generates Y.

(e) Describe the sensation of a person in S^3 being acted upon by a general one-parameter subgroup of $\mathrm{SO}(4)$.

Problem 2.7.10. Give a geometric description of the homomorphism $\mathrm{SO}(4) \to \mathrm{SO}(3) \times \mathrm{SO}(3)$. (Hint: Consider the space of right-handed Hopf flows and the space of left-handed Hopf flows.)

Chapter 3

Geometric Manifolds

In chapter 1 we looked at a good number of manifolds. In doing this, we relied more on intuition and common sense than on definitions. It is now time to study manifolds a bit more systematically.

Manifolds come to us in nature and in mathematics by many different routes. Very frequently, they come naturally equipped with some special pattern or structure, and to understand the manifold we need to "see" the pattern. At other times, a manifold may come to us naked; by finding structures that fit it, we can gain new insight, relate it to other manifolds, and take better care of it.

The fact that there are all these different grades, or flavors, of manifolds was not clearly understood during the early development of topology. Different constructions were perceived more as alternative technical contexts for doing topology than as building blocks for essentially different structures. One of the remarkable achievements of topologists over the past forty years has been to come to grips with these distinctions, which are, contrary to intuition, substantive: for example, topological, piecewise linear and differentiable manifolds are inequivalent in dimensions four and higher, although in dimensions two and three the distinctions collapse.

Most of the myriad other possible structures—complex structures, foliations, hyperbolic structures, and so on—are considerably more restrictive than differentiable structures. These more restrictive structures can have great power in dimensions two and three.

3.1. Basic Definitions

A *manifold* is a topological space that is locally modeled on $\mathbf{R}^n$.

What it means to be locally modeled on $\mathbf{R}^n$ depends on what property, or pattern, of $\mathbf{R}^n$ we want to capture. The idea is to patch the manifold together seamlessly from small pieces of fabric having the given pattern. A pattern is described operationally, in terms of the transformations that preserve it; by allowing chunks of $\mathbf{R}^n$ to be glued together only according to these transformations, we get a manifold with the desired pattern. The set of allowed gluing maps should satisfy some natural properties:

Definition 3.1.1 (pseudogroup). A *pseudogroup* on a topological space X is a set $\mathcal{G}$ of homeomorphisms between open sets of X satisfying the following conditions:

(a) The domains of the elements $g \in \mathcal{G}$ cover X.

(b) The restriction of an element $g \in \mathcal{G}$ to any open set contained in its domain is also in $\mathcal{G}$.

(c) The composition $g_1 \circ g_2$ of two elements of $\mathcal{G}$, when defined, is in $\mathcal{G}$.

(d) The inverse of an element of $\mathcal{G}$ is in $\mathcal{G}$.

(e) The property of being in $\mathcal{G}$ is *local*, that is, if $g : U \to V$ is a homeomorphism between open sets of X and U is covered by open sets U_α such that each restriction $g|_{U_\alpha}$ is in $\mathcal{G}$, then $g \in \mathcal{G}$.

It follows from these conditions that every pseudogroup contains the identity map on any open set; the *trivial pseudogroup* is the one that contains these maps and no others. At the other extreme, the largest pseudogroup on $\mathbf{R}^n$ is the pseudogroup Top of all homeomorphisms between open subsets of $\mathbf{R}^n$. A *topological manifold* is one for which the gluing homeomorphisms are in Top—they need satisfy no further conditions. Such a space has the local topological structure of $\mathbf{R}^n$, but no more.

More generally, for any pseudogroup $\mathcal{G}$ on $\mathbf{R}^n$, a $\mathcal{G}$-*manifold* is a manifold for which the gluing maps lie in $\mathcal{G}$. Here is a more precise statement:

Definition 3.1.2 ($\mathcal{G}$-manifold). Let $\mathcal{G}$ be a pseudogroup on $\mathbf{R}^n$. An n-dimensional $\mathcal{G}$-*manifold* is a topological space M with a $\mathcal{G}$-*atlas* on it. A $\mathcal{G}$-atlas is a collection of $\mathcal{G}$-*compatible coordinate charts* whose domains cover M. A coordinate chart, or *local coordinate system*, is a pair (U_i, ϕ_i), where U_i is open in M and $\phi_i : U_i \to \mathbf{R}^n$ is a homeomorphism onto its image. Compatibility means that, whenever two charts (U_i, ϕ_i) and (U_j, ϕ_j) intersect, the *transition map* or *coordinate change*

$$\gamma_{ij} = \phi_i \circ \phi_j^{-1} : \phi_j(U_i \cap U_j) \to \phi_i(U_i \cap U_j)$$

is in $\mathcal{G}$.

Example 3.1.3 (differentiable manifolds). If $\mathcal{C}^r$, for $r \geq 1$, is the pseudogroup of C^r diffeomorphisms between open sets of $\mathbf{R}^n$ a $\mathcal{C}^r$-manifold is called a differentiable manifold (of class C^r), or C^r manifold. A $\mathcal{C}^r$-isomorphism is called a diffeomorphism. C^∞ manifolds are also called *smooth manifolds*.

Convention 3.1.4 (manifolds are Hausdorff and have countable basis). We make a standing assumption that manifolds are Hausdorff except where otherwise noted. Non-Hausdorff manifolds do sometimes arise naturally: see Example 3.5.5, for example. We also assume that manifolds have a countable basis of open sets. Then, for topological manifolds and many other types of manifolds where $\mathcal{G}$ is large enough, the same $\mathcal{G}$-structure on M can be described by a finite $\mathcal{G}$-atlas (where each chart's domain may have countably many connected components). See [Mun66, 20–21], for example.

Two $\mathcal{G}$-atlases on a topological space M define the same $\mathcal{G}$-structure if they are *compatible*, in the sense that their union is also a $\mathcal{G}$-atlas. Compatibility between atlases is an equivalence relation (cf. Exercise 3.1.8(b)). Mathematicians have a loathing for ambiguity and indefiniteness, so that often a manifold is defined by one of two polite fictions: an equivalence class of $\mathcal{G}$-atlases, or a maximal $\mathcal{G}$-atlas. Either object is so huge and complicated as to be unimaginable, in contrast with individual $\mathcal{G}$-atlases, which are generally not so hard to construct and deal with. But it is good to keep in mind that the choice of a particular atlas is not an essential part of the structure of a manifold.

Just as important as the class of manifolds with a $\mathcal{G}$-structure is the class of maps that preserve that structure. The simplest case is when two $\mathcal{G}$-manifolds are taken to one another by a homeomorphism that, when expressed in terms of local charts, is given by an element of $\mathcal{G}$ (on a small enough neighborhood of each point). Such a map is called a $\mathcal{G}$-*isomorphism* (also a $\mathcal{G}$-*automorphism* if the domain and image are the same).

More generally, if the map is a local homeomorphism (for example, a covering map) locally expressible as an element of $\mathcal{G}$, we call it a *local* $\mathcal{G}$*-isomorphism*, or a local homeomorphism preserving the $\mathcal{G}$-structure.

Exercise 3.1.5 (induced structure). Let $\pi : N \to M$ be a local homeomorphism from a topological space N into a $\mathcal{G}$-manifold M. Give N a natural $\mathcal{G}$-structure so that π preserves the $\mathcal{G}$-structure. (Compare Figure 1.4.)

We will encounter many examples where two pseudogroups of interest are contained in one another. For instance, every pseudogroup is contained in Top. If $\mathcal{H} \subset \mathcal{G}$, an $\mathcal{H}$-atlas is automatically also a $\mathcal{G}$-atlas, so an $\mathcal{H}$-manifold also has a $\mathcal{G}$-structure. The $\mathcal{G}$-structure is the $\mathcal{G}$*-relaxation* of the $\mathcal{H}$-structure; conversely, the $\mathcal{H}$-structure is an $\mathcal{H}$*-stiffening* of the $\mathcal{G}$-structure—the idea is that there is more flexibility in constructing $\mathcal{G}$-structures.

We will often want to analyze all the ways in which a given $\mathcal{G}$-structure can be stiffened into an $\mathcal{H}$-structure. Generally, however, we'll want to view as equivalent two stiffenings that differ in an uninteresting way, in the sense that the gap can be bridged by a $\mathcal{G}$-automorphism. More precisely, suppose that M is a $\mathcal{G}$-manifold and that $g : M \to M$ is a $\mathcal{G}$-automorphism of M. If N is an $\mathcal{H}$-stiffening of M, given by some atlas $\{(U_i, f_i)\}$, we can form another $\mathcal{H}$-stiffening gN of M by taking the atlas $\{(g^{-1}(U_i), f_i \circ g)\}$. Thus the group of $\mathcal{G}$-automorphisms of M acts on the set of $\mathcal{H}$-stiffenings of M. By identifying N with gN for all g and N, we get the set of $\mathcal{H}$-stiffenings of M *up to* $\mathcal{G}$*-automorphisms*.

Example 3.1.6 (smoothings). Let $\mathcal{G} = \mathcal{C}^r$, with $r < \infty$, and $\mathcal{H} = \mathcal{C}^\infty$. The question then is to study the $\mathcal{C}^\infty$-stiffenings, or *smoothings*, of a given C^r manifold M. It was shown by Whitney [Whi36] that any M has only one smoothing up to C^r diffeomorphism—informally, any differentiable manifold has exactly one smooth structure. It is for this reason that one can, in many situations, blur the distinction between smooth and differentiable.

The situation is not so simple for smoothings of topological manifolds ($\mathcal{G} = \mathsf{Top}$ and $\mathcal{H} = \mathcal{C}^\infty$). Milnor [Mil56] proved the surprising result that there are several inequivalent smoothings of S^7 considered as a topological manifold. After these *exotic spheres*, other manifolds having inequivalent differentiable structures, or no differentiable structure, were found.

In low dimensions, as we have mentioned, those distinctions collapse: every two- or three-dimensional topological manifold has a differentiable structure unique up to diffeomorphism. See Section 3.10 and [Mun60].

We will sometimes consider on the set of $\mathcal{H}$-stiffenings other equivalence relations, arising from the action of a proper subgroup of the group of $\mathcal{G}$-automorphisms and therefore finer than the one just described. Definition 4.6.4 (moduli space and Teichmüller space) gives an important example. But unless we say otherwise, whenever we discuss stiffenings we will have in mind the equivalence relation coming from the full group of $\mathcal{G}$-automorphisms.

We won't give a complete treatment of $\mathcal{G}$-structures and the variety of beautiful results and questions relating to them: in fact, we will soon focus on a small class of rigid pseudogroups. But here is a list of examples, most of which will play a part later in the book. For a more thorough discussion of pseudogroups, you can consult [Hae58].

Example 3.1.7 (real analytic manifolds). Let $\mathcal{C}^\omega$ be the pseudogroup of real analytic diffeomorphisms between open subsets of $\mathbf{R}^n$. A $\mathcal{C}^\omega$-manifold is called a *real analytic manifold*. Real analytic diffeomorphisms are uniquely determined by their restriction to any open set; this will be essential in the study of the developing map in Section 3.4. It is a deep theorem that every smooth manifold admits a unique real analytic stiffening [Whi36].

Differentiability and analyticity are local properties, so checking that $\mathcal{C}^r$ and $\mathcal{C}^\omega$ are pseudogroups is immediate. Many other interesting properties that one might think of for transitions maps are not local, however. Defining a pseudogroup based on such a property requires a bit of care to ensure that condition 3.1.1(e) is satisfied.

Exercise 3.1.8. (a) Given a set $\mathcal{G}_0$ of homeomorphisms between open subsets of X, show that there is a unique minimal pseudogroup $\mathcal{G}$ on X that contains $\mathcal{G}_0$. We say that $\mathcal{G}$ is *generated* by $\mathcal{G}_0$. (Typically, we will start with a $\mathcal{G}_0$ that satisfies all conditions in Definition 3.1.1 except for locality.)

(b) Why is condition 3.1.1(e) necessary? (Hint: let $\mathcal{G}_0$ consist of translations of $\mathbf{R}$ restricted to open subsets of $\mathbf{R}$. Show that compatibility between $\mathcal{G}_0$-atlases on S^1 is not an equivalence relation.)

Example 3.1.9 (foliations). Write $\mathbf{R}^n$ as the product $\mathbf{R}^{n-k} \times \mathbf{R}^k$ and let $\mathcal{G}$ be the pseudogroup generated by diffeomorphisms ϕ (between open subsets of $\mathbf{R}^n$) that take horizontal factors to horizontal factors, that is, diffeomorphisms that have the form

$$\phi(x,y) = (\phi_1(x,y), \phi_2(y)),$$

for $x \in \mathbf{R}^{n-k}$ and $y \in \mathbf{R}^k$. The pseudogroup $\mathcal{G}$ consists of all diffeomorphisms between open sets of $\mathbf{R}^n$ whose Jacobian at every point is an $n \times n$ matrix such that the lower left $(n-k) \times k$ block is 0. A $\mathcal{G}$-structure is called a *foliation* of codimension k (or dimension $n-k$). To visualize a foliation one should think not of local coordinate charts, but of the inverse images of factors $\mathbf{R}^{n-k} \times \{y\}$ under the charts, which piece together globally to give the *leaves* of the foliation.

One-dimensional foliations exist on a great many manifolds: any nowhere vanishing vector field has an associated foliation, obtained by following the flow lines. One example is the Hopf fibration of S^3 illustrated in Figure 2.31. In this case all the leaves close upon themselves, but in general the leaves may spiral around the manifold in a complicated way:

Exercise 3.1.10 (irrational foliation on the torus). For a torus $T^2 = \mathbf{R}^2/\mathbf{Z}^2$, consider the constant vector field $X(x,y) = (1,\alpha)$. The flow lines of X are the images of the straight lines $y = \alpha x + y_0$ on T^2, under the projection map. Show that these leaves are topological circles if α is rational, and dense subsets of T^2 if α is irrational. What value of α gives the circles of Figure 2.31?

The torus has many other foliations. For example, the quotient of $\mathbf{R}^2 \setminus \{0\}$ by a homothety centered at the origin is a topological torus. Since the foliation of $\mathbf{R}^2 \setminus \{0\}$ by horizontal lines is preserved by homothety, we get a foliation on the torus by passing to the quotient: see Figure 3.1.

Exercise 3.1.11 (the Reeb foliation). Apply the process of figure 3.1 in one higher dimension, that is, starting from a foliation by horizontal planes in $\mathbf{R}^3 \setminus \{0\}$. What is the quotient manifold? Describe the quotient foliation qualitatively; show that one of its leaves is a torus. Cutting the manifold open along this torus gives two solid tori, which can be glued back to form the three-sphere, as shown in Figure 2.31. How smooth is the resulting foliation of S^3?

The discovery of this foliation [Ree52] was something of a surprise; before then, no foliation of S^3 by surfaces was known. Later it became clear that codimension-one foliations are very common: Lickorish [Lic65] and Novikov [Nov65] showed that they exist for every

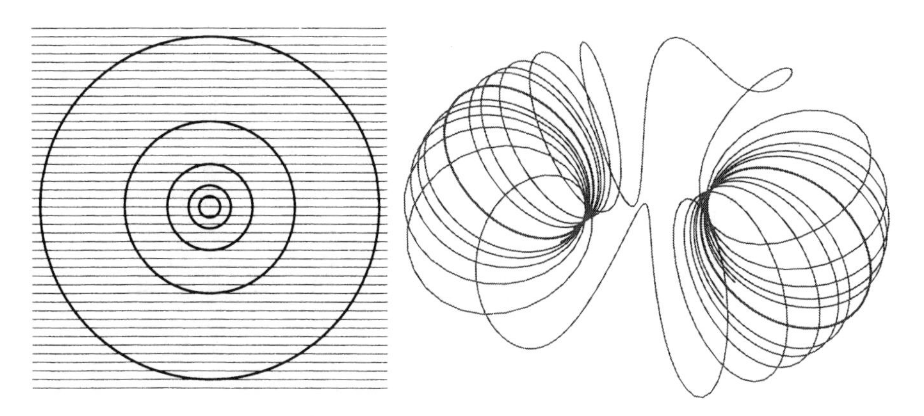

Figure 3.1. A foliation on the torus. (a) The horizontal foliation is preserved by the homothety $x \mapsto 2x$. (b) A foliation is obtained on the quotient manifold, a torus. Two of the leaves are circles, hence compact; these are darker in the picture. All other leaves, of which two are shown, spiral around, accumulating onto the circles.

oriented three-manifold, and Wood [Woo69] extended this to non-orientable three-manifolds. More generally, a manifold of any dimension has a codimension-one foliation if and only if its Euler number is zero [Thu76]. (See [Hae58] and [Thu74] for related results for foliations of codimension greater than one.)

On the other hand, Haefliger [Hae58] proved that no foliation of S^3 by surfaces can be real analytic, and Novikov [Nov65] proved that every codimension-one foliation of S^3, as well as many other three-manifolds, has a closed leaf that is a torus. Gabai [Gab83, Gab87] has constructed foliations with no closed torus leaves on many three-manifolds outside the scope of Novikov's theorem. These results have interesting consequences for the topology of three-manifolds.

Example 3.1.12 (measured manifolds). Diffeomorphisms between open subsets of $\mathbf{R}^n$ that preserve n-dimensional Lebesgue measure, or volume, form a pseudogroup Vol. A manifold with a Vol-structure is a *measured manifold*, since it comes equipped with a measure.

Exercise 3.1.13 (area-preserving vector fields in the plane). A vector field is called area-preserving if the flow obtained by integrating it is area-preserving. Show that the vector field obtained by rotating clockwise by 90° the gradient of a smooth function $\mathbf{R}^2 \to \mathbf{R}$ preserves area, and that every area-preserving vector field has this form.

Problem 3.1.14. Generalize the criterion for an area-preserving vector field given in the preceding exercise to an arbitrary Riemannian surface. (Hint: The criterion works locally. To find a global criterion, rather than working with a smooth function f, you need to use df, which pieces together globally to make a closed one-form on the surface. If the surface is non-orientable, you get a closed one-form twisted by the orientation bundle of the surface.)

Exercise 3.1.15 (existence of measure stiffening). Show that every closed smooth manifold M has a measure stiffening. (Hint: choose a measure on M by using a partition of unity to average Lebesgue measures in different coordinate charts. Show that for any bounded positive smooth function f on an open subset of $\mathbf{R}^n$, there is a diffeomorphism of the open subset into another open subset of $\mathbf{R}^n$ whose Jacobian has determinant f. Use this to find a measure atlas.)

Problem 3.1.16 (classification of measure stiffenings). Show that two measure stiffenings on a compact connected manifold are differentiably equivalent if and only if they have the same total volume. This depends on some knowledge of differential forms, and on de Rham's theorem. Here is an outline:

(a) The total volume of a measured manifold is invariant up to Vol equivalence.

(b) Assume that M is oriented, so the measure is represented by an n-form α. If ω is a differential form and X is a vector field, the *Lie derivative* $L_X\omega$ measures how much ω changes under the flow of X. More precisely,

$$L_X\omega = \frac{\partial}{\partial t}(\phi_X^t)^*\omega,$$

where ϕ_X^t is the diffeomorphism obtained by integrating the vector field X over time t. Use the formula

$$L_X\omega = i_X d\omega + d i_X\omega,$$

where i_X indicates contraction with X (that is, $i_X\omega(X_1, \ldots, X_n) = \omega(X, X_1, \ldots, X_n)$) to show that for any $(n-1)$-form β, there is a vector field X such that $L_X\alpha = d\beta$.

(c) For any two n-forms α_0 and α_1 such that

$$\int_M \alpha_0 = \int_M \alpha_1,$$

realize a homotopy $\alpha_t = (1-t)\alpha_0 + t\alpha_1$ between them by constructing a suitable one-parameter family of vector fields X_t.

(d) The space of measure stiffenings of any compact connected manifold M, up to differentiable equivalence, is parametrized by positive real numbers.

Example 3.1.17 (complex manifolds). When n is even, $\mathbf{R}^n$ can be identified with $\mathbf{C}^{n/2}$. Let Hol be the pseudogroup of biholomorphic maps between open subsets of $\mathbf{C}^{n/2}$. (A *biholomorphic map* is a holomorphic map having a holomorphic inverse; it turns out that any holomorphic homeomorphism is biholomorphic.) A Hol-manifold is called a *complex manifold* of complex dimension $n/2$, and also a *Riemann surface* when $n = 2$.

When $n = 2$, a homeomorphism is holomorphic if and only if it is conformal and preserves orientation. Therefore an orientation-preserving isometry of the Poincaré disk model for $\mathbf{H}^2$ is biholomorphic, and every orientable hyperbolic surface inherits the structure of a complex manifold.

Stereographic projection from the unit sphere to $\mathbf{C}$ is a conformal map. A collection of maps obtained by rotating the sphere and then mapping by stereographic projection to $\mathbf{C}$ constitutes an atlas for a complex structure on S^2 (provided they don't all omit the same point); this is the complex structure of $\mathbf{CP}^1$, the Riemann sphere.

Similarly, orientation-preserving isometries of $\mathbf{E}^2$ are holomorphic. It follows that a hyperbolic, Euclidean, or elliptic structure on any orientable surface can be relaxed to a complex structure.

This easy observation has a converse, the celebrated *uniformization theorem*: every complex structure on a surface S can be stiffened to a hyperbolic, Euclidean or elliptic structure that makes S into a complete metric space. This is a very strong statement, as it applies to *any* Riemann surface—even one that is not closed, or has infinitely many holes. The uniformization theorem is closely related to the Riemann mapping theorem, and was the subject of much attention (and contention) by Poincaré, Klein and others, in the latter part of the nineteenth century.

Moreover, any Riemannian metric on an orientable surface S gives rise to a complex structure on S, because around each point in S one can find *isothermal coordinates*, that is, coordinates in which the metric has the form $ds^2 = f(u,v)\,(du^2 + dv^2)$; the transition between two such sets of coordinates is clearly conformal. (For an "elementary" proof of the existence of isothermal coordinates, see [sC55]; for a much more general treatment, see [Hör90].) Since every differentiable surface can be given a Riemannian metric—using partitions of unity, for example—we see that any orientable surface has a complete Riemannian metric of constant curvature. As in the case of closed surfaces (Problem 1.3.15), the three possibilities are mutually exclusive within each homeomorphism class, except that surfaces homeomorphic to $\mathbf{R}^2$, $S^1 \times \mathbf{R}$ or the Möbius strip can be given both Euclidean and hyperbolic structures.

We close this section with terminology that, although standard, is potentially quite confusing.

A *manifold-with-boundary* is, in general, not a manifold; it is a space locally modeled on the half-space $\mathbf{R}^n_+ = \{(x_1, \ldots, x_n) \in \mathbf{R}^n : x_n \geq 0\}$. This means that each point p in a manifold-with-boundary has a neighborhood that looks like a neighborhood of some point in $\mathbf{R}^n_+$, either in the interior or on the boundary. The points that don't have neighborhoods like those of an interior point of $\mathbf{R}^n_+$ form the *boundary*.

A manifold (with no boundary) that is compact is called a *closed manifold*, and a non-compact one is sometimes called *open*. This is sometimes in conflict with the usage of "open" and "closed" in general topology, but there isn't much we can do about it.

3.2. Triangulations and Gluings

As with most raw definitions, direct use of Definition 3.1.2 is rare. Atlases are an inconvenient way to describe a manifold, and normally one uses some "higher-level" operation, such as gluing together pieces of $\mathbf{R}^n$ or taking the quotient of an existing manifold by a group, to construct new manifolds. Thus, in Section 1.3 we glued polygons together to get surfaces, and in Section 1.4 we described three-manifolds by gluing polyhedra.

As long as we were dealing with surfaces, our intuition was a pretty sound guide. However, some of the conclusions we might reach from this experience break down in high dimensions. In three dimensions, most of these difficulties can still be circumvented, so that throughout most of this book, we will discuss polyhedra and gluings and three-manifolds in a rather intuitive way. But in this section and the next we will work through some of the relevant technical issues so that we can comfortably ignore them subsequently.

We consider first gluings of simplices, then gluings of convex polyhedra, which are more common in practice. Recall that an n-*simplex* σ is the convex hull of $n+1$ affinely independent points

$v_0, \ldots, v_n$ (in some affine space, necessarily of dimension at least n). The convex hull of a subset of $\{v_0, \ldots, v_n\}$ is a *face*, or *subsimplex*, of σ. As usual, a one-dimensional face is called an *edge* of σ, and a zero-dimensional face (or the unique point in it) is a *vertex*. A face of dimension $n - 1$ will be called a *facet*. The *interior* of a simplex is what's left when you take away its proper faces; it can also be defined as the topological interior of the simplex considered as a subset of its affine hull.

A *simplicial complex* is a locally finite collection Σ of simplices (in some affine space), satisfying the following two conditions: any face of a simplex in Σ is also in Σ, and the intersection of two simplices in Σ is either empty or a face of both. The union of all simplices in Σ is called the *polyhedron* of Σ, and denoted by $|\Sigma|$. The k*-skeleton* of Σ is the subcomplex consisting of simplices of dimension k or lower.

If a simplicial complex has finitely many simplices, whose union forms a convex set, we recover the familiar notion of a *convex polyhedron*, defined, for instance, as the convex hull of a finite set of points. When we talk about a convex polyhedron we'll be referring to this notion, even if finiteness is not explicitly mentioned.

Given a map from the vertices of a simplicial complex to an affine space, there is a unique way to extend it to the complex's polyhedron so that the map is affine within each simplex. Such a map is called *simplicial*. Naturally, we want to consider two complexes isomorphic if there is a simplicial homeomorphism between the two.

A *subdivision* of a complex Σ is any complex Σ' having the same polyhedron as Σ, and such that every simplex in Σ' is contained in some simplex in Σ. The importance of simplicial complexes as a technical tool is that an arbitrary continuous map from a polyhedron (into, say, an affine space) can be approximated by a simplicial map in the same homotopy class, if we subdivide the domain appropriately (see, for example, [Mun84, 89–95]).

Barycentric subdivision is the most common subdivision procedure. It is defined by induction on the k-skeleton, for increasing values of k. To subdivide a k-simplex σ, we take as a new vertex the average of the vertices of σ, and cone off the previously defined subdivision of the boundary of σ.

The polyhedron of a simplicial complex, or a space homeomorphic to it, is said to be *triangulated* by the complex. More exactly, a *triangulation* of a topological space X is a simplicial complex Σ, together with a homeomorphism $|\Sigma| \to X$; the various concepts defined for simplicial complexes can be transferred to X by the homeomorphism. We've already used triangulated surfaces, in Section 1.3 (see especially Figure 1.16).

Among triangulated spaces are spaces obtained by gluing simplices.

Definition 3.2.1 (gluing). An *n-dimensional (rectilinear) gluing* consists of a finite set of n-simplices, a choice of pairs of simplex facets such that each facet appears in exactly one of the pairs, and an affine identification map between the facets of each pair.

We are interested in the quotient space of the union of the simplices by the equivalence relation generated by the identification maps. This quotient space itself is sometimes called a gluing. We make it into a simplicial complex by embedding it in some affine space in such a way that each simplex of the barycentric subdivision of the component simplices is affinely embedded.

Exercise 3.2.2. Show that there is such an embedding.

Exercise 3.2.3. In a gluing of three-dimensional simplices, each edge enters into exactly two gluings, one for each of its faces. Composing these, one gets a cycle of gluings that eventually must return to the original edge. Suppose that the composition of gluings around the cycle reverses the edge's orientation. Describe a neighborhood of the fixed point of the return map of the edge to itself, in the resulting identification space.

When is a gluing (or, more generally, a triangulated space) a manifold? Exercise 1.3.2 answered the question in two dimensions. For higher dimensions, we need the notions of the *link* and *star* of a simplex. If σ is a simplex in a simplicial complex Σ, let $\tau_1, \ldots, \tau_k$ be the simplices of Σ containing σ. For each τ_i, let σ_i be the simplex opposite σ in τ_i, in the sense that $\sigma \cap \sigma_i = \varnothing$ and τ_i is the convex hull of $\sigma \cup \sigma_i$. The link of σ, denoted $\operatorname{lk} \sigma$ or $\operatorname{lk}_\Sigma \sigma$, is the simplicial complex consisting of the σ_i. The star $\operatorname{st} \sigma$ is the union of the interiors of the τ_i.

Exercise 3.2.4 (cone on link). The *cone* on a topological space X, which we denote by CX, is the product $X \times [0,1]$ with $X \times \{1\}$ collapsed into a point. Show that, if σ is a p-simplex in a simplicial complex Σ, a point in the interior of σ has $\operatorname{st} \sigma$ as a neighborhood in $|\Sigma|$, and $\operatorname{st} \sigma$ is homeomorphic to $D^p \times C\,|\operatorname{lk} \sigma|$.

Proposition 3.2.5 (spherical links imply manifold). *Let X be a triangulated space of dimension n. If the link of every simplex* *of dimension p is homeomorphic to an $(n-p-1)$-sphere, X is a topological manifold.*

In fact, the conclusion is true whenever the links of vertices are $(n-1)$-spheres.

Proof of 3.2.5. By Exercise 3.2.4, every point in X has a neighborhood of the form $D^p \times CS^{n-p-1}$, which is homeomorphic to $D^p \times D^{d-p}$, since the cone on the sphere is a ball. These neighborhoods cover X, so X is a manifold. $\boxed{3.2.5}$

Exercise 3.2.6. Show that a manifold obtained by gluing is orientable if and only if, when the faces of each simplex are oriented consistently, all face identifications are orientation-reversing (cf. Exercise 1.3.2).

It would be natural to suppose that the converse of Proposition 3.2.5 is true as well, but this is not the case if $n \geq 5$. The first counterexamples were found by R. D. Edwards. Example 3.2.11 below gives a somewhat different counterexample, found by Cannon [Can79].

(The right criterion is this: the polyhedron of a simplicial complex is a topological manifold if and only if the link of each cell has the homology of a sphere, and the link of every vertex is simply connected. The proof is far beyond the scope of the current discussion.)

In three dimensions this sort of thing can't happen:

Proposition 3.2.7 (manifolds have spherical links in dimension three). *A three-dimensional gluing is a three-manifold if and only if the link of every vertex is homeomorphic to S^2.*

Proof of 3.2.7. The "if" part is true for any triangulated space, as mentioned after the statement of Proposition 3.2.5. Alternatively, for three-dimensional gluings, it is easy to see that the link of an edge must be a circle, so we can apply Proposition 3.2.5 itself.

For the converse, we need some test that non-manifolds will fail, and we choose local simple connectivity. A topological space X is *(locally) simply connected* at $x \in X$ if for any neighborhood U of x there is a smaller neighborhood $V \subset U$ of x such that $V \setminus \{x\}$ is simply connected. Manifolds of dimension greater than two clearly are simply connected everywhere.

If X is a three-dimensional gluing and v is a vertex of X, the link of v is a two-dimensional gluing and therefore a manifold, by Exercise 1.3.2(b). Clearly X is simply connected at v if and only if $\operatorname{lk} v$ is simply connected. Since the only simply connected closed surface is S^2, we conclude that if the link of v is not a sphere, X is not a manifold. 3.2.7

Even when a three-dimensional gluing is not a manifold, it can be made into one by removing the vertices whose links are not spheres. This is what we did in Example 1.4.8. Alternatively, we can remove an open neighborhood of each bad vertex, to obtain a compact manifold-with-boundary.

Proposition 3.2.8 (gluing is manifold if and only if $\chi = 0$). *A three-dimensional gluing X is a three-manifold if and only if its Euler number is zero. In general, if X has k vertices $v_1, \dots, v_k$, we have*

$$\chi(X) = k - \frac{1}{2}\sum_{i=1}^{k} \chi(\operatorname{lk} v_i).$$

Proof of 3.2.8. Let e, f and t be the number of edges, faces and tetrahedra in X. Then $f = 2t$, since each face lies on two tetrahedra and each tetrahedron has four faces. We also have

$$\sum_{i=1}^{k} \chi(\operatorname{lk} v_i) = 2e - 3f + 4t,$$

since each edge accounts for two vertices in links of vertices, each face accounts for three edges and each tetrahedron for four faces. The desired equality follows.

Since the Euler number of the two-sphere is 2, and the Euler number of every other closed surface is less than 2, we get $\chi(X) \geq 0$, with equality if and only if X is a manifold. 3.2.8

So far we've only considered gluings of simplices, and only by affine maps. What happens when we allow gluings of arbitrary convex polyhedra, by arbitrary maps? Is the topology of the identification space determined by just the gluing pattern, or does it depend on the particular choice of a gluing?

The set of *vertices* of a convex polyhedron is the unique minimal set of points whose convex hull is the polyhedron; a *face* is an intersection of the convex polyhedron with a half-space, having dimension less than the dimension of the polyhedron. A *facet* is a face of codimension one.

Two convex polyhedra are *combinatorially equivalent* if there is a one-to-one correspondence of their faces of every dimension that preserves incidence. Thus the vertices are in one-to-one correspondence; each edge of one polyhedron corresponds to an edge of the other polyhedron with corresponding vertices; and so on.

If α is a combinatorial equivalence between polyhedra P and Q, a homeomorphism $h : P \to Q$ is in the *combinatorial class* of α if h sends each face of every dimension to the face specified by α. It is easy to see that every combinatorial equivalence can be realized by a homeomorphism in its combinatorial class, but of course not all homeomorphisms are in the combinatorial class of a combinatorial equivalence.

Now consider a finite set of combinatorial classes of n-dimensional convex polyhedra. A *gluing pattern* is a choice of pairs of polyhedron facets (each facet appearing in exactly one of the pairs), together with a combinatorial equivalence between the facets in each pair. A *gluing* realizing this pattern is a choice of actual convex polyhedra of the given combinatorial types, and a choice of actual homeomorphisms between the faces in the given combinatorial equivalence classes.

Not any choice of homeomorphisms will do, however: there is a compatibility condition. Any face β of dimension $n-2$ or less is a face of two or more facets, so it enters into two or more pairings. Compositions of the various identifications will result in the identification of β with possibly many other faces. A bit of reflection shows that there will always be chains of these identifications that take β to itself. The compatibility condition is this: For any face β and any chain of identifications whose composition takes β back to itself in the combinatorial class of the identity, the composed identification must actually be the identity.

Exercise 3.2.9 (standard gluing). A *barycentric subdivision* of a polyhedron is defined like the barycentric subdivision of a simplex, with new vertices being taken arbitrarily in the interior of the corresponding faces.

(a) Given a combinatorial equivalence between two polyhedra, use barycentric subdivision to realize a homeomorphism in its combinatorial class.

(b) Show that, for a given gluing pattern, the identification maps built in this way satisfy the compatibility condition.

(c) Make the quotient space of the gluing thus defined into a simplicial complex, as in Definition 3.2.1 and the subsequent exercise.

(d) Extend Propositions 3.2.7 and 3.2.8 to this context.

Problem 3.2.10 (uniqueness of gluings). (a) Show that a homeomorphism of the unit ball in $\mathbf{R}^n$ that is the identity on its boundary is isotopic to the identity. (Hint: comb all the tangles to a single point. This is called *the Alexander trick*.)

(b) Show that any two gluings realizing a given gluing pattern yield homeomorphic identification spaces.

Example 3.2.11 (double suspension of Poincaré space). The *suspension* ΣX of a topological space X is the product $X \times [0,1]$ with $X \times \{0\}$ and $X \times \{1\}$ each collapsed into a point, called a *suspension point*. For example, the suspension of the n-sphere is the $(n+1)$-sphere.

Let P be the Poincaré dodecahedral space of Example 1.4.4, made into a simplicial complex by the recipe in Exercise 3.2.9. The suspension ΣP is not a manifold, because the link of a suspension point is P itself, which is not simply connected (see the proof of Proposition 3.2.7). But, amazingly, the *double suspension* $\Sigma^2 P = \Sigma\Sigma P$ is homeomorphic to S^5.

This shows that the converse of Proposition 3.2.5 is false, because along the *suspension circle*—the suspension of the suspension points—the links of vertices and edges are not spheres. The links of vertices on the suspension circle are homeomorphic to ΣP, which is simply connected, so this is not an obstruction to $\Sigma^2 P$ being a manifold. The links of the edges on the suspension circle are homeomorphic to P, which is not simply connected; this might seem like an obstruction, but turns out not to be.

The number of possible gluings for polyhedra grows very fast with the number of facets.

Exercise 3.2.12. (a) In how many ways can one pair each face of a cube with its opposite, so as to produce an oriented three-manifold?

(b) Show that the number of gluing patterns for an octahedron is $8! \cdot 3^4/(4! \cdot 2^4) = 8505$. Can you estimate how many of these yield manifolds? How many yield orientable manifolds?

(c) Same question for an icosahedron.

Because of symmetries, many gluing patterns give obviously homeomorphic results—for example, in the case of an icosahedron, this reduces the number of different gluing diagrams by a factor of about 120. (Since some gluings have a certain amount of symmetry, the reduction is not actually quite this much.) This still leaves a huge number which are not obviously homeomorphic. It seems unlikely that the same phenomenon occurs as in dimension two, when very large numbers of different gluing patterns give homeomorphic manifolds, but how can we tell? We need more tools to be able to name and recognize a three-manifold when it is described in different ways.

3.3. Geometric Structures on Manifolds

It is convenient to broaden slightly the definition of a $\mathcal{G}$-manifold (3.1.2), by allowing $\mathcal{G}$ to be a pseudogroup on any connected manifold X, not just on $\mathbf{R}^n$. As long as $\mathcal{G}$ acts transitively, this does not give any new types of manifolds:

Exercise 3.3.1 (locality of $\mathcal{G}$-structures). Let $\mathcal{G}$ be a pseudogroup on any manifold X and $U \subset X$ any open subset such that for every $x \in X$ there is some element of $\mathcal{G}$ taking x into U. Let $\mathcal{G}_U$ be the subpseudogroup consisting of elements whose domain and range are contained in U. Show that every $\mathcal{G}$-manifold has a $\mathcal{G}_U$-stiffening unique up to $\mathcal{G}$-isomorphism.

Many important pseudogroups come from group actions on manifolds. Given a group G acting on a manifold X, let $\mathcal{G}$ be the pseudogroup generated by restrictions of elements of G: see Exercise 3.1.8(a). Thus every $g \in \mathcal{G}$ agrees locally with elements of G: the domain of g can be covered with open sets U_α such that $g|_{U_\alpha} = g_\alpha|_{U_\alpha}$ for $g_\alpha \in G$. A $\mathcal{G}$-manifold is also called a (G, X)-*manifold*.

Example 3.3.2 (Euclidean manifolds). If G is the group of isometries of Euclidean space $\mathbf{E}^n$, a $(G, \mathbf{E}^n)$-manifold is called a *Euclidean*, or *flat*, manifold; the structure of these manifolds is what we discussed informally in Section 1.1. As we saw in Section 1.3, the torus and the Klein bottle are the only compact two-dimensional manifolds that can be given Euclidean structures, but they have many such structures. We'll return to this question in Example 4.6.11.

It is an altogether non-trivial result of Bieberbach [Bie11, Bie12] that, in any dimension, there are only finitely many compact Euclidean manifolds up to homeomorphism, and that any such manifold can be finitely covered by a torus of the same dimension. We will consider Euclidean manifolds in more detail in Section 4.2, when these results will be proved.

Figure 3.2 shows a gluing construction for a three-dimensional example, which we'll encounter again in Section 4.3, where three-dimensional Euclidean manifolds are classified (see Figure 4.7).

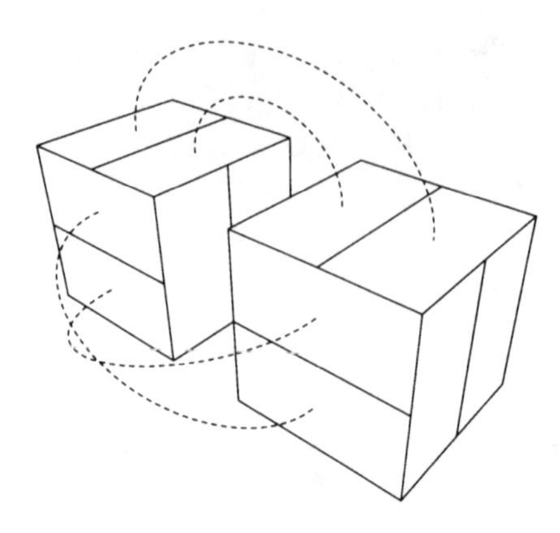

Figure 3.2. A Euclidean three-manifold. Starting with two identical cubes, mark each facet with a bisecting altitude, in such a way that no two altitudes meet. Then pair facets that face the same way by means of a reflection in a line parallel to the facets' chosen altitudes. The result is a Euclidean manifold.

Exercise 3.3.3. Prove the last assertion in the caption of Figure 3.2.

(a) Show that the identification space is a manifold M.

(b) Find a group of isometries of $\mathbf{E}^3$ whose quotient is M. The existence of this group shows that M has a Euclidean structure.

(c) Find a subgroup of finite index isomorphic to $\mathbf{Z}^3$. By looking at the quotient of $\mathbf{E}^3$ by this subgroup, show that M can be finitely covered by a torus (as expected by Bieberbach's theorem).

Example 3.3.4 (affine manifolds). If G is the group of affine transformations of $\mathbf{R}^n$, a $(G, \mathbf{R}^n)$-manifold is called an *affine manifold*.

As an example, consider again the homothety of Figure 3.1. As the quotient of $\mathbf{R}^2 \setminus \{0\}$ by a group of affine transformations (the group generated by this homothety), the torus has an affine structure (why?).

Here is another method, due to John Smillie, for constructing affine structures on T^2 from any quadrilateral Q in the Euclidean plane. Identify the opposite edges of Q by the orientation-preserving similarities that carry one to the other. Since similarities preserve angles, the sum of the angles about the vertex in the resulting complex is 2π, so a neighborhood of the vertex has an affine structure (why?). We will see in Part II how such structures on T^2

are intimately connected with questions concerning Dehn surgery in three-manifolds.

J. P. Benzecri (see [Mil58]) showed that the only closed two-dimensional affine manifolds are tori and Klein bottles. An important open question about affine manifolds is whether a compact affine manifold must have Euler number zero.

A Euclidean structure on a manifold automatically gives an affine structure. Bieberbach proved that closed Euclidean manifolds with the same fundamental group are equivalent as affine manifolds. This is proved in Theorem 4.2.2.

Example 3.3.5 (elliptic manifolds). If G is the orthogonal group $O(n+1)$ acting on the sphere S^n, a (G, S^n)-manifold is called *spherical*, or *elliptic*. The Poincaré dodecahedral space (Example 1.4.4) and lens spaces (Example 1.4.6) are spherical manifolds.

Example 3.3.6 (hyperbolic manifolds). If G is the group of isometries of hyperbolic space $\mathbf{H}^n$, a $(G, \mathbf{H}^n)$-manifold is a *hyperbolic manifold*. We discussed hyperbolic surfaces in Section 1.2 and a three-dimensional example, the Seifert–Weber dodecahedral space, in Example 1.4.5.

In each of the three preceding examples, showing that a manifold has the specified geometric structure amounts to showing that each point has a neighborhood isometric to the appropriate ball. (You should justify this in the light of Definition 3.1.2.) This is certainly the case if the manifold is a quotient of $\mathbf{E}^3$, S^3 or $\mathbf{H}^3$ by a group of isometries.

Alternatively, if the manifold is defined by gluing *convex polyhedra* in $\mathbf{E}^3$, S^3 or $\mathbf{H}^3$ (that is, sets of non-empty interior that can be expressed as intersections of closed half-spaces), the condition can be verified by checking that the dihedral angles add up to 360° around each edge, and that corners fit together exactly to form a spherical neighborhood of a point in the model space. (Actually, the second condition follows from the first.) We checked the edge condition in Examples 1.4.4, 1.4.5 and 1.4.6. Likewise, we can see that the manifold of Figure 3.2 is Euclidean simply by observing that the edges of the cubes are identified in groups of four, with dihedral angles of 90°, and the corners are identified in groups of eight octants.

We use the preceding remark to give a hyperbolic structure to several open manifolds:

Example 3.3.7 (the figure-eight knot complement). We saw in Example 1.4.8 how a certain gluing of two tetrahedra (minus their vertices) yields a space homeomorphic to the complement of a figure-eight knot in S^3. We now give this space a hyperbolic structure.

Let the two tetrahedra be regular ideal tetrahedra in hyperbolic space, that is, regular tetrahedra whose vertices are at infinity. Combinatorially, a regular ideal tetrahedron is a simplex with its vertices deleted; geometrically, it can be modeled on a regular Euclidean tetrahedron inscribed in the unit sphere, interpreted in the projective model. The dihedral angles of this polyhedron are 60°, as can be seen from Figure 3.3.

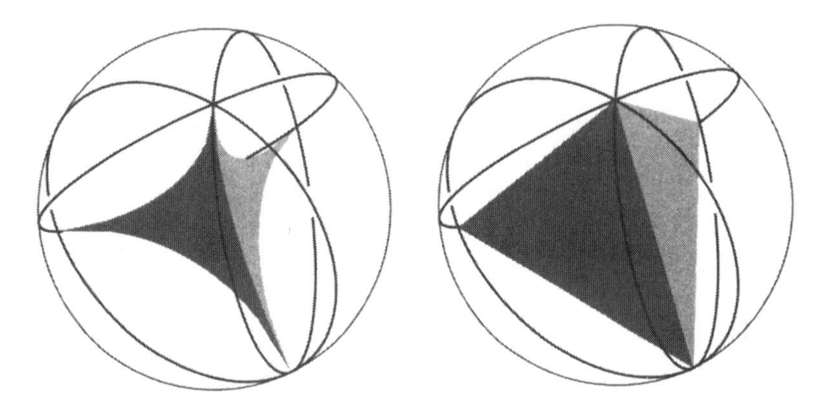

Figure 3.3. A regular ideal tetrahedron. The dihedral angles of a polyhedron can be read off from the angles between the circles at infinity determined by the faces in the Poincaré model (a); for an ideal polyhedron, the circles are the same as in the projective model, and in this case they equal 60°, by symmetry (b).

Exercise 3.3.8. (a) Justify the caption of Figure 3.3, by showing that the angle between two planes in hyperbolic space is the same as the angle of their bounding circles on the sphere at infinity, seen in the Poincaré ball model.

(b) Show that the dihedral angles of any ideal hyperbolic tetrahedron—not just a regular one—add up to 360°. (Hint: project Figure 3.3 stereographically from one vertex, so three of the four circles become lines.)

Now we glue the faces of the two tetrahedra using hyperbolic isometries, following the combinatorial pattern of Example 1.4.8. As we discussed there, edges are identified six at a time, so the dihedral angles around each edge add up to 360°. We don't have to check the vertex condition, because there are no vertices.

Fix two adjacent ideal regular tetrahedra T and T' in hyperbolic space and label their faces and edges as in Figure 1.24, making sure that their common face is labeled consistently (call it D). Then there is a unique orientation-preserving hyperbolic isometry taking face A of T to face A of T' (and matching the arrows), and likewise for the B and C pairs. The quotient of $\mathbf{H}^3$ by the group G generated by these isometries is our manifold, and G is a representation of the manifold's fundamental group.

This example is closely linked to a family of groups studied by Bianchi [Bia92] in connection with number theory: indeed, by choosing T to have vertices 0, 1, ω and $\bar{\omega}$, where $\omega = -\frac{1}{2} + \frac{\sqrt{3}}{2}i$ is a cube root of unity, we see that all the generators can be chosen to lie in $\mathrm{PSL}(2, \mathbf{Z}[\omega]) \subset \mathrm{PSL}(2, \mathbf{C})$. (In fact, G has index 12 in $\mathrm{PSL}(2, \mathbf{Z}[\omega])$). A different representation for G [Mag74, 153–155] was known to Gieseking, in a thesis published in 1912.

The connection with the figure-eight knot, however, is much more recent. Riley [Ril82] and Jørgensen [Jor77] independently showed that the complement of the figure-eight knot has a hyperbolic structure, the former by finding a representation for its fundamental group, and the latter by looking at the space as a punctured torus bundle over the circle.

The technique of Example 3.3.7 may seem very special, but it actually applies to many different knot and link complements. One divides the complement into a union of ideal polyhedra, then attempts to realize these polyhedra as ideal hyperbolic polyhedra and glue them together to form a hyperbolic manifold. Several people have implemented this procedure on computers, including Daryl Cooper, Colin Adams, and Jeff Weeks. See also [Hat83a]. Of course, not every way of gluing simplices is related to a geometric construction.

Example 3.3.9 (Whitehead link). Figure 3.4 shows two views of the *Whitehead link*. This link may be spanned by a two-complex, which cuts the complement into an octahedron with vertices deleted: see Figures 3.5 and 3.6. As in Example 1.4.8, the two-cells of the complex give the faces of the octahedron, once we open up at the arrows and collapse each piece of the knot to a point (corresponding to a deleted vertex).

A hyperbolic structure for the complement is obtained by making the octahedron an ideal regular octahedron in hyperbolic space,

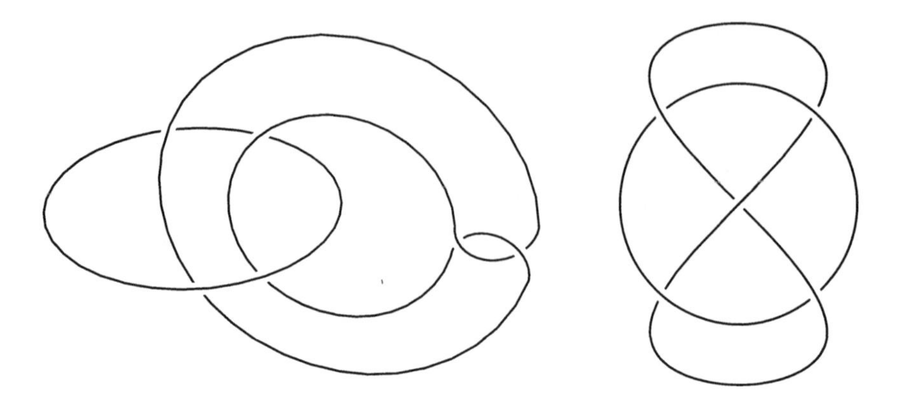

Figure 3.4. The Whitehead link.

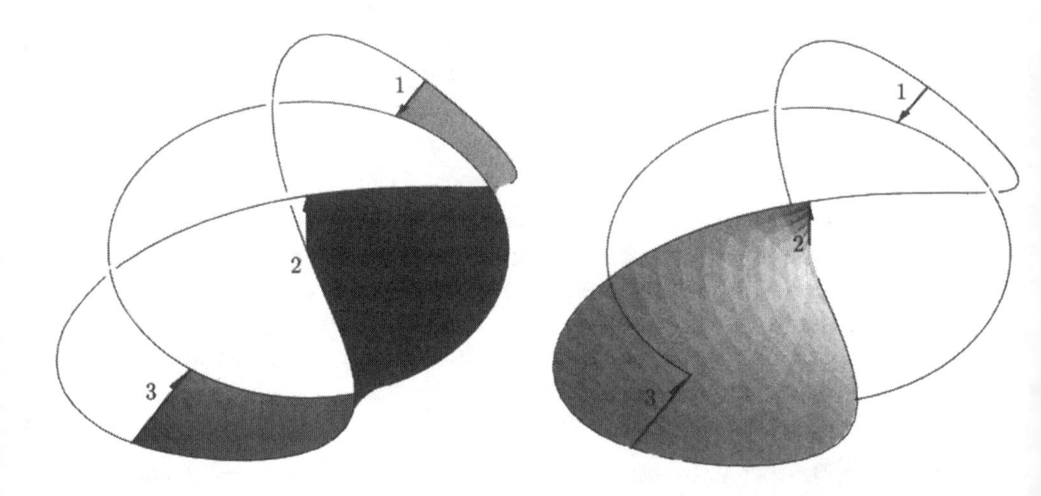

Figure 3.5. Cutting the Whitehead link complement. A two-complex spanning the Whitehead link cuts its complement so as to form an octahedron with vertices deleted. There are four two-cells, two of which are shown here; the other two are symmetrically placed. The three one-cells are indicated by the numbered arrows. The pattern of identification on the surface of the octahedron is shown in Figure 3.6.

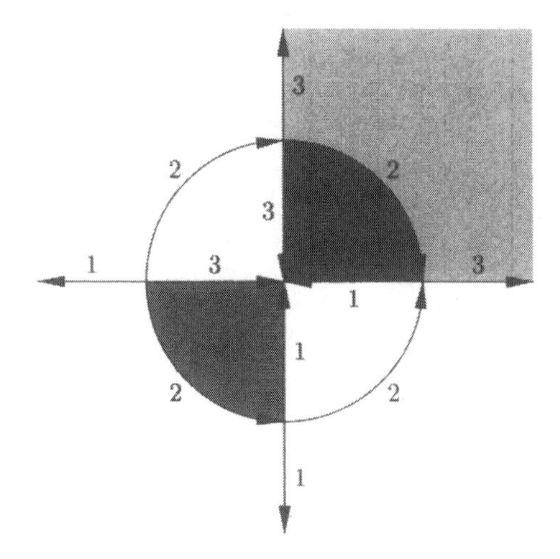

Figure 3.6. The Whitehead link: gluing the octahedron. This shows the surface of a spherical octahedron, projected stereographically from one vertex. (Imagine that the solid occupies the half-space behind the page.) Gluing the octahedron according to this pattern gives the complement of the Whitehead link. The arrows correspond to Figure 3.5, and the surfaces visible in that figure are shown here with the same shading.

which appears like a Euclidean regular octahedron in the Klein model. The dihedral angles of a regular octahedron are $90°$, by the same reasoning as in Figure 3.3, so neighborhoods of the edges fit four to an edge. The gluing is done with hyperbolic isometries of the octahedron's faces. Again, we don't have to worry about vertices.

Example 3.3.10 (Borromean rings). Figure 3.7 shows the Borromean rings, a link with lots of interesting properties—for example, if you cut open one of the circles, the other two are unlinked. The complement of this link can be given a hyperbolic structure coming from gluing two ideal octahedra. To see this, place a short stick connecting two loops at each of the crossings; these will be the edges of the octahedra (Figure 3.7, right). The faces will map to two-cells spanning the eight plane regions (one of them unbounded) determined by the three circles in the top view (Figure 3.7, left). Notice that the pattern of these regions is the same as the stereographic projection of a spherical octahedron from the center of a face. We therefore imagine each octahedron blown up into a half-space, with this pattern on the boundary.

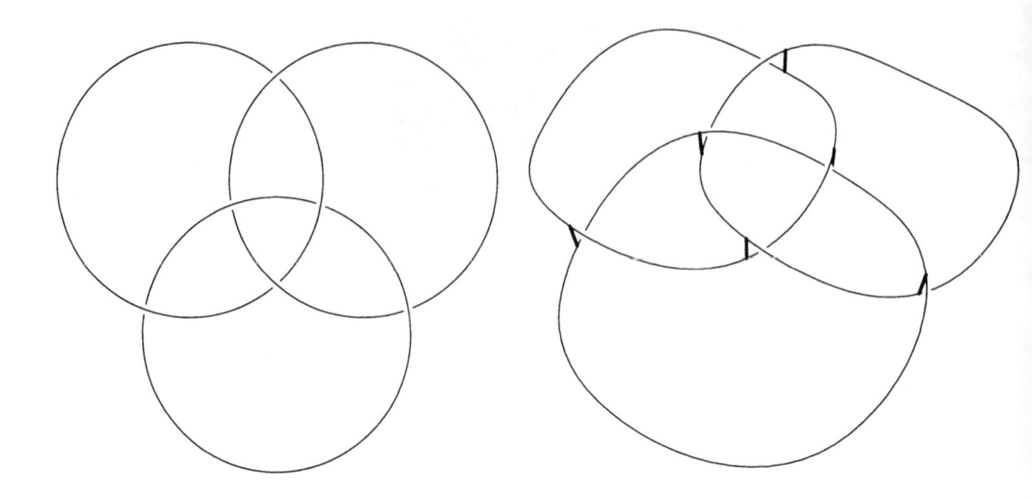

Figure 3.7. The Borromean rings. Top view and slanted view of the Borromean rings, showing the one-skeleton of a spanning two-complex.

One octahedron approaches the link from below, the other from above. Before coupling with the link, each face turns into a hexagon: the vertices are ripped open and alternate edges of the hexagon are shortened and turned in order to match the vertical sticks. The other edges of the hexagons are shrunk to points so that each face again becomes a triangle. Each face of the upper octahedron is glued to the face directly below it with a 120° rotation; the sense of the rotation is alternately clockwise and counterclockwise, as in intermeshed gears.

To give this manifold a hyperbolic structure we can use two octahedra with 90° dihedral angles, as in the previous example of the Whitehead link, since four faces are glued to each edge in the resulting complex.

Definition 3.3.11 (manifolds with geodesic boundary). We can also put geometric structures on manifolds-with-boundary (see the end of Section 3.1), by considering appropriate pseudogroups on a model manifold-with-boundary X. For example, let $X = \mathbf{E}^+_n$ be a Euclidean half-space. A Euclidean manifold with geodesic boundary is a $\mathcal{G}$-manifold, where $\mathcal{G}$ is the pseudogroup generated by isometric homeomorphisms between open subsets of X. Hyperbolic manifolds with geodesic boundary and elliptic manifolds with geodesic boundary are defined similarly.

A Euclidean (or hyperbolic, or elliptic) manifold with geodesic boundary can be doubled to give a manifold (without boundary) having the same structure: one just takes two copies of the manifold-with-boundary and glues them together by the identity map on the boundary.

Example 3.3.12 (hyperbolic manifold with geodesic boundary). We will again glue two tetrahedra together, but in a pattern different from the one used in Example 3.3.7: see Figure 3.8.

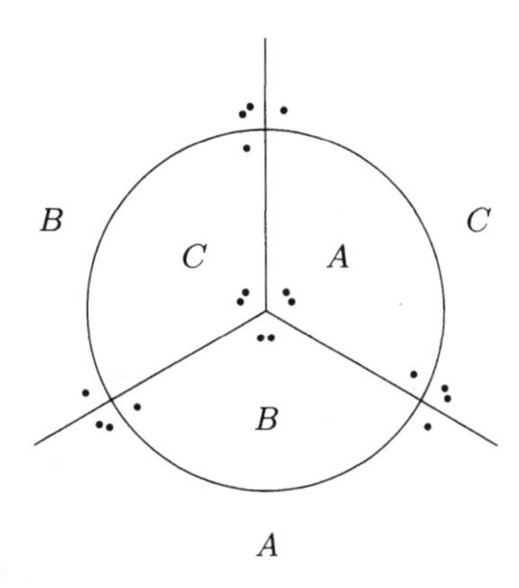

Figure 3.8. A gluing of two tetrahedra. Two tetrahedra glued base to base make a hexahedron, whose boundary is shown here in stereographic projection (compare Figure 3.6). The six faces are then glued in pairs with matching letters, oriented in such a way that the dots also match. (Since all edges collapse to one, we cannot use numbers as in Figure 3.6 to specify the gluing.)

Exercise 3.3.13. Express the gluing of Example 3.3.7 according to the conventions of Figure 3.8. How different are the two gluings?

The complex K obtained from this gluing has only one vertex, one edge, four faces, and two tetrahedra, so its Euler number is 2. The link of the vertex is a surface of genus two, as can be seen by hand or by using Proposition 3.2.8. As in Example 3.3.7, we can remove from K a cone-like closed neighborhood of the vertex, obtaining a manifold M. But we cannot give M a (complete) hyperbolic structure using the technique of Example 3.3.7, for this would require finding two ideal hyperbolic tetrahedra whose dihedral angles together add up to 360°, and this is impossible by Exercise 3.3.8(b).

Instead, we will look at the manifold-with-boundary $\bar{M}$ obtained by removing from K an *open* neighborhood of the vertex. We will give $\bar{M}$ a hyperbolic structure so that the boundary, which is a surface of genus two, is geodesic. To do this, consider again an ideal regular tetrahedron, centered at the origin in the projective model. If we move the vertices away from the origin, parts of the resulting tetrahedron are outside the sphere at infinity. To obtain a finite object again, we chop off the tips, cutting along planes dual to the vertices. Each such plane is perpendicular to the edges converging at the dual vertex, so we're left with a polyhedron that looks combinatorially like Figure 3.9, so long as the edges of the tetrahedron are not entirely outside infinity (in which case the truncating planes intersect one another within $\mathbf{H}^3$, and carve out a finite tetrahedron).

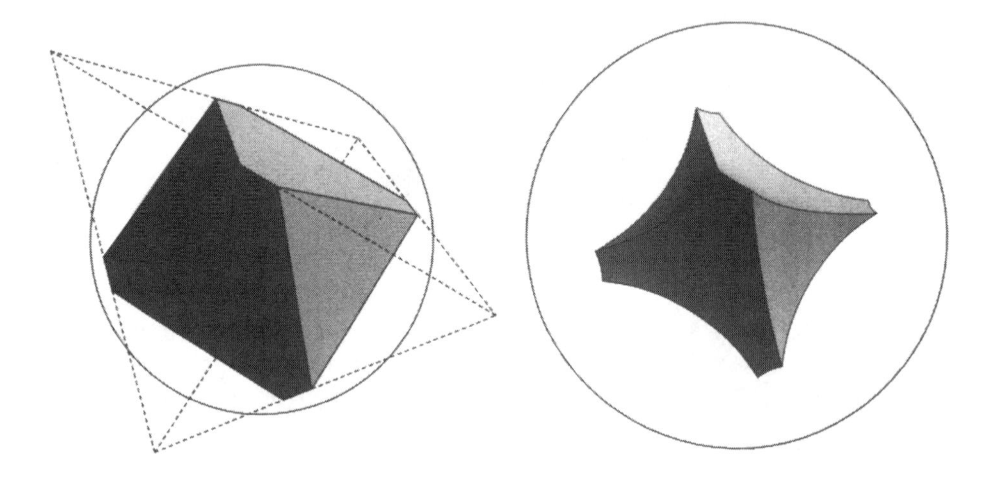

Figure 3.9. Truncated tetrahedron. A regular hyperbolic tetrahedron with vertices outside infinity, truncated by the planes dual to vertices, is shown in the projective model (left) and in the Poincaré model (right). All dihedral angles are either 30° (between hexagonal faces) or 90° (between a hexagonal and a triangular face).

The dihedral angles for the tetrahedron in this construction can be chosen at will between 60° (for the ideal tetrahedron) and 0° (for a tetrahedron with edges tangent to the sphere at infinity). Figure 3.9 shows the truncated tetrahedron with 30° angles, which is the one we need in order to fit together the twelve edges of the two tetrahedra. Gluing two of these in the pattern of Figure 3.8 gives a hyperbolic structure with geodesic boundary for $\bar{M}$: the faces of the original tetrahedra meet the boundary at right angles.

The manifold M of Example 3.3.12 turns out to be homeomorphic to the subset of $\mathbf{R}^3$ shown in Figure 3.10, whose complement in S^3 (the holes plus the outside of the ball, including the point at infinity) is a genus-two handlebody. In general, for any gluing diagram involving

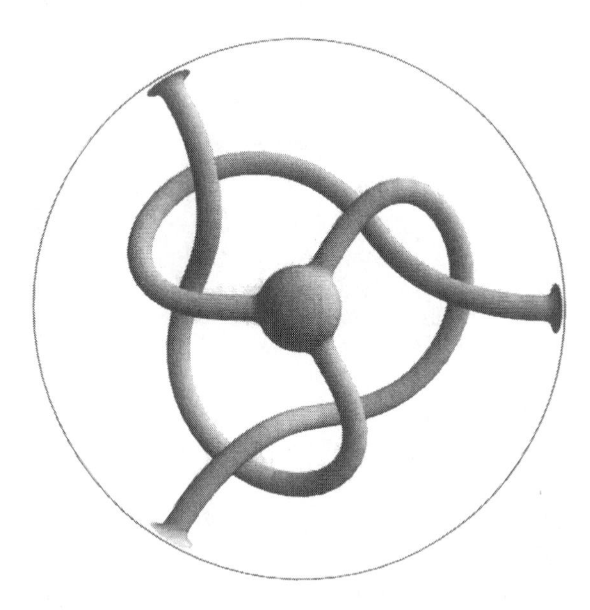

Figure 3.10. The tripus or knotted Y. The manifold of Example 3.3.12 is like a worm-eaten apple, with three holes that tangle and meet at the center as shown. The holes form a handlebody in the shape of a knotted Y or "tripus." See also Figure 3.14.

a single polyhedron, there is a method to try to realize the manifold resulting from the gluing (minus the vertices) as a submanifold of S^3. The method always works if each pair of faces is glued across a vertex of the polyhedron—under the conventions of Figure 3.8, this means that faces with the same letter meet at a vertex that has the same marking on both sides.

We illustrate the method for the gluing of Example 3.3.12 with a sequence of figures, following George Francis. Starting from Figure 3.8, we remove regular neighborhoods of edges and vertices, including the vertex at infinity. This leaves the surface gouged by furrows, which come together at pits (Figure 3.11, left). Next the pits grow into craters, while the furrows shorten and widen, becoming strips separating the craters (Figure 3.11, middle). The faces become small T-shaped mesas, the only elements to rise above the

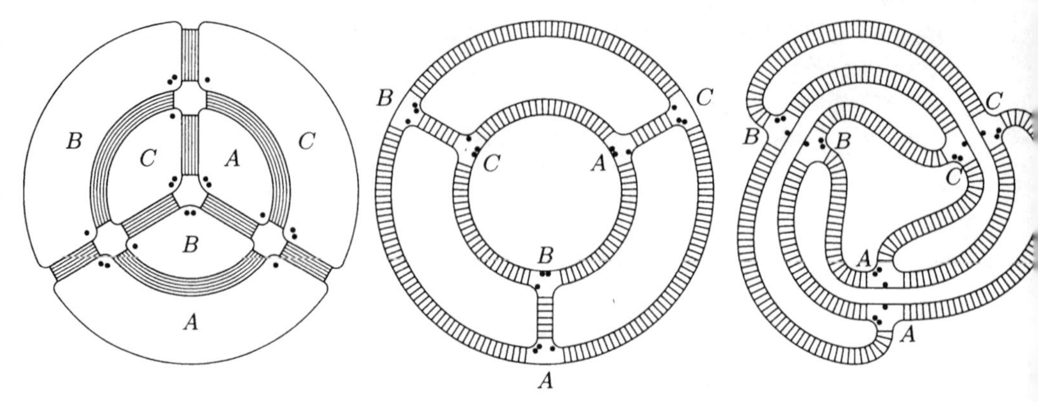

Figure 3.11. Preparing for the gluing. Starting from Figure 3.8, we remove neighborhoods of edges and vertices, then shrink the faces and rearrange so that matching faces are directly in front of one another.

level of the craters. By turning the inner circle 120° clockwise we can arrange for matching tees to face each other (Figure 3.11, right). It is here that we use the fact that each pair of matching faces is separated solely by a vertex (crater). Now the faces are in a convenient position for gluing.

Each gluing is performed as in Figure 3.12, by erecting an arc that bridges between facing tees. As we push down on a bridge and flatten it, the space below becomes an underground tube. The overall picture is now as in Figure 3.13, left. The nine original strips have combined into a single closed one, which we straighten out as

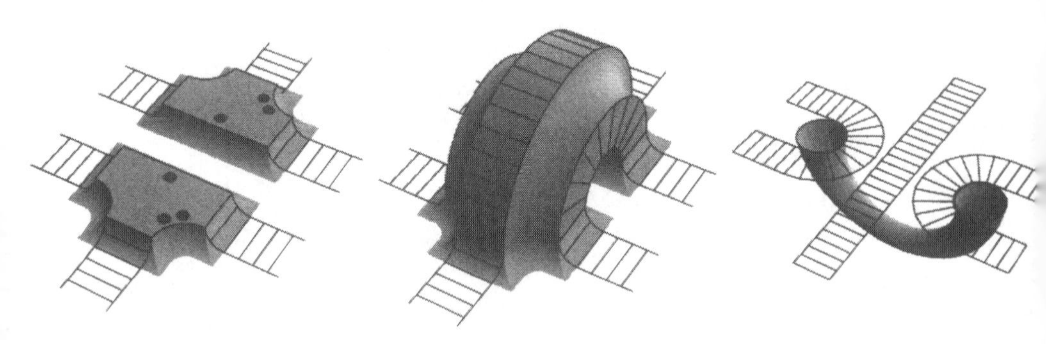

Figure 3.12. Gluing the faces. When two tees are glued together, they create a tunnel (middle), which we can push underground (right).

much as possible so that it will be easier to glue on the deleted edge neighborhoods. As we straighten out the strip, the underground tubes get stretched and entangled, since we're pulling them by their vents, which must remain on the same side of the strip where they start (Figure 3.13, middle).

Figure 3.13. Gluing the edge neighborhoods. After the face identifications have been performed, we are left with a single strip and three underground tubes under the (transparent) surface of the earth. We make the strip into an annulus, and glue onto it a thickened disk coming from the deleted edge neighborhoods. The final object is like an igloo on the surface of the earth, with three tunnel entrances to the igloo.

To conclude the construction, we must glue the deleted edge neighborhoods back on. They started in the shape of solid half-cylinders, but we distort them into wedges so that they fit together around the single edge of the identification space. The result is a cylindrical rod, or, equivalently, a thickened disk. We glue it back on in the form of a dome whose circumference matches the strip (Figure 3.13, right). Finally, we shrink the dome and press it down, pushing underground the air bubble enclosed by it, with its three vents. The bubble becomes the inner ball in Figure 3.10, where the earth is drawn round instead of flat.

Exercise 3.3.14. The method above sometimes works even when it is not the case that each pair of faces is to be glued across a vertex of the gluing diagram. For instance, only one pair of tees may be initially in a position to be glued, but, after this pair is glued, the change in the topology of the strips may allow another pair to be put in the right position, and so on. Make diagrams to illustrate this process for the gluings of Exercise 3.3.13 and Figure 3.6. (The first of these is given in [Fra87, 153–155].)

Figure 3.14. Ferguson's *Knotted Wye*. Helaman Ferguson's marble rendition of the knotted Y (see Figure 3.13) adorns the lobby of the University of Minnesota's Geometry Center. The three intertwining limbs of the Y meet in two masses, the bulge and the base, which should be imagined as separate (in the sculpture, the bulge is supported by the base, and the limbs are supported by one another). This arrangement is inspired by a trefoil knot: to go from the trefoil to the knotted Y, cut open the three outer loops, gather the six free ends in two triplets, and thicken.

3.4. The Developing Map and Completeness

Several times so far we have used the intuitive idea of unrolling a geometric manifold—for instance, in Section 1.1, we unrolled a Euclidean torus out on the plane, thus getting a tiling of the plane. In this section we will give a proper definition for the notion of unrolling, or *developing*, and generalize it to its natural context.

Let X be a connected real analytic manifold, and G a group of real analytic diffeomorphisms acting transitively on X. An element of G is then completely determined by its restriction to any open subset of X. We will look at a (G, X)-manifold M.

Let $U_1, U_2, \ldots$ be coordinate charts for M, with maps $\phi_i : U_i \mapsto X$ and transition functions

$$\gamma_{ij} = \phi_i \circ \phi_j^{-1} : \phi_j(U_i \cap U_j) \to \phi_i(U_i \cap U_j).$$

By the definition of a (G, X)-manifold, each γ_{ij} agrees locally with elements of G, and so can be seen as a locally constant map into G. By composing with ϕ_j we get a locally constant map $U_i \cap U_j \to G$, which we also call γ_{ij}.

Now suppose that (U_i, ϕ_i) and (U_j, ϕ_j) are two charts whose domains contain the same point x. Then we can modify ϕ_j by composing with $\gamma_{ij}(x) \in G$, so that it coincides with ϕ_i around the point x. In fact, if $U_i \cap U_j$ is connected, the modified ϕ_j agrees with ϕ_i in the whole intersection, so we get a map $U_i \cup U_j \to X$ that extends ϕ_i. But in general we run into inconsistencies if we try to extend the coordinate charts too far. The way to avoid the inconsistencies is to pass to the universal cover.

Fix a basepoint $x_0 \in M$ and a chart (U_0, ϕ_0) whose domain contains the basepoint, and let $\pi : \tilde{M} \to M$ be the universal covering map of M. We think of $\tilde{M}$ as the space of homotopy classes of paths in M that start at the basepoint, and we take a path α representing a point $[\alpha] \in \tilde{M}$, so that $\alpha(1) = \pi([\alpha])$. We subdivide α at points

$$x_0 = \alpha(t_0),\ x_1 = \alpha(t_1),\ \ldots, x_n = \alpha(t_n),$$

where $t_0 = 0$ and $t_n = 1$, so that each subpath is contained in the domain of a single coordinate chart (U_i, ϕ_i). We go along α, successively adjusting each chart ϕ_i so it agrees with (the previously adjusted) ϕ_{i-1} in a neighborhood of $x_i \in U_{i-1} \cap U_i$. These adjusted charts form the *analytic continuation* of ϕ_0 along the path (Figure 3.15). The last adjusted chart is

$$\psi = \gamma_{01}(x_1)\gamma_{12}(x_2) \ldots \gamma_{n-1,n}(x_{n-1})\phi_n.$$

Exercise 3.4.1 (analytic continuation is well defined). Verify the preceding formula. Show that the germ of ψ at $\alpha(1)$ does not depend on the choices made during the process of analytic continuation, or even on the path α, so long as α varies within the same homotopy class. Therefore we can set $\phi_0^{[\alpha]} = \psi$, and the notation is well defined. How does $\phi_0^{[\alpha]}$ change if we change the basepoint x_0 or the initial chart ϕ_0?

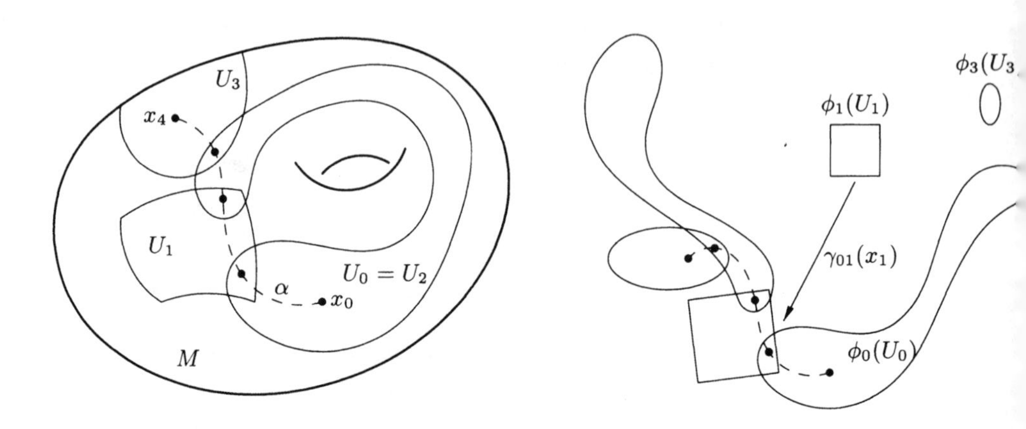

Figure 3.15. Analytic continuation. Here M is an affine torus (left). The path α (dashed) is subdivided at points $x_0, x_1, \ldots, x_4$ (marked by dots) so that each segment $[x_i, x_{i+1}]$ lies entirely in a coordinate patch U_i. The analytic continuation of ϕ_0 along α is ϕ_0 itself on a neighborhood of the first segment, $\gamma_{01}(x_1)\phi_1$ on a neighborhood of the second, and so on. The analytic continuation can be thought of as a multivalued map from M to $X = \mathbf{E}^2$, or as a map from the universal cover $\tilde{M}$ to X.

Definition 3.4.2 (developing map). For a fixed basepoint and initial chart ϕ_0, the *developing map* of a (G, X)-manifold M is the map $D : \tilde{M} \to X$ that agrees with the analytic continuation of ϕ_0 along each path, in a neighborhood of the path's endpoint. In symbols,

$$D = \phi_0^\sigma \circ \pi$$

in a neighborhood of $\sigma \in \tilde{M}$. If we change the initial data (the basepoint and the initial chart), the developing map changes by composition in the range with an element of G (Exercise 3.4.1).

If we give $\tilde{M}$ the (G, X)-structure induced by the covering map π (Exercise 3.1.5), the developing map is a local (G, X)-diffeomorphism between $\tilde{M}$ and X.

Although G acts transitively on X in the cases of primary interest, this condition is not necessary for the definition of D. For example, if G is the trivial group and X is closed, then closed (G, X)-manifolds are precisely the finite-sheeted covers of X, and D is the covering projection.

Now let σ be an element of the fundamental group of M. Analytic continuation along a loop representing σ gives a germ ϕ_0^σ that is comparable to ϕ_0, since they are both defined at the basepoint (Figure 3.16). Let g_σ be the element of G such that $\phi_0^\sigma = g_\sigma \phi_0$;

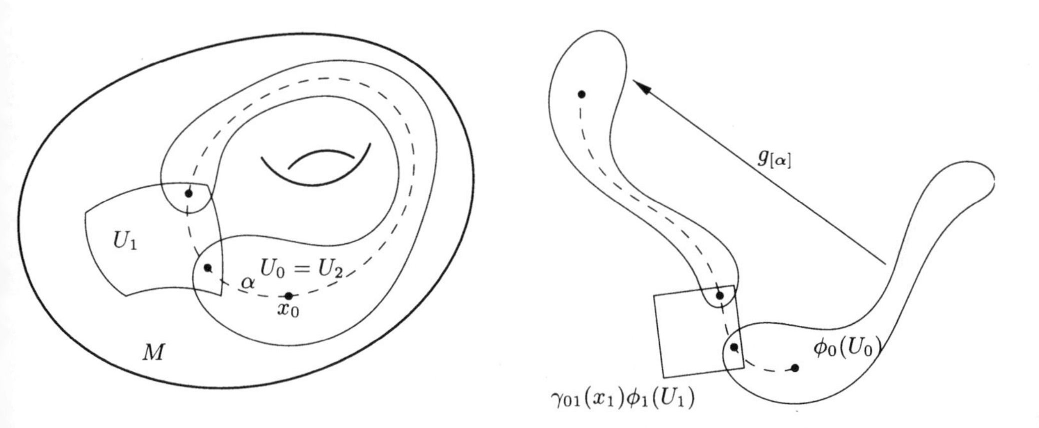

Figure 3.16. The holonomy around a path. For the torus of Figure 3.15, analytic continuation around the loop α requires two coordinate changes: from ϕ_0 to $\gamma_{01}(x_1)\phi_1$ to $\gamma_{01}(x_1)\gamma_{12}(x_2)\phi_2$. Therefore the holonomy around α is $g_{[\alpha]} = \gamma_{01}(x_1)\gamma_{12}(x_2) = \gamma_{01}(x_1)\gamma_{01}^{-1}(x_2)$.

we call g_σ the *holonomy* of σ. It follows easily from Definition 3.4.2 that

$$D \circ T_\sigma = g_\sigma \circ D,$$

where $T_\sigma : \tau \mapsto \sigma\tau$ is the covering transformation associated with σ. Applying this equation to a product, we see that the map $H : \sigma \mapsto g_\sigma$ from $\pi_1(M)$ into G is a group homomorphism, which we call the *holonomy* of M. Its image is the *holonomy group* of M. Note that H depends on the choices involved in the construction of D (Definition 3.4.2): when D changes, H changes by conjugation in G.

Exercise 3.4.3. What are the developing maps and holonomies of

(a) a Euclidean torus, as in Example 3.3.2, and

(b) an affine torus, as in Example 3.3.4 (see Figure 3.17).

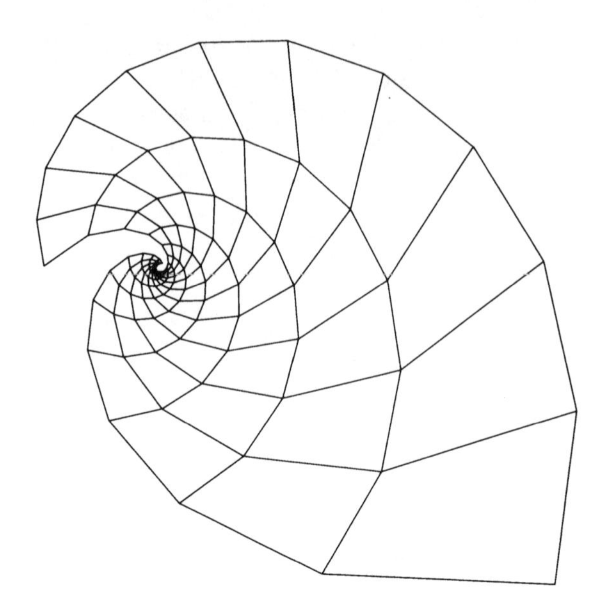

Figure 3.17. Developing the affine torus. The developing map of an affine torus constructed from a quadrilateral (Example 3.3.4) generally omits a single point of the plane.

Exercise 3.4.4 (free homotopy classes). Recall that two loops in a path-connected space are *freely homotopic* if they are homotopic as maps from S^1 into the space (without reference to a basepoint).

(a) Free homotopy classes of loops are in one-to-one correspondence with conjugacy classes of elements of the fundamental group.

(b) Let $D : \tilde{M} \to X$ be the developing map of a (G, X)-manifold M. To a free homotopy class of loops in M is associated a conjugacy class in the holonomy group of M. Given a representative $\alpha : S^1 \to M$ of this class, the choice of a lift $\tilde{\alpha} : \mathbf{R} \to \tilde{M}$ of α pinpoints the holonomy of α.

In general, the holonomy of M need not determine the (G, X)-structure on M, but there is an important special case in which it does. We say that M is a *complete* (G, X)*-manifold* if $D : \tilde{M} \to X$ is a covering map. Since a covering of a simply connected space is a homeomorphism, we see that if M is complete and X is simply connected, we can identify $\tilde{M}$ with X by the developing map, and often will do so tacitly. The identification is canonical up to composition in the range with an element of G. We also get the following result:

Proposition 3.4.5 (holonomy group characterizes manifold). *If G is a group of analytic diffeomorphisms of a simply connected space X, any complete (G, X)-manifold may be reconstructed from its holonomy group Γ, as the quotient space X/Γ.*

Convention 3.4.6. Since Γ acts on X from the left, it would be more correct to write the quotient as $\Gamma\backslash X$. However, for our purposes the distinction is so seldom important that for esthetic reasons we will continue to write X/Γ.

Exercise 3.4.7. The torus of Figure 3.17 is plainly not complete. Find conditions for an affine torus obtained by Smillie's procedure (Example 3.3.4) to be complete.

Exercise 3.4.8 (similarity manifolds). A *similarity structure* is a $(\mathrm{Sim}(n), \mathbf{E}^n)$-structure, where $\mathrm{Sim}(n)$ is the group of similarities of $\mathbf{E}^n$. What are the possible similarity structures on S^1? Which of them are complete?

Because of the significant relation implied by Proposition 3.4.5, it is often worthwhile replacing a model space X that is not simply connected by its universal cover $\tilde{X}$. There is a covering group $\tilde{G}$ acting on $\tilde{X}$, namely the group of homeomorphisms of $\tilde{X}$ that are lifts of elements of G. Then $\tilde{G}$ can be described in the form of an extension

$$1 \to \pi_1(X) \to \tilde{G} \to G \to 1.$$

There is a one-to-one correspondence between (G, X)-structures and $(\tilde{G}, \tilde{X})$-structures, but the holonomy for a $(\tilde{G}, \tilde{X})$-structure contains more of the information.

Exercise 3.4.9 (analysis of stabilizers). We investigate more about the structure of $\pi_1(X)$ and $\tilde{G}$ when X is not simply connected.

(a) Show that the fundamental group of any topological group is abelian.

(b) If G is a topological group and $H \subset G$ a closed subgroup, we say that G has a *local cross-section* (with respect to H) if there is a set S of unique coset representatives in a neighborhood of the identity in G/H, such that the map $x \mapsto Hx$ is a homeomorphism on S.

Let X be a manifold on which G acts transitively by homeomorphisms, and take a point $x \in X$, with stabilizer G_x. Assume that G has a local cross-section with respect to G_x, and that the map $p : g \mapsto gx$ is open.

Construct a homomorphism from the fundamental group of X to the group of path components of G_x, and show that the kernel of this homomorphism is central. In particular, if G_x is path-connected, the fundamental group of X is abelian.

(c) (Harder.) Show that the local cross-section and openness conditions are satisfied if G is a Lie group, for example.

Here is a useful sufficient condition for completeness.

Proposition 3.4.10 (compact stabilizers imply completeness). *Let G be a Lie group acting analytically and transitively on a manifold X, and such that the* †*stabilizer G_x of x is compact, for some (hence all) $x \in X$. Then every closed (G, X)-manifold M is complete.*

Proof of 3.4.10. Transitivity implies that the given condition at one point x is equivalent to the same condition everywhere. So fix $x \in X$, and let T_xX be the tangent space to X at x. There is an analytic homomorphism of G_x to the linear group of T_xX, whose image is compact. We now use the following fact, which is important in its own right:

Lemma 3.4.11 (existence of invariant metric). *Let G act transitively on an analytic manifold X. Then X admits a G-invariant Riemannian metric if and only if, for some $x \in X$, the image of G_x in* $\mathrm{GL}(T_xX,)$ *has compact closure.*

Proof of 3.4.11. One direction is clear: If G preserves a metric, G_x maps to a subgroup of $O(T_xX)$, which is compact. To prove the converse, fix x and assume that the image of G_x has compact closure H_x. Let Q be any positive definite form on T_xX. Using Haar measure on H_x (Exercise 2.5.20), average the set of transforms g^*Q, for $g \in H_x$, to obtain a quadratic form on T_xX invariant under H_x. Propagate this to the tangent space of every other point $y \in X$, by pulling back by any element $g \in G$ that takes y to x; the pullback is independent of the choice of g. The resulting Riemannian metric is invariant under G. $\boxed{3.4.11}$

Exercise 3.4.12. Using a local cross-section (see Exercise 3.4.9) for G with respect to H_x, prove that any G-invariant Riemannian metric is analytic, and in particular the one found at the end of the above lemma.

If M is any (G, X)-manifold, we can use charts to pull back the invariant Riemannian metric from X to M; the result is a Riemannian metric on M invariant under any (G, X)-map. Now in a Riemannian manifold, we can find for any point y a ball $B_\varepsilon(y)$ of radius $\varepsilon > 0$ that is ball-like (that is, a homeomorphic image of the round ball under the exponential map), and convex (any two points in it are joined by a unique geodesic inside the ball). If M is closed, we can choose ε uniformly, by compactness. We may also assume that all ε-balls in X are contractible and convex, since G is a transitive group of isometries.

Then, for any $y \in \tilde{M}$, the ball $B_\varepsilon(y)$ is mapped homeomorphically by D, for, if $D(y) = D(y')$ for $y' \neq y$ in the ball, the geodesic connecting y to y' maps to a self-intersecting geodesic, contradicting the convexity of ε-balls in X. Furthermore, D is an isometry between $B_\varepsilon(y)$ and $B_\varepsilon(D(y))$, by definition.

Now take $x \in X$ and $y \in D^{-1}(B_{\varepsilon/2}(x))$. The ball $B_\varepsilon(y)$ maps isometrically, and thus must properly contain a homeomorphic copy of $B_{\varepsilon/2}(x)$. The entire inverse image $D^{-1}(B_{\varepsilon/2}(x))$ is then a disjoint union of such homeomorphic copies. Therefore D †evenly covers X, so it is a covering projection, and M is complete. 3.4.10

Example 3.4.13. Proposition 3.4.10 says that the universal cover of a closed elliptic three-manifold M is S^3. Since S^3 is compact, $\pi_1(M)$ is finite. In particular, if M is simply connected it is homeomorphic to S^3.

Example 3.4.14. A topological space X whose universal cover is contractible is called an *Eilenberg–MacLane space*, and is called a $K(G, 1)$, where G is the fundamental group of X. (The 1 means that X satisfies the condition that $\pi_k(X)$ is trivial for $k \neq 1$; there are analogous definitions for $K(G, n)$ for arbitrary n.) A space which is a $K(G, 1)$ is determined by G up to homotopy equivalence. From Proposition 3.4.10 we see that closed hyperbolic manifolds and closed Euclidean manifolds are $K(G, 1)$'s. This means that their fundamental groups are extremely important.

For manifolds that are not closed, checking completeness is a more subtle business. The following criterion is often helpful, and it justifies the use of the word "complete" for (G, X)-manifolds.

Proposition 3.4.15 (meanings of completeness). *Let G be a transitive group of real analytic diffeomorphisms of X with compact stabilizers G_x. Fix a G-invariant metric on X (whose existence is guaranteed by Lemma* 3.4.11*), and let M be a (G, X)-manifold with the metric inherited from X. The following conditions are equivalent:*

(a) *M is a complete (G, X)-manifold.*

(b) *For some $\varepsilon > 0$, every closed ε-ball in M is compact.*

(c) *For every $a > 0$, every closed a-ball in M is compact.*

(d) *There is some family of compact subsets S_t of M, for $t \in \mathbf{R}_+$, such that $\bigcup_{t\in\mathbf{R}^+} S_t = M$ and S_{t+a} contains the neighborhood of radius a about S_t.*

(e) *M is complete as a metric space.*

Proof of 3.4.15. (a) $\Rightarrow$ (b). If $p : Y \to Z$ is a covering map between Riemannian manifolds and p preserves the Riemannian metric, we have $\bar{B}_\varepsilon(p(y)) = p(\bar{B}_\varepsilon(y))$ for any $y \in Y$ and any $\varepsilon \in \mathbf{R}$, because distances are defined in terms of path lengths, and paths in Z can be lifted to paths in Y. So the compactness of ε-balls in Y implies the same in Z, and conversely. Now we choose $\varepsilon > 0$ so that the closed ε-ball about some point in X is compact; since G acts transitively on X, the same ε works for all points. Therefore closed ε-balls are compact in X, and likewise in $\tilde{M}$ and in M.

(b) $\Rightarrow$ (c). We show this by induction. Suppose that (c) is true for a certain value of $a \geq \varepsilon$. Then $\bar{B}_a(x)$ can be covered with finitely many $\varepsilon/2$-balls, and therefore $\bar{B}_{a+\varepsilon/2}(x)$ can be covered with a finite number of ε-balls, which are compact. This shows that $\bar{B}_{a+\varepsilon/2}(x)$ is compact.

(c) $\Rightarrow$ (d). Let S_t be the ball of radius t about some fixed point.

(d) $\Rightarrow$ (e). Any Cauchy sequence in M must be contained in S_t, for some sufficiently high t, hence it converges.

(e) $\Rightarrow$ (a). Suppose M is metrically complete. We will show that the developing map $D : \tilde{M} \to X$ is a covering map by proving that any path α_t in X can be lifted to $\tilde{M}$ (since local homeomorphisms with the path-lifting property are covering maps). First we need to see that $\tilde{M}$ is metrically complete. In fact, the projection to M of any Cauchy sequence in $\tilde{M}$ has a limit point $x \in M$. Since x has a compact neighborhood which is evenly covered in $\tilde{M}$, and whose components are definitely separated in the metric of $\tilde{M}$, the Cauchy sequence converges also in $\tilde{M}$.

Consider now any path α_t in X. If it has a lifting $\tilde{\alpha}_t$ for t in a closed neighborhood $[0, t_0]$, then it has a lifting for $t \in [0, t_0 + \varepsilon)$, for some $\varepsilon > 0$. On the other hand, if it has a lifting for t in a half-open interval $[0, t_0)$, the lifting extends to $[0, t_0]$ by the completeness of $\tilde{M}$. Hence, M is complete as a (G, X)-manifold. 3.4.15

Conditions (b)–(e) are equivalent for any Riemannian manifold, not just a (G, X)-manifold. This is part of a fundamental theorem in Riemannian geometry, due to Hopf and Rinow [HR31], which also says that metric completeness is equivalent to geodesic completeness, in the sense that the exponential map at any point is globally defined. See [dC92, Chapter 7] and [GHL90], for example.

For hyperbolic manifolds obtained by gluing polyhedra, condition (d) seems to be the most useful for showing a structure is complete, and condition (e) for showing a structure is not complete (see Example 3.4.17, for instance).

Example 3.4.16 (completeness of knot complements). The hyperbolic structure that we constructed for the figure-eight knot complement in Example 3.3.7 is complete. Recall that we glued two copies T and T' of an ideal regular tetrahedron. The gluing maps look like Euclidean isometries when T and T' are symmetrically embedded in the projective model, as in Figure 3.3(b). Now an ideal regular tetrahedron has a finite symmetry group, and therefore a unique center of symmetry (Corollary 2.5.19), coinciding with the center of the ball in Figure 3.3(b). The balls of radius t about the centers of T and T', intersected with the respective tetrahedra, are matched nicely by the gluing maps, so they yield sets S_t meeting criterion (d) of Proposition 3.4.15.

The same argument shows that the hyperbolic structures that we gave to the complements of the Whitehead link (Example 3.3.9) and of the Borromean rings (Example 3.3.10) are complete.

Example 3.4.17 (completeness of plane gluings). If we glue two ideal hyperbolic triangles in the pattern indicated by Figure 3.18, we get a sphere with three punctures. Unlike the previous exam-

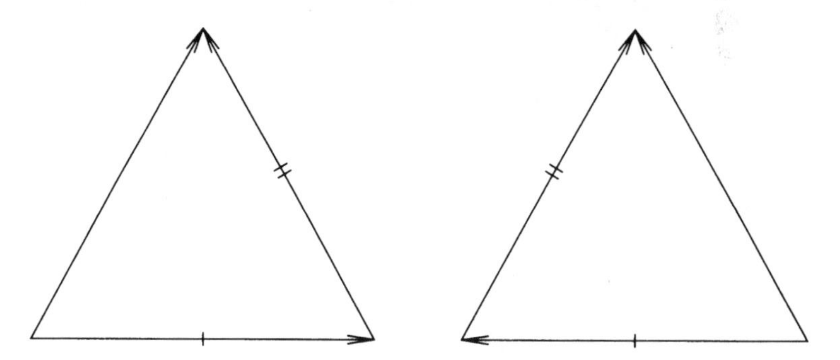

Figure 3.18. Gluing diagram for the thrice punctured sphere.

ple, here there is freedom in the choice of the gluing maps: the identification between paired sides may be modified by an arbitrary translation. Therefore, we have not just one, but a family of hyperbolic structures on the thrice punctured sphere, parametrized by $\mathbf{R}^3$. (They need not be all distinct, and in fact they are not.) We are interested in knowing which among these structures are complete.

More generally, suppose S is an oriented hyperbolic surface obtained by gluing ideal hyperbolic triangles, leaving no free edges. An *ideal vertex* of S is an equivalence class of vertices of the component triangles, identified according to the gluing pattern. Associated with each ideal vertex v of S there is an invariant $d(v)$, defined as follows:

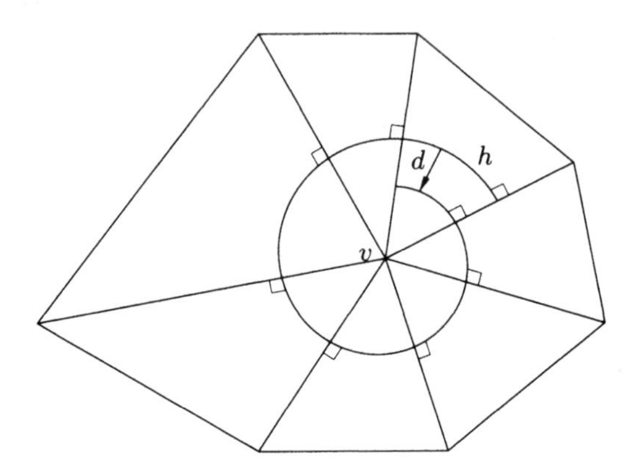

Figure 3.19. Extending a horocycle. The distance $d = d(v)$ associated with an ideal vertex v encodes the holonomy of a path going once around v, counterclockwise.

In one of the triangles incident on v, let h be a horocycle segment "centered at v", that is, orthogonal to the two sides of the triangle that go off to v. Extend h counterclockwise about v by horocycle segments that meet successive edges orthogonally. Eventually this horocyclic curve will reenter the original triangle as a horocycle concentric with h, at a signed distance $d(v)$ from h—by convention, the sign is positive in the situation of Figure 3.19. It is easy to see that $d(v)$ does not depend on the initial choice of h.

Proposition 3.4.18 (completeness criterion for plane gluings). *The surface S is complete if and only if $d(v) = 0$ for all ideal vertices v.*

Proof of 3.4.18. Suppose that $d(v) > 0$ for some v. Extending the horocyclic curve further around the vertex, we see that it has bounded length, because each successive circuit is shorter than the previous one by a constant factor < 1. The sequence of intersections of the curve with one of the edges is a non-convergent Cauchy sequence, so S fails condition 3.4.15(e). The case $d(v) < 0$ is analogous: we just extend the horocycle clockwise.

If, on the other hand, $d(v)$ is zero for all vertices, we can remove small horoball neighborhoods of each vertex to obtain a compact subset $S_0 \subset S$. Let S_t be the set obtained by removing from S smaller horoball neighborhoods, bounded by horocycles a distance t from the original ones. The subsets S_t satisfy 3.4.15(d), so S is complete. 3.4.18

Exercise 3.4.19. In the gluing of Figure 3.18, show that the conditions $d(v) = 0$ give three linearly independent linear equations in three variables, so there is only one way to get a complete surface. Show that, conversely, any complete thrice punctured sphere can be decomposed into two ideal triangles. In the language of Section 4.6, the Teichmüller space of a thrice punctured sphere is a single point. See also Figure 3.20.

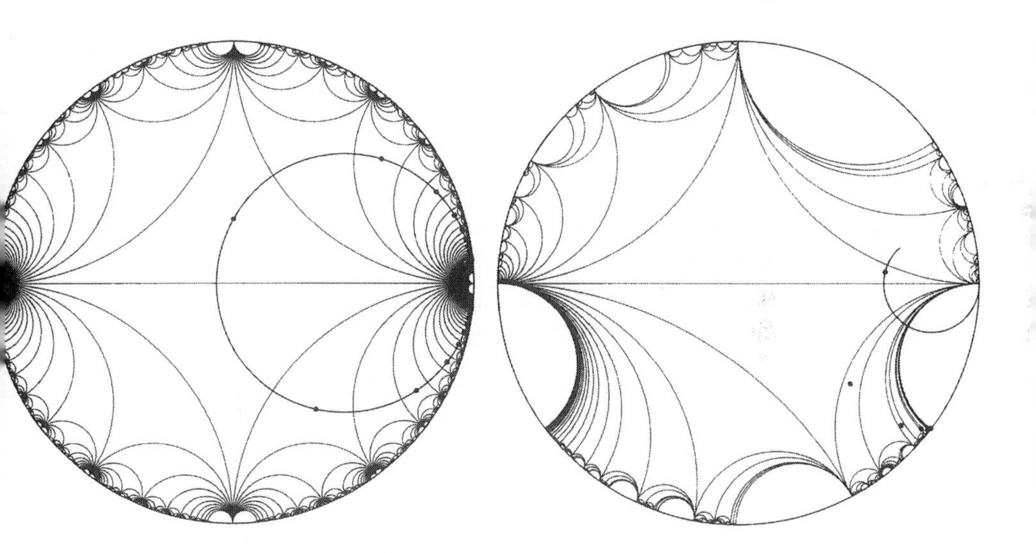

Figure 3.20. Development of the thrice punctured sphere. A complete (left) and an incomplete (right) hyperbolic surface obtained from the gluing of Figure 3.18 are developed onto the hyperbolic plane. In the first case, horocyclic continuation gives a closed curve, always passing through the same reference point in the triangle; on the right, it gives a curve that moves away from the reference point. Also note, on the right, the regions of $\mathbf{H}^2$ that are not in the image of the developing map.

Exercise 3.4.20. Show that the holonomy around an ideal vertex such that $d(v) = 0$ is parabolic, and that the corresponding end of the surface (a "neighborhood" of the ideal vertex) is isometric to the thin end of the pseudosphere. We call such ideal vertices *cusps*—they will play a prominent role in Section 4.5 (The Thick-Thin Decomposition). The next proposition describes the situation around a vertex with $d(v) \neq 0$.

Proposition 3.4.21 (completion of a hyperbolic surface). *The metric completion $\bar{S}$ of S is a hyperbolic surface with geodesic boundary. It has one boundary component of length $|d(v)|$ for each ideal vertex v where $d(v) \neq 0$.*

Proof of 3.4.21. For a given ideal vertex v with $d(v) \neq 0$, each horocyclic spiral has an endpoint in the metric completion. Distinct spirals are uniformly spaced, so they have distinct endpoints. The family of spirals is parametrized by their intersection with some fixed edge heading toward v, with intersections that are $d(v)$ apart corresponding to the same spiral. Thus we must adjoin a circle of endpoints for each ideal vertex v with $d(v) \neq 0$. This circle is the set of accumulation points of a geodesic (for instance, an edge incident on v), so is itself geodesic. After adjoining these endpoints, one readily verifies that every Cauchy sequence converges, so $\bar{S}$ is a surface with boundary. 3.4.21

It is easier to visualize this situation by turning Figure 3.19 around: instead of geodesics heading straight toward v and horocycles that spiral toward v, the more natural picture is one of horocycles heading straight toward the boundary circle, and geodesics orthogonal to them spiraling around the circle, as in Figure 3.21.

Clearly, the analysis in Example 3.4.17 applies not only when S is a gluing of ideal triangles, but also when it is a gluing of triangles with some or all vertices finite, so long as finite vertices are matched to finite vertices and the sum of angles at each finite vertex of the identification space is 2π.

We now reformulate Proposition 3.4.18 in a way that admits of an easy generalization to higher dimensions. Let v be an ideal vertex of a surface S obtained by gluing hyperbolic triangles. Then the link $\operatorname{lk} v$, which is a circle, has a canonical similarity structure (Exercise 3.4.8).

To see this, we identify the portion of $\operatorname{lk} v$ lying in a single triangle incident on v with an arbitrarily chosen horocycle segment centered at v, as in Figure 3.19. By considering the upper half-plane

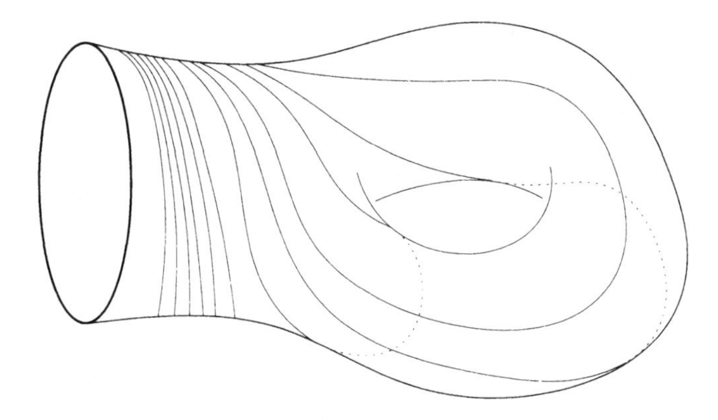

Figure 3.21. Geodesics accumulate on the boundary. An ideal vertex for which $d(v)$ is not zero yields a circle of length $|d(v)|$ in the compactification. The edges of the glued-up triangles spiral around the boundary circle.

model, with v at infinity, we see that this horocyclic cross-section has a Euclidean structure, and that projection between two such cross-sections is a Euclidean similarity, by Exercise 2.2.10. Thus $\operatorname{lk} v$ has a similarity structure within each triangle. These local structures piece together to give a similarity structure on all of $\operatorname{lk} v$, because when we go around the vertex we end up with a cross-section parallel to the one we started with, and therefore differing from it by a similarity. In fact, it is easy to see that the similarity is the contraction by $e^{d(v)}$. Thus, saying that $d(v) = 0$ is equivalent to saying that the holonomy of the similarity structure on $\operatorname{lk} v$ consists only of isometries. This, in turn, is equivalent to the existence of a Euclidean stiffening for the similarity structure:

Exercise 3.4.22 (criterion for stiffening). If G is a group acting analytically on a manifold X and H is a subgroup of G, a G-structure has an H-stiffening if and only if its holonomy group is conjugate to a subgroup of H.

Here then is the promised generalization of Proposition 3.4.18. Consider a gluing of n-dimensional polyhedra (according to the definition on page 123, this means all faces are paired). Suppose the polyhedra are realized in $\mathbf{H}^n$, possibly with some vertices at infinity; that the gluing maps are isometries; and that the resulting space M (not including the vertices at infinity) is a hyperbolic manifold. An *ideal vertex* of M is, of course, an equivalence class of vertices at infinity of the component polyhedra. If v is an ideal vertex, we give its link $\operatorname{lk} v$ a similarity structure coming from horospherical cross-sections of the polyhedra incident on v, just as in the case of a surface.

Theorem 3.4.23 (completeness criterion for gluings). *The following conditions are equivalent:*

(a) *M is complete.*

(b) *For each ideal vertex v of M, the holonomy of* $\operatorname{lk} v$ *consists of isometries.*

(c) *For each ideal vertex v, the similarity structure on* $\operatorname{lk} v$ *can be stiffened to a Euclidean structure.*

(d) *For each ideal vertex v, the link* $\operatorname{lk} v$ *is complete as a similarity manifold.*

Proof of 3.4.23. Saying that the holonomy of $\operatorname{lk} v$ consists of isometries is saying that by following any loop in the link we get back to the same horospherical cross-section we started with. Thus the partial cross-sections match up to form a global cross-section, which bounds a "horoball neighborhood" of v. Deleting horoball neighborhoods of all the ideal vertices, we get a compact subset M_0. As in the proof of Proposition 3.4.18, we can then shrink the deleted neighborhoods to get a sequence of compact subsets M_t that satisfy Proposition 3.4.15(d).

On the other hand, if the holonomy of some loop in $\operatorname{lk} v$ is a contraction rather than an isometry, the horospherical cross-section is pushed toward v by a fixed distance every time we go around this loop. This means that the distance between corresponding points on successive cross-sections decreases exponentially, so these points form a Cauchy sequence. Again as in the proof of Proposition 3.4.18, this sequence has no limit point in M, so M is not complete. This proves the equivalence of (a) and (b).

The equivalence between (b) and (c) follows from Exercise 3.4.22, since $\operatorname{Isom} \mathbf{E}^n$ is normal in $\operatorname{Sim}(n)$.

To see that (c) implies (d), observe that $\operatorname{lk} v$ is a closed manifold (since there are no free facets in the gluing), and apply Proposition 3.4.10.

Finally, (d) implies (b): if $\operatorname{lk} v$ is complete as a similarity manifold, then $\mathbf{E}^n$ is its universal cover and its holonomy has no fixed points. Therefore, the holonomy consists of isometries. 3.4.23

Exercise 3.4.24. When M is not complete, the similarity structure on $\operatorname{lk} v$ may depend on the decomposition of M into ideal polyhedra.

Exercise 3.4.25. Show that the only closed oriented surface that has a similarity structure is the torus. (Compare with Problem 1.3.15.) Thus the existence of a similarity structure on the links imposes topological constraints on a hyperbolic manifold obtained by gluing, even when the manifold is not complete.

Exercise 3.4.26. Use Theorem 3.4.23 to prove that the structures in the complements of the figure-eight knot (Example 3.3.7), the Whitehead link (Example 3.3.9) and the Borromean rings (Example 3.3.10) are complete. (This was done by a different method in Example 3.4.16).

Exercise 3.4.27. What are the analogs of Proposition 3.4.21 and Figure 3.21 in three dimensions? We will return to this point in a later volume, when we discuss Dehn surgery and rigidity of geometric structures. We will see then that we can deduce non-trivial topological information by studying the completions.

3.5. Discrete Groups

According to Proposition 3.4.5, when G is a group of analytic diffeomorphisms of a simply connected manifold X, complete (G, X)-manifolds (up to isomorphism) are in one-to-one correspondence with certain subgroups of G (up to conjugacy by elements of G). There are certain traditional fallacies concerning the characterization of the groups that are holonomy groups for complete (G, X)-manifolds, so it is worth going through the definitions carefully.

Definition 3.5.1 (group action properties). Let Γ be a group acting on a topological space X by homeomorphisms. We will normally consider the action to be *effective*; this means that the only element of Γ that acts trivially is the identity element of Γ, so in effect we can see Γ as a group of homeomorphisms of X. Here are other properties that the action might have:

(i) The action is *free* if no point of X is fixed by an element of Γ other than the identity.

(ii) The action is *discrete* if Γ is a discrete subset of the group of homeomorphisms of X, with the †compact-open topology.

(iii) The action *has discrete orbits* if every $x \in X$ has a neighborhood U such that the set of $\gamma \in \Gamma$ mapping x inside U is finite.

(iv) The action is *wandering* if every $x \in X$ has a neighborhood U such that the set of $\gamma \in \Gamma$ for which $\gamma U \cap U \neq \varnothing$ is finite.

(v) Assume X is locally compact. The action of Γ is *properly discontinuous* if for every compact subset K of X the set of $\gamma \in \Gamma$ such that $\gamma K \cap K \neq \varnothing$ is finite.

Exercise 3.5.2. Show that Γ acts properly discontinuously if and only if the map $\Gamma \times X \to X \times X$ given by $(\gamma, x) \mapsto (\gamma x, x)$ is proper. (A *proper map* is one under which the inverse image of any compact set is compact.)

Exercise 3.5.3 (property hierarchy). Show that each of the last three properties in 3.5.1 is (strictly) stronger than the previous one. (Hint for the inequivalence of (iii) and (iv): Construct an action of $\mathbf{Z}$ on $S^1 \times \mathbf{R} \times [-1, 1] \setminus \{s_0\} \times \mathbf{R} \times \{0\}$, with discrete orbits, by choosing a family of diffeomorphisms of the cylinders $S^1 \times \mathbf{R} \times \{t\}$ whose orbits spiral with slope t.)

Proposition 3.5.4 (wandering and free implies manifold quotient). *Let Γ be a group acting freely on a connected (Hausdorff) manifold X, and assume the action is wandering. Then the quotient space X/Γ is a (possibly non-Hausdorff) manifold and the quotient map is a covering map.*

Proof of 3.5.4. Given $x \in X$, take a neighborhood U of x that intersects only finitely many of its translates γU. Using the Hausdorffness of X and the freeness of the action, choose a smaller neighborhood of x whose translates are all disjoint. Then each translate maps homeomorphically to its image in the quotient, so the image is evenly covered. Since x was arbitrary, this implies the claim. $\boxed{3.5.4}$

Example 3.5.5 (non-Hausdorff quotient). Let $X = \mathbf{R}^2 \setminus 0$ and let $\Gamma = \mathbf{Z}$ be the group of diffeomorphisms generated by $(x, y) \mapsto (2x, \frac{1}{2}y)$. The conditions of Proposition 3.5.4 are met, but the quotient $M = X/\Gamma$ is not Hausdorff (Figure 3.22).

If we modify X slightly, by removing the negative y-axis, we get a free and wandering action of $\mathbf{Z}$ on $\mathbf{R}^2$. (In fact, the remarkable *plane translation theorem*, due to Brouwer [Bro12], says among other things that every free action of $\mathbf{Z}$ on $\mathbf{R}^2$ is wandering.) The quotient is, again, a non-Hausdorff manifold. In particular, there are non-homeomorphic non-Hausdorff two-manifolds with fundamental group $\mathbf{Z}$, and universal covering space homeomorphic to $\mathbf{R}^2$.

Exercise 3.5.6. Describe the quotients of Example 3.5.5 as a gluing of a small number of (Hausdorff) surfaces.

We now reinstate the convention that all manifolds are Hausdorff unless we state otherwise.

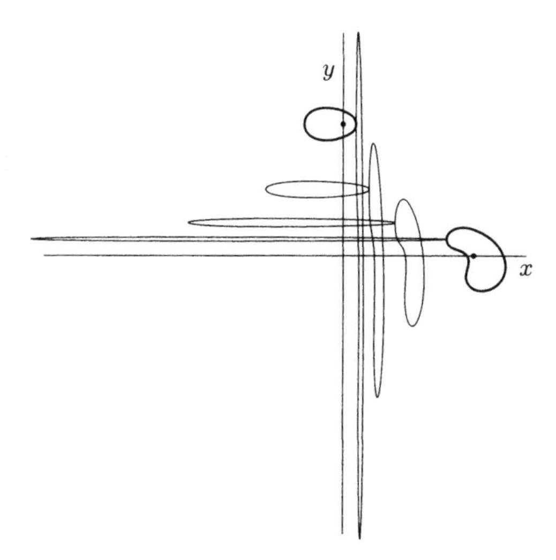

Figure 3.22. Action with non-Hausdorff quotient. The quotient of $\mathbf{R}^2 \setminus 0$ by the action generated by the linear map $(x, y) \mapsto (2x, \frac{1}{2}y)$ is a manifold, because the action is wandering and free, but it is not Hausdorff: every neighborhood of the image of $(1, 0)$ intersects every neighborhood of the image of $(0, 1)$. This is because the iterates of any neighborhood of a point on the x-axis accumulate along the y-axis, and vice versa.

Proposition 3.5.7 (free and properly discontinuous gives good quotient). *Let Γ be a group acting on a manifold X. The quotient space X/Γ is a manifold with $X \to X/\Gamma$ a covering projection if and only if Γ acts freely and properly discontinuously.*

Proof of 3.5.7. If the action is free and properly discontinuous, we just have to check that the quotient is Hausdorff, by Proposition 3.5.4 and Exercise 3.5.3. Suppose x and y are points in X on distinct orbits. Let K be a union of two disjoint compact neighborhoods of x and y which contain no translates of x or y. Then $K \setminus \bigcup_{\gamma \neq 1} \gamma K$ (where 1 is the identity element) is still a union of a neighborhood of x with a neighborhood of y, and these neighborhoods project to disjoint neighborhoods in X/Γ.

For the converse, suppose that X/Γ is Hausdorff and that $p : X \to X/\Gamma$ is a covering projection. For any pair of points $(x_1, x_2) \in X \times X$, we will find neighborhoods U_1 of x_1 and U_2 of x_2 such that $\gamma(U_1) \cap U_2 \neq \varnothing$ for at most one $\gamma \in \Gamma$. If x_2 is not on the orbit of x_1, this follows from the Hausdorff property of X/Γ, because $p(x_1)$ and $p(x_2)$ have disjoint neighborhoods. If x_2 has the form γx_1, this follows from the fact that p is a covering projection—we take for U_1 a neighborhood of x_1 that projects homeomorphically to the quotient space, and let $U_2 = \gamma U_1$.

Now let K be any compact subset of X. Since $K \times K$ is compact, there is a finite covering of $K \times K$ by product neighborhoods of the form $U_1 \times U_2$, where U_1 has at most one image under Γ intersecting U_2. Therefore, the set of elements $\gamma \in \Gamma$ such that $\gamma K \cap K \neq \varnothing$ is finite, and Γ acts freely and properly discontinuously. $\boxed{3.5.7}$

The next criterion is often convenient for checking proper discontinuity. Recall that a map $f : X \to Y$ is *equivariant* with respect to a group G acting on X and Y if, for any $\gamma \in G$, the map f conjugates the action of γ on X to the action of γ on Y.

Proposition 3.5.8 (proper map preserves proper discontinuity). *Suppose that Γ acts on the spaces X and Y, and that $f : X \to Y$ is a proper, surjective, equivariant map. Then the action on X is properly discontinuous if and only if the action on Y is.*

Proof of 3.5.8. Images and inverse images of compact sets under f are compact. Every compact subset of X is contained in a set of the form $f^{-1}(K)$, where $K \subset Y$ is compact, so it suffices to consider such sets to check proper discontinuity in X. The proposition now follows because $K \cap \gamma K \neq \varnothing$ if and only if $f^{-1}(K) \cap \gamma f^{-1}(K) \neq \varnothing$. $\boxed{3.5.8}$

Exercise 3.5.9 (nondiscrete with Hausdorff quotients). Find a group G of homeomorphisms of $\mathbf{R}^2$ that is not discrete, but has a Hausdorff quotient $\mathbf{R}^2/G$.

Exercise 3.5.10 (discrete subgroups of Lie groups are properly discontinuous). Let G be a Lie group and $\Gamma \subset G$ a discrete subgroup. Consider the action of Γ on G by left translation.

(a) Show that Γ has discrete orbits.

(b) Show that the action of Γ is wandering. (Hint: Let U be a neighborhood of the identity 1 such that $\Gamma \cap U = \{1\}$. Find a neighborhood V of the identity such that $V = V^{-1}$ and $V^2 \subset U$. Show that an orbit of Γ can intersect the neighborhood Vx of any element $x \in G$ in at most one point.)

(c) Show that the action of Γ is properly discontinuous. (Hint: given $x, y \in G$, find neighborhoods U of x and V of y such that $UV^{-1}VU^{-1}$ intersects Γ in the identity element. Then there is at most one element $\gamma \in \Gamma$ such that γV intersects U.)

An alternate approach to this problem is to use a left invariant metric.

Corollary 3.5.11. *Suppose G is a Lie group and X is a manifold on which G acts transitively with compact stabilizers G_x. Then any discrete subgroup of G acts properly discontinuously on X.*

Proof of 3.5.11. The map $G \to X = G/G_x$ is proper. Apply Proposition 3.5.8 and Exercise 3.5.10. 3.5.11

Thus, in the cases of most interest to us, conditions (ii)–(v) in Definition 3.5.1 are in fact equivalent. Putting this together with Proposition 3.4.5 (holonomy group characterizes manifold), Proposition 3.4.10 (compact stabilizers imply completeness) and Proposition 3.5.7 (free and properly discontinuous gives good quotient), we get an important corollary:

Corollary 3.5.12 (classification of complete (G, X)-structures on a manifold). *Suppose G is a Lie group acting transitively, analytically and with compact stabilizers on a simply connected manifold X. If M is a closed differentiable manifold, (G, X)-structures on M (that is, (G, X)-stiffenings of M up to diffeomorphism) are in one-to-one correspondence with conjugacy classes of discrete subgroups of G that are isomorphic to $\pi_1(M)$ and act freely on X with quotient M. If M is not closed, we get the same correspondence if we look only at complete (G, X)-structures on M.*

Note the condition "with quotient M", which is necessary because $\pi_1(M)$ does not determine the diffeomorphism class of M. For example, both the punctured torus and the three-punctured sphere have as fundamental group the free group $\mathbf{Z} * \mathbf{Z}$ on two generators.

When a group Γ acts properly discontinuously on a space X in such a way that the quotient space X/Γ is compact, we say that the action (or Γ itself) is *cocompact*. If X is a Riemannian manifold and Γ is a group of isometries such that the quotient space X/Γ has finite volume, Γ is *cofinite*. Obviously cocompact groups are cofinite, but the converse is not true: Example 3.3.7 (the figure-eight knot complement) provides a counterexample.

Exercise 3.5.13. Explain how to define the volume of X/Γ when the action of Γ is not free.

Sometimes cofinite and cocompact groups are also called *lattices* and *uniform lattices*, respectively, but we will avoid this terminology, since it conflicts with the more common use of the term lattice, namely, a discrete subgroup of $\mathbf{R}^n$ isomorphic to $\mathbf{Z}^n$.

Problem 3.5.14 (proper discontinuity of topological groups). Work out a theory generalizing the definitions and results of this section to actions of topological groups on locally compact spaces, subsuming results of this section under the case that G has the discrete topology.

3.6. Bundles and Connections

Geometric structures on manifolds are closely related to another concept we have been using before discussing formally: fiber bundles, or bundles for short.

Definition 3.6.1 (fiber bundle). If (G, X) is a topological group G acting on a topological space X, a (G, X)-*bundle* (or, more formally, a *fiber bundle* with *structure group* G and *fiber* X) consists of the following data: a *total space* E, a *base space* B, a continuous map $p : E \to B$, called the *bundle projection*, and a *local trivialization*, explained below. We also say that the space E *fibers* over B with fiber X.

A local trivialization is like an atlas: it gives a way to describe the part of the bundle over a small neighborhood V as a product $V \times X$, up to adjustments by the action of G on fibers. Different elements of G may apply to different fibers, but they must depend continuously on the base.

More formally, a local trivialization is a covering of B by a collection of open sets U_i, and for each U_i a homeomorphism $\phi_i : p^{-1}(U_i) \to U_i \times X$ which gives p when composed with the projection $U_i \times X \to U_i$. The ϕ_i are required to be such that, for each intersecting U_i and U_j, the composition

$$\psi_{ij} = \phi_i \circ \phi_j^{-1} : (U_i \cap U_j) \times X \to (U_i \cap U_j) \times X$$

has the form

$$\psi_{ij}(u, x) = (u, \gamma_{ij}(u)x),$$

where $\gamma_{ij} : U_i \cap U_j \to G$ is continuous.

(The word "fibration" is not a synonym for fiber bundle. It includes fiber bundles, but it also encompasses more general maps that behave homotopically like fiber bundles, but don't have the same geometric description as locally modeled on a projection of a cartesian product to one of its factors.)

Just as for an atlas defining a $\mathcal{G}$-structure on a manifold, a local trivialization is not an essential part of the structure, just a tool used to define it. Two local trivializations are equivalent if their union is also a local trivialization.

Example 3.6.2 (product bundle). The *product* (G, X)-bundle over a space B is $B \times X$.

Example 3.6.3 (covering space). Any covering space of a connected space B is a bundle, where the fiber is some set S with the discrete topology, and the structure group is the group of permutations of S.

Example 3.6.4 (Möbius strip). The Möbius strip is a $(C_2, \mathbf{R})$-bundle over S^1, where C_2 is the cyclic group of order two acting non-trivially. Figure 3.23 shows a possible set of local trivializations.

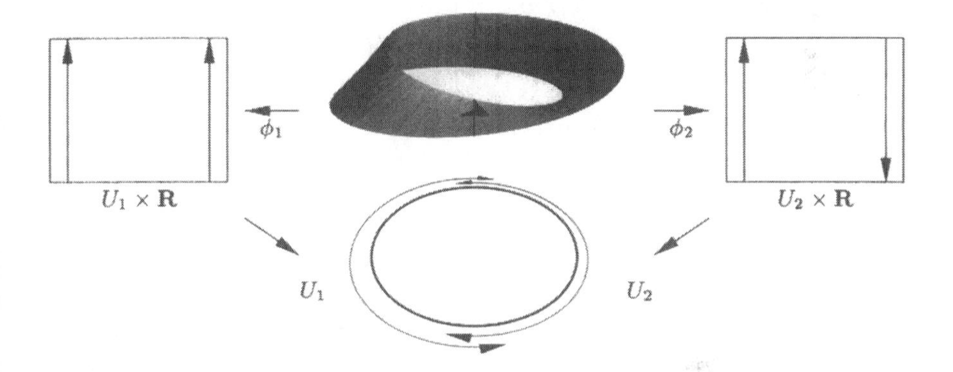

Figure 3.23. A non-trivial line bundle over the circle. The Möbius strip is one of the two $\mathbf{R}$-bundles over S^1 (the other being the product $S^1 \times \mathbf{R}$). Divide the base into two open intervals U_1 and U_2, and glue together the local trivializations $U_1 \times \mathbf{R}$ and $U_2 \times \mathbf{R}$ so they agree over one component of their overlap, and differ by a flip of the fiber over the other component.

Example 3.6.5 (mapping torus). A useful source of manifolds is the construction known as the *mapping torus* of a diffeomorphism. If M is a smooth manifold and $\phi : M \to M$ is a diffeomorphism, the mapping torus M_ϕ is obtained from the cylinder $M \times [0,1]$ by identifying the two ends via the map ϕ. Clearly, M_ϕ is an M-bundle over the circle. The Möbius strip is a particular case of this.

Exercise 3.6.6. Describe to your satisfaction (using local trivializations or otherwise) the following fiber bundles:

(a) The Klein bottle as a circle bundle over the circle.

(b) The Hopf fibration (Figure 2.31).

(c) $\mathbf{RP}^3$ as a circle bundle over S^2. (Hint: The Hopf flow commutes with the antipodal map. For another relationship, see Example 3.7.6.)

Example 3.6.7 (vector bundle). An important special case of fiber bundles is when X is a vector space and $G = \mathrm{GL}(X)$ is its group of linear automorphisms. In this case, a (G, X)-bundle is called a *vector bundle*.

Example 3.6.8 ((G, X)-foliation). If the base space B is an n-manifold and the fiber X is an m-manifold, the total space E is an $(n+m)$-manifold which has a foliation whose leaves are the fibers. It has a geometric structure which is a stiffening of a codimension-n foliation: a foliation with (G, X)-structure on its leaves, or a *tangentially (G, X) foliation*. Not every tangentially (G, X) foliation is a bundle, though, because each fiber in a bundle must be a copy of X, not just locally modeled on X.

The only data really needed to define a bundle is the base space, the covering by U_i, and the *cocycle* or collection of transition maps γ_{ij}, which satisfy the *cocycle condition* $\gamma_{ij}\gamma_{jk} = \gamma_{ik}$ wherever this composition is defined (on $U_i \cap U_j \cap U_k$). The total space is obtained by pasting together the $U_i \times X$ using the transition maps. Any set of continuous functions $\gamma_{ij} : U_i \cap U_j \to G$ satisfying the cocycle condition will work to form a bundle. (If G is abelian and has the discrete topology, the γ_{ij} form a one-cocycle in the sense of Čech cohomology.)

Since the cocycle makes no reference to the fiber X, but only to the group G, it follows that if Y is any other space on which G acts, we can construct for every (G, X)-bundle over a base B an *associated* (G, Y)-bundle over B. Moreover, if $\rho : G \to H$ is a group homomorphism and H acts on the space Z, we can form an associated (H, Z)-bundle, where the transition maps are the images under ρ of the transition maps for the original bundle.

Exercise 3.6.9. Show that, if G is abelian, there is a canonical action of G on each fiber E_x of a (G, X)-bundle $E \to B$, which extends to an action of G on the total space E.

A *bundle map* between two (G, X)-bundles is a continuous map between the base spaces covered by a continuous map between the total spaces which acts on each fiber as a G-isomorphism that depends continuously on the base point. A bijective bundle map whose inverse is also a bundle map is a *bundle isomorphism*. A (G, X)-bundle isomorphic to a product bundle $B \times X$ is also called *trivial*.

It frequently happens that when $\mathcal{G}$ is a pseudogroup acting on a manifold Y, the elements of $\mathcal{G}$ give rise to bundle isomorphisms of a certain type of bundle over (subsets of) Y. In such a case, there is an *associated bundle* over any $\mathcal{G}$-manifold, obtained by pasting together with this action. Here is the most important example of this phenomenon:

Example 3.6.10 (tangent bundle). Let $\mathcal{G} = \mathcal{C}^r$ be the pseudogroup of C^r diffeomorphisms between subsets of $Y = \mathbf{R}^n$ (Example 3.1.3), and consider the product bundle $\mathbf{R}^n \times \mathbf{R}^n$ to $\mathbf{R}^n$, where the fiber has the structure of a vector space, with group $\mathrm{GL}(n, \mathbf{R})$. For any diffeomorphism $f : U \to V$ in $\mathcal{G}$, f together with its derivative is a bundle isomorphism from $U \times \mathbf{R}^n$ over $V \times \mathbf{R}^n$. Given a differentiable n-manifold M with charts (U_i, ϕ_i), we can paste together the product bundles $U_i \times \mathbf{R}^n$ using the derivatives of the transition maps. This defines the *tangent bundle* TM of M. The fiber at a point $x \in M$ is called the *tangent space* $T_x M$ at x.

Example 3.6.11 (tangent sphere bundle). The *tangent sphere bundle* (or *tangent circle bundle*, in dimension two) of a differentiable n-manifold M is obtained by collapsing each ray in TM to a point, so the fiber becomes the $(n-1)$-sphere S^{n-1}. If M has a Riemannian metric, we can think of the tangent sphere bundle as the subset of TM consisting of tangent vectors of unit length. In this case we often call it the *unit tangent bundle* to M, and denote it UTM.

A *principal bundle* is one in which the fiber is the structure group itself, and the action is by left translations.

Example 3.6.12 (frame bundle). Let $G = \mathcal{C}^r$ and $Y = \mathbf{R}^n$ as in 3.6.10, and consider the principal bundle $\mathbf{R}^n \times \mathrm{GL}(n, \mathbf{R})$ with structure group and fiber $\mathrm{GL}(n, \mathbf{R})$. We think of the fiber as the set of *frames*, or ordered bases for $\mathbf{R}^n$, since $\mathrm{GL}(n, \mathbf{R})$ acts simply transitively on this set. The associated principal bundle over a differentiable n-manifold M is called the *frame bundle* of M. An element of the fiber over $x \in M$, called a *frame* at x, can be be thought of as a choice of a basis for the tangent space $T_x M$.

A *section* of a bundle $p : E \to B$ is a left inverse to the bundle projection, that is, a continuous map $s : B \to E$ such that $p \circ s = \mathrm{Id}$.

A section singles out, in a continuous way, a point in each fiber. If the fiber X is contractible and the base is reasonable (for example if it has a countable basis), every (G, X)-bundle admits a section, but otherwise sections are likely not to exist.

Whenever $H \subset G$ is a subgroup, any (H, X) bundle is also a (G, X)-bundle. When a (G, X)-bundle is expressed as an (H, X)-bundle, it is called a *reduction* of the bundle to structure group H. A reduction of the structure group is analogous to a stiffening of a $\mathcal{G}$-structure.

In general, if H is a closed subgroup of G and G has a local cross-section with respect to H (Exercise 3.4.9), the quotient map $G \to G/H$ is a principal H-bundle over G/H. Then any principal G-bundle $E \to B$ can be written as a two-stage bundle $E \to E/H \to B$, where E/H is the $(G, G/H)$-bundle associated with E, and $E \to E/H$ is a principal H-bundle. If E can be reduced to structure group H, all the transition maps γ_{ij} are in H, so the pasting maps ψ_{ij} for the total space preserve the subgroup H. In this case $E/H \to B$ admits a section: in each fiber, pick the point that maps to the coset of the identity under local trivializations.

The converse is also true: if E/H has a section, E can be reduced to structure group H. To show this, we introduce another construction for making new bundles out of old ones. If $E \to B$ is a (G, X)-bundle and $f : C \to B$ is a continuous map, the *pullback* of E by f is the (G, X)-bundle $f^*E \to C$, with total space

$$f^*E = \{(c, e) \in C \times E : f(c) = p(e)\},$$

bundle projection $(c, e) \mapsto c$, and local trivializations $(c, e) \mapsto (c, \pi_X \circ \phi_i(e))$, where ϕ_i is a local trivialization of E and π_X denotes projection to the X-factor.

Proposition 3.6.13 (reduction to subgroup). *Let G be a topological group and H a closed subgroup, and assume that G has a local cross-section with respect to H. (This happens, for example, if G is a Lie group.) A principal G-bundle $E \to B$ can be reduced to structure group H if and only if the associated $(G, G/H)$-bundle over B has a section.*

Proof of 3.6.13. We've already proved the "only if" part. If $E/H \to B$ has a section s, the pullback of $E \to E/H$ by s is a principal H-bundle over B. The associated (H, G)-bundle is E (check this).

3.6.13

Example 3.6.14 (every manifold has a Riemannian metric). The tangent bundle of a differentiable manifold provides a good example of this phenomenon, in fact one that we have already used (Lemma 3.4.11.) Its natural structure group is $\mathrm{GL}(n, \mathbf{R})$, which has the subgroup $O(n) \subset \mathrm{GL}(n, \mathbf{R})$. The quotient space $\mathrm{GL}(n, \mathbf{R})/O(n)$ has a nice interpretation as the space P of positive definite quadratic forms on $\mathbf{R}^n$. Since P is a convex subset of a vector space, it is contractible. Therefore, the $(\mathrm{GL}(n, \mathbf{R}), \mathrm{GL}(n, \mathbf{R})/O(n))$-bundle associated to the tangent bundle of a manifold has a section. It follows that the tangent bundle can always be reduced to structure group $O(n)$. Such a reduction is equivalent to the choice of a Riemannian metric.

An interesting special case of bundles is when the structure group G has the discrete topology (compare Example 3.6.3), so the transition maps γ_{ij} are locally constant. Such a bundle is called *flat*. If G is not discrete, we still use the expression *flat G-bundle* to refer to a G-bundle that has a reduction to structure group $\hat{G}$, where $\hat{G}$ denotes G with the discrete topology.

If the fiber and base of a flat bundle are both manifolds, the gluing maps ψ_{ij} for the total space preserve a "horizontal" foliation, transverse to the fibers, as well as the foliation by fibers of Example 3.6.8. The transverse foliation is called a *flat connection* for the bundle. (More general connections are discussed below; they are a little harder to visualize.)

The leaves of the transverse foliation are covering spaces of the base. For any path in the base between points x and y, the lifts to leaves of the transverse foliation give a homeomorphism between the fibers over x and y, called a *holonomy map*, and which depends only on the homotopy class of the path relative to the endpoints. Expressed in terms of local trivializations, this map corresponds to the action of an element of G, which of course depends on the choice of local trivializations. But if $x = y$, we get a well-defined element of G, called the holonomy around the loop. If the base B is path-connected and we pick a basepoint b for it, the holonomy around loops beginning and ending at b gives a homomorphism $\pi_1(B) \to G$, called the holonomy of the bundle.

Proposition 3.6.15 (classification of flat bundles). *Let G be a topological group and B a path-connected space with a basepoint b, and let X be a space on which G acts effectively. Flat (G, X)-bundles over B equipped with an identification of X with the fiber over b, up* *to isomorphisms that are the identity on B and preserve the given identifications on the fibers over b, are in one-to-one correspondence with homomorphisms $\pi_1(B) \to G$.*

Proof of 3.6.15. This is really a statement about $(\hat{G}, X)$-bundles, since G and $\hat{G}$ have the same algebraic structure. In other words, we can assume that G is discrete.

We already know that a (G, X)-bundle gives a holonomy homomorphism $\pi_1(B) \to G$, which clearly does not change under isomorphisms of the specified type.

Given a homomorphism $\rho : \pi_1(B) \to G$, first form the trivial (G, X)-bundle over the universal cover $\tilde{B}$, then divide it by $\pi_1(B)$ acting by covering transformations on the base and by ρ in the direction of the fibers. This gives a bundle over B with holonomy ρ.

Two bundles with the same holonomy are isomorphic: you just map one to the other by taking leaves of the horizontal foliation to leaves of the horizontal foliation. $\boxed{3.6.15}$

Using this result, the problem of finding a (G, X)-stiffening of a smooth manifold can be subdivided nicely into two steps. Consider the product bundle $X \times X$, with its obvious flat connection and the diagonal section $\{(x, x) : x \in X\}$, which is transverse to the horizontal foliation. Over a (G, X)-manifold M, with charts (U_i, ϕ_i), we can define a (G, X)-bundle by piecing together locally trivial neighborhoods $U_i \times X$. The transition maps γ_{ij} are defined as follows: $\gamma_{ij}(u)$ is the element of G that agrees with $\phi_i \circ \phi_j^{-1}$ near $\phi_j(u)$. The gluing maps ψ_{ij} preserve the flat connection and diagonal section of $X \times X$, so the resulting (G, X)-bundle over M has a flat connection and a section transverse to the foliation. The space of leaves of the foliation restricted to a small neighborhood has a canonical (G, X)-structure, coming from projection along the leaves to a fiber, so any manifold of complementary dimension transverse to the foliation also inherits a (G, X)-structure. In particular, the (G, X)-structure on M can be reconstructed from the flat (G, X)-bundle, together with the section transverse to the foliation.

The problem of constructing a (G, X)-structure on a manifold M therefore breaks into two steps. First, find a homomorphism $\pi_1(M) \to G$, which will serve as the holonomy for the (G, X)-structure. Then construct a section of the associated flat bundle, transverse to the foliation. This method works with particular success when M is non-compact.

Here is another construction that associates with a (G, X)-manifold M a canonical flat bundle over M, this time with fiber G. An element of the total space consists of a point $x \in M$, together with the germ at x of a (G, X)-map into X. Any two such germs differ by an element of G. A local trivialization for this bundle is determined by any atlas for M. The leaves of the horizontal foliation are covering spaces of M equipped with (G, X)-maps to X: they are all the possible developing maps for M.

Example 3.6.16 (Euclidean manifold has flat tangent bundle). For a Euclidean manifold M, we get a flat principal $(\operatorname{Isom} \mathbf{E}^n)$-bundle over M. $\operatorname{Isom} \mathbf{E}^n$ has an obvious homomorphism to $O(n)$, and $O(n)$ acts on $\mathbf{R}^n$, so we can form the associated flat $(O(n), \mathbf{R}^n)$-bundle. This can be identified with the reduction to structure group $O(n)$ of the tangent bundle of M (see Example 3.6.14). Therefore, the tangent bundle of any Euclidean manifold (also known as a *flat manifold*) has a natural flat orthogonal connection.

Example 3.6.17 (hyperbolic manifold has flat sphere bundle). The previous example has an interesting analog for hyperbolic manifolds. Each fiber of the tangent sphere bundle of $\mathbf{H}^n$ can be identified with the sphere at infinity S^{n-1}_∞, by extending geodesic rays to infinity (page 56). The group Möb_{n-1} of isometries of hyperbolic space acts on the sphere at infinity, so there is an associated flat $(\text{Möb}_{n-1}, S^{n-1}_\infty)$-bundle over any hyperbolic manifold. This is canonically identified with the tangent sphere bundle of the manifold, which therefore has a flat connection. In particular, the tangent circle bundle of a hyperbolic surface, which is a three-manifold, has a codimension-one foliation. These foliations have very interesting geometric and dynamical properties. Notice, however, that the structure group for this bundle is the Möbius group, rather than the orthogonal group or the linear group. The tangent space of a hyperbolic manifold cannot in general be reduced to a discrete orthogonal, or even linear, structure group.

As promised, we now turn to connections in general. A connection can be thought of as an infinitesimal form of a local trivialization. Let $p : E \to B$ be a smooth (G, X)-bundle: that is, E, B, and X are all smooth manifolds, of dimensions $m + n$, n and m, say; p is a smooth map, G is a Lie group acting smoothly on X, and the cocycle γ_{ij} defining E is smooth. A *connection* for E is an n-plane field τ transverse to the fibers, and satisfying an additional

(G, X)-compatibility condition: for any fiber E_x, there must exist some smooth local coordinate chart for E such that τ is tangent to the horizontal directions.

Another way to express the compatibility condition is this: Given any smooth n-plane field τ transverse to the fibers and given a path α in the base between points x and y, just as in the case of a flat connection, there is a map between some subset of the fiber E_x to some subset of E_y. This map, called the *holonomy* of τ along α, or *parallel translation* along α, is obtained by lifting α to a path $\tilde{\alpha}$ always tangent to τ: given an initial lift $\tilde{x}$ of x, the lift tangent to τ is uniquely determined until and unless it shoots off to infinity. For any compact subset of E_x, the lift is defined at least for a sufficiently short initial segment of α. If X is connected, the (G, X)-compatibility condition for τ is that the holonomy along any path α preserves the (G, X)-structure on the fiber.

Given two connections τ and σ, one can take an affine combination $t\tau + (1 - t)\sigma$: the plane τ_e at any point $e \in E$ is like the graph of a linear function, and the set of linear functions is an affine space. Connections always exist locally (in any coordinate chart for a local trivialization), and therefore they can be pieced together using a partition of unity to give a connection globally.

The most important example of a connection is the *Levi-Civita connection*, an orthogonal connection defined on the tangent bundle of any Riemannian manifold M. Given a path α on M, there is an infinitesimally best fit of a neighborhood of α to a neighborhood of some path α' in $\mathbf{E}^m$, where m is the dimension of M. Under this best fitting, parallel translation for the Levi-Civita connection maps to parallel translation in $\mathbf{E}^m$, that is, moving the tangent space of $\mathbf{E}^m$ along α' while keeping every vector parallel to the original.

The Levi-Civita connection can be described more directly if M is a submanifold of some Euclidean space $\mathbf{E}^n$, with its inherited Riemannian metric. It is easier to first describe a related connection, Euclidean rather than orthogonal. Given two nearly parallel m-planes $P, P' \subset \mathbf{E}^n$, the orthogonal projection from one to the other is nearly an isometry: it distorts the metric by a factor no greater than the cosine of the angle between the planes. If we take a one-parameter family of m-planes that mediate between P and P', and look at the composition of orthogonal projections from $P = P_0$ to P_1 to P_2 and so on to $P_N = P'$, where $P_1, \ldots, P_{N-1}$ are elements of the family taken in order, the distortion decreases as the subdivision gets finer, and in the limit we get an actual isometry. Applying this to the family of tangent planes to M along a path α, we get a flow of Euclidean isometries, with trajectories orthogonal to the planes. (Physically, this amounts to rolling the tangent plane to M along α, without slipping.) This flow defines parallel translation for a certain Euclidean connection on TM. The Levi-Civita connection is obtained by normalizing the Euclidean connection by translations to keep the origin fixed, converting it into an orthogonal connection.

The transverse foliation of a flat bundle—in fact, any foliation of a manifold—determines a *tangent plane field*. Conversely, given a plane field on a manifold, we can ask when it is the tangent plane field of a foliation. Such a plane field is called *integrable*.

Saying that the plane field for a connection is integrable is saying that the connection is flat, that is, it pieces together to give a transverse foliation. For a flat connection, the holonomy along a path depends only on the homotopy class of the path, and in particular it is the identity for any small loop. This can fail to be the case for a general connection, and the extent of the failure is measured by the *curvature* of the connection. In order to define it, we recall some basic facts about Lie algebras (see [Hoc65]).

A *Lie algebra* is a vector space V equipped with a bilinear map $[\,,\,] : V \times V \to V$ that is skew-symmetric (so that $[y,x] = -[x,y]$) and satisfies the *Jacobi identity*

3.6.18. $$[[x,y],z] + [[y,z],x] + [[z,x],y] = 0$$

for all $x, y, z \in V$. The tangent space T_0G of a Lie group G at the identity has the structure of a Lie algebra: T_0G can be identified with the space of left-invariant vector fields on G (extend a tangent vector to all of G by left translation), and given two such fields X and Y, their Lie bracket is also one (Exercise 3.6.19). This is called the Lie algebra of G. For every finite-dimensional Lie algebra there is at least one Lie group with that Lie algebra. Two Lie groups have isomorphic Lie algebras if and only if they have isomorphic universal covers.

Exercise 3.6.19 (Lie bracket). The *Lie bracket* $[X, Y]$ of two differentiable vector fields X and Y on a manifold M measures by how much the flows that they generate fail to commute. Starting from a point p, follow the flow of X for a time s, then follow Y for a time t, then $-X$ for a time s, then $-Y$ for a time t. Most likely the quadrilateral won't close exactly.

(a) Show that the size of the gap decreases at least linearly with the product st. Therefore we can take the second derivative with respect to s

and t to get a vector at p. This is, by definition, the value of $[X, Y]$ at p.

(b) Show that the Lie bracket operation on vector fields satisfies the conditions for a Lie algebra.

Curvature is a map from two-vectors on M to the Lie algebra $\mathfrak{g}$ of G, that is, a $\mathfrak{g}$-valued two-form: given two vectors v and w at a point $p \in M$, map the unit square into M so that the vectors $(1,0)$ and $(0,1)$ at the origin are taken to v and w. For each s and t, let $H(s,t)$ be the holonomy of the Levi-Civita connection around the boundary of the rectangle $[0,s] \times [0,t]$. Then the second derivative of this holonomy, with respect to s and t, is the curvature of $v \wedge w$ (compare the definition of the Lie bracket in Exercise 3.6.19). Note that if $v = w$ the holonomy is 0, so the second derivative really is an alternating function.

In the case of the Levi-Civita connection for a Riemannian surface, there is only a one-dimensional vector space of tangent two-vectors, and a one-dimensional space of infinitesimal automorphisms of the fiber, or infinitesimal rotations (that is, the Lie algebra of $O(2)$ is one-dimensional). Given a local orientation, each of these vector spaces has a canonical basis: the area form, and the unit-speed counterclockwise infinitesimal rotation. Gaussian curvature of a surface is the curvature of the Levi-Civita connection expressed in this basis.

More generally, for an n-dimensional Riemannian manifold, the *sectional curvature* at a point p with respect to a tangent plane P is obtained by restricting the domain of the curvature map to two-vectors in P, and projecting the image to the Lie subalgebra of infinitesimal rotations of P.

3.7. Contact Structures

Just as foliations are related to flat connections and integrable plane fields, *contact structures* in three dimensions are related to a different kind of plane field, one that is as non-integrable as possible. We will take some time to develop the basic picture for contact structures, because they give an interesting example of a widely occuring pattern for manifolds that is hard to see until your mind and eyes have been attuned.

Let τ be the plane field in $\mathbf{R}^3$ that associates to $(x, y, z) \in \mathbf{R}^3$ the plane spanned by the vectors $(1,0,0)$ and $(0,1,x)$ (Figure 3.24). To understand τ, it helps to think about *Legendrian curves*, which are curves in $\mathbf{R}^3$ whose tangent vectors are always contained in τ. Pause a minute to visualize how it is possible to get from any point in $\mathbf{R}^3$ to any other along a Legendrian curve. This property immediately distinguishes τ from the tangent plane field of a foliation.

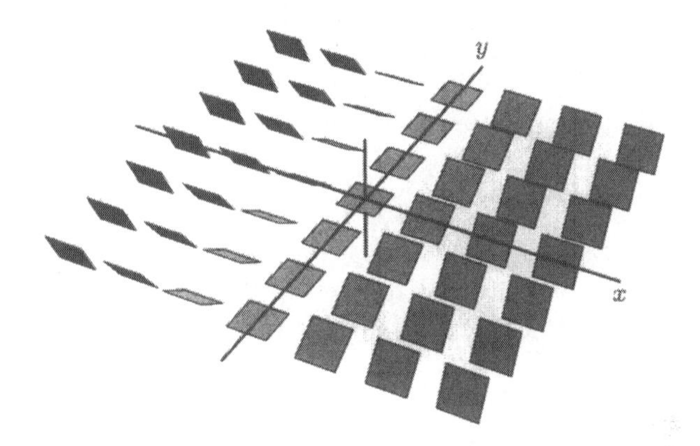

Figure 3.24. The plane field τ. The field is invariant under vertical translations, so the figure shows only the plane $z = 0$.

Exercise 3.7.1 (Legendrian lifts). Given a piecewise differentiable curve $\gamma : [0,1] \to \mathbf{R}^2$ in the xy-plane, and a point in $\mathbf{R}^3$ that projects to $\gamma(0)$, there is a unique lift of γ to a Legendrian curve $\tilde{\gamma} : [0,1] \to \mathbf{R}^3$ that projects to γ.

In the language of Section 3.6, τ is a connection for the $(\mathbf{R}, \mathbf{R})$-bundle $\pi_z : \mathbf{R}^3 \to \mathbf{R}^2$, where π_z is projection along the z-axis. The Legendrian lift of a curve is parallel translation around the curve. The lift of a closed curve γ is not necessarily closed; in fact, the difference in height of its endpoints is equal to the signed area enclosed by γ. One way to see this is to express τ in a dual way, as the kernel of the one-form $\omega = -x\,dy + dz$. Then $d\omega = -dx \wedge dy = \pi_z^* dA$, where dA is the area form in the plane, and π_z^* indicates the pullback under π_z. If $\tilde{\gamma}$ is completed to a closed curve by means of a vertical segment, the integral of ω around the closed curve reduces to the integral along the vertical segment, which is minus the difference in heights of its endpoints. By Stokes' theorem, this equals the integral of $-\pi_z^* dA$ over a surface bounded by the closed curve, which reduces to minus the signed area enclosed by γ.

Thus, the connection τ has constant curvature 1. Parallel translation around a curve γ is the identity if and only if the total curvature bounded by γ adds up to zero; this happens if and only if γ encloses a signed area of 0.

A *contact diffeomorphism* between subsets of $\mathbf{R}^3$ is one that preserves τ. Clearly, contact diffeomorphisms map Legendrian curves to Legendrian curves. The *contact pseudogroup* Con is the pseudogroup of contact diffeomorphisms between open sets of $\mathbf{R}^3$, and a *contact structure* on a three-manifold is a Con-structure. The pullback of τ under the charts is the *contact plane field* on the manifold.

There are many contact automorphisms of $\mathbf{R}^3$. As a start, the map

$$(x, y, z) \mapsto (x + x_0, y + y_0, z + z_0 + x_0 y)$$

preserves τ. Note that this map preserves the foliation of $\mathbf{R}^3$ by vertical lines, and also the foliation by lines parallel to the x-axis (see also Exercise 3.8.6(c)). Here are two generalizations:

Example 3.7.2 (area-preserving automorphisms lift). Let ϕ be a diffeomorphism of the xy-plane that preserves area (or multiplies it by a constant factor). Then there is a contact automorphism $\tilde{\phi}$ of $\mathbf{R}^3$ that preserves the foliation by vertical lines, and projects to ϕ under π_z. Moreover, any two such maps differ by a vertical translation.

For given an arbitrary point $p \in \mathbf{R}^3$, we can connect p to a fixed point $q \in \mathbf{R}^3$ by a smooth Legendrian curve γ. If $\tilde{\phi}$ is to map Legendrian curves to Legendrian curves, the only possible candidate for $\tilde{\phi}(p)$ is the endpoint of the Legendrian lift of $\phi \circ \pi_z \circ \gamma$ that starts at $\tilde{\phi}(q)$. This endpoint does not depend on the choice of γ: the lifts of two curves with the same endpoints have the same endpoints if and only if the signed area enclosed by the curves is zero, and this area property is preserved by ϕ.

This shows the existence and uniqueness of $\tilde{\phi}$, and also that every contact automorphism of $\mathbf{R}^3$ that preserves the foliation by vertical lines is of this type.

Example 3.7.3 (automorphisms of the yz-plane lift). Another good way to think about the standard contact structure in $\mathbf{R}^3$ is to use the projection π_x to the yz-plane along lines parallel to the x-axis. These lines are Legendrian curves, and they form a *Legendrian foliation*. Consider an arbitrary Legendrian curve γ, and its projection $\pi_x(\gamma)$ to the yz-plane. At any time when the derivative $d\pi_x(\gamma(t))/dt$ is non-zero, you can reconstruct the x-coordinate from the projection: it is the slope of this tangent vector in the yz-plane.

Now let ϕ be any diffeomorphism of the yz-plane. Then there is a unique contact diffeomorphism $\tilde{\phi}$, defined on most of $\mathbf{R}^3$, that preserves the foliation by lines parallel to the x-axis, and projects to ϕ under π_x.

For given a point $p = (x, y, z) \in \mathbf{R}^3$, the projection of τ to the yz-plane is a line of slope x. The derivative of ϕ maps the vector $(1, x)$ at (y, z) to some other vector; we set the x-coordinate of $\tilde{\phi}(p)$ to the slope of this vector. If the slope is infinite, $\tilde{\phi}$ is undefined at p. By construction, $\tilde{\phi}$ preserves τ. If ϕ maps vertical lines to vertical lines, the slope is never infinite, and we get a contact automorphism of $\mathbf{R}^3$.

Clearly, the contact automorphisms obtained by this construction are precisely those that preserve the foliation of $\mathbf{R}^3$ by lines parallel to the x-axis.

In this construction, we can avoid the awkwardness of having $\tilde{\phi}$ undefined at certain points by adding a point at infinity to the x-axis to complete the circle of directions, and extending τ to $\mathbf{R}^2 \times \mathbf{RP}^1$ with vertical planes for $x = \infty$. (Formally, we define a contact structure on $\mathbf{R}^2 \times \mathbf{RP}^1$ with two charts: one, the identity, is defined for x finite; the other, $(x, y, z) \mapsto (-1/x, z, -y)$, is defined for $x \neq 0$.) Then the lift $\tilde{\phi}$ is a contact automorphism of $\mathbf{R}^2 \times \mathbf{RP}^1$.

In this construction, it's best to think of $\mathbf{R}^2 \times \mathbf{RP}^1$ as the projectivized tangent space to $\mathbf{R}^2$, so that an element of $\mathbf{R}^2 \times \mathbf{RP}^1$ is a point in the plane together with a line through that point. A Legendrian path corresponds to a motion of a pair (point, line through point) such that the line is tangent to the motion of the point.

Exercise 3.7.4 (lifting curves in the yz-plane). Suppose we're given a differentiable map β of some open interval to the yz-plane. Let π_x denote projection along the x-axis onto the yz-plane.

(a) Show that if the derivative of β is never zero, so that β is an immersion, there is a unique Legendrian curve γ immersed in $\mathbf{R}^2 \times \mathbf{RP}^1$ such that $\pi_x \circ \gamma = \beta$, where π_x is projection along the x-axis onto the yz-plane. (Write down the formula.)

(b) A typical Legendrian curve is likely to have a tangent parallel to the x-axis at some point, so that its projection to the yz-plain fails to be an immersion. For example, the projection of $(\frac{3}{2}t, t^2, t^3)$ has a cusp at $t = 0$. Show that the conclusion of part (a) still holds if the first and second derivatives of β are never simultaneously zero. (Hint: do first the case where the tangent to β is never vertical.)

(c) Show that if for every t there is some r such that the r-th derivative of β is non-zero, the conclusion still holds, except that the lift may not be immersed.

(d) What can you say in general?

Proposition 3.7.5 (tangent circle bundle has contact structure). *The tangent circle bundle of a smooth surface has a canonical contact structure, which is preserved by the derivative of any diffeomorphism.*

Proof of 3.7.5. We first look at $\mathbf{R}^2 \times S^1$, the tangent circle bundle of $\mathbf{R}^2$. This space is a double cover of the contact manifold $\mathbf{R}^2 \times \mathbf{RP}^1$ that we defined after Example 3.7.3—the fiber of the tangent circle bundle is a circle of unit tangent vectors, rather than a circle of directions. We give $\mathbf{R}^2 \times S^1$ the induced contact structure. Diffeomorphisms of $\mathbf{R}^2$ lift to contact automorphisms of $\mathbf{R}^2 \times S^1$, just as they lift to automorphisms of $\mathbf{R}^2 \times \mathbf{RP}^1$ in Example 3.7.3. Therefore, given an atlas for a surface, the contact structures on the tangent circle bundles of the chart domains patch together to give a contact structure on the tangent circle bundle of the surface. 3.7.5

Example 3.7.6. As a corollary, T^3, $\mathbf{RP}^3$ and S^3 have contact structures. T^3 is diffeomorphic to the tangent circle bundle of the two-torus. $\mathbf{RP}^3$ is diffeomorphic to SO(3) (Exercise 2.7.7), which in turn can be identified with the tangent circle bundle of S^2 (why?). Therefore $\mathbf{RP}^3$ has a contact structure. Since S^3 is the double cover of SO(3), it also has a contact structure.

In fact, Martinet [Mar71] proved that every orientable compact three-manifold has a contact structure.

You can get a good physical sense for the contact structure on the tangent circle bundle of a surface by thinking about ice skating, or bicycling. A skate that is not scraping sideways describes a Legendrian curve in the tangent circle bundle to the ice. It can turn arbitrarily, but any change of position is in the direction it points. Likewise, as you cycle along, the direction of the bicycle defines a ray tangent to the earth at the point of contact of the rear wheel. Assuming you are not skidding, the rear wheel moves in the direction of this ray, and its motion describes a Legendrian curve in the tangent circle bundle of the earth.

Young children are sometimes given bicycles with training wheels, some distance off to the side of the rear wheel. The training wheel also traces out a Legendrian curve—in fact, for any real number t, the diffeomorphism ϕ_t of $\mathbf{R}^2 \times S^1$ that takes a tangent ray a signed distance t to the left of itself is a contact automorphism. The training wheel path is the image of the rear wheel path under such a transformation. Note that this transformation applied to curves in the plane often creates or removes cusps (Figure 3.25). The same thing happens when you mow a lawn, if you start by making a big circuit around the edge of the lawn and move inward.

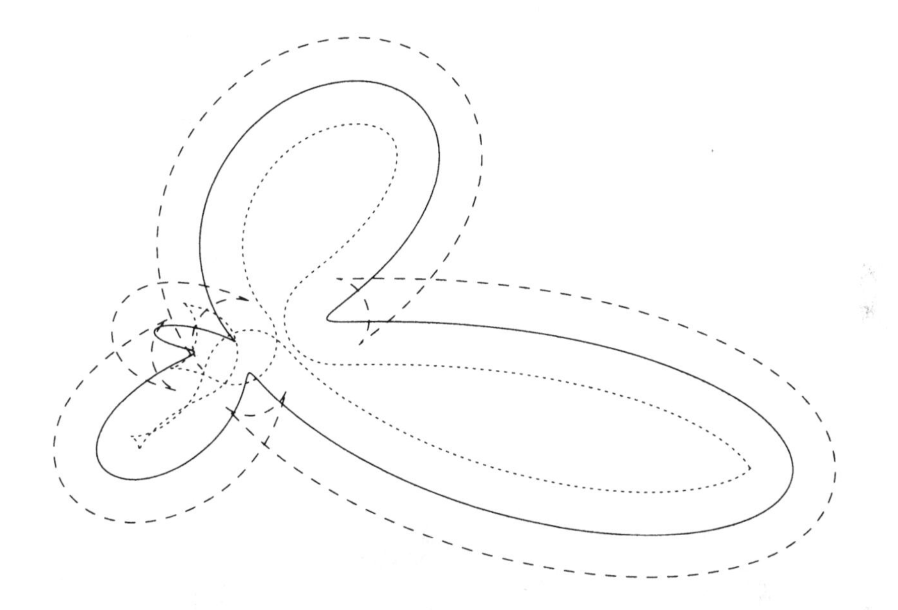

Figure 3.25. A family of parallel curves. The transformation ϕ_t of the tangent circle bundle of the plane which translates a ray a distance t to its left is a contact automorphism. The figure shows the projections to the plane of the images of a single curve, for various values of t. Notice how cusps develop.

The rear wheels of an automobile also follow Legendrian curves, related by a transformation ϕ_t. (The two wheels turn at slightly different speeds, thanks to the differential gear; otherwise one or both would skid whenever the car went around a curve.) Parallel parking implements the fact that any two points are connected by a Legendrian path. The front wheels are more complicated, since the line joining them is not perpendicular to their path unless the car is moving straight. The relation between the front and back wheels of a bicycle or a car is also complicated: the rear wheels depend not just on the current position and direction of the front wheels, but on their history.

The contact automorphism ϕ_t can be defined not only for $\mathbf{R}^2$, but for any oriented surface with a Riemannian metric. It sends each tangent ray to a tangent ray at a point obtained by translating it along the perpendicular geodesic, toward the left. (The surface needs to be oriented so we can decide consistently which way is left.) As t varies, the automorphisms ϕ_t form a flow ϕ. This is closely related to the surface's *geodesic flow*. The geodesic flow of a Riemannian manifold is the flow ψ that moves each unit tangent vector along its geodesic at unit speed. Really ϕ and ψ are the same flow in slightly different guises, identified by the map that rotates each unit tangent vector counterclockwise by 90°.

Exercise 3.7.7. (a) Proposition 3.7.5 gives one canonical plane field for the tangent circle bundle of a smooth surface. If the surface has a Riemannian metric, there is another: the Levi-Civita connection, restricted from the tangent bundle to the tangent sphere bundle. Compare the two, and explain the relation between the Levi-Civita connection and the geodesic flow (or the training-wheel flow) on the surface. See also Corollary 3.7.16.

(b) Describe the contact structures of Example 3.7.6 geometrically.

(c) What is the relationship among the contact structure on S^3, the Hopf flow, and Levi-Civita connection on S^2?

Problem 3.7.8 (Legendrian foliations). In a contact three-manifold, it is easy to construct, at least locally, a one-dimensional Legendrian foliation, that is, a foliation by Legendrian curves: in a small neighborhood, any nonzero section of τ is a vector field whose flow lines give such a foliation.

(a) Show that every Legendrian foliation is locally equivalent to the foliation by lines parallel to the x-axis in our model contact structure in $\mathbf{R}^3$.

(b) Work out an interpretation of contact three-manifolds with Legendrian foliations as three-manifolds locally modeled on the tangent circle bundle of $\mathbf{R}^2$, up to derivatives of diffeomorphisms.

(c) Find and analyze some interesting examples.

Let's return to our model plane field τ in $\mathbf{R}^3$. Let X be any smooth vector field tangent to τ. We claim that unless X is identically zero, the flow it generates cannot consist of contact automorphisms. For consider a short Legendrian arc α transversal to X at a point p. Under the flow, α sweeps out a quadrilateral, three sides of which are Legendrian arcs. If the remaining side—the translate of α under the flow—were also Legendrian, we would have a closed Legendrian curve, which is impossible because the quadrilateral's projection to the xy-plane has non-zero area. In fact, we see that the flow of X tilts τ along the axis through X, changing its slope in the transverse direction in proportion to the length of X. It follows that:

Proposition 3.7.9 (horizontal contact perturbations). *For each infinitesimal perturbation of τ, there is a unique vector field X tangent to τ which achieves the perturbation.*

Corollary 3.7.10 (contact local rigidity). *Let π be a contact plane field on a three-manifold M. There is a neighborhood U of π in the C^2 topology such that any plane field π' in U also defines a contact structure, and is in fact C^∞-equivalent to π by a diffeomorphism close to the identity in the C^2 topology.*

Proof of 3.7.10. Given a plane field π' near π, there is a smooth homotopy from π to π'. By Proposition 3.7.9, there is a vector field X_0 whose action on π matches the homotopy to first order at $t = 0$. The value of X_0 at any point depends on the value and first time derivative of π at that point. Since the perturbation is C^2-small, solutions X_t also exist for all times $0 \leq t \leq 1$, and we get a time-dependent vector field, which is C^1 as a function on $M \times [0,1]$. By integrating this vector field we obtain a C^2-small diffeomorphism taking π to π'. 3.7.10

It would appear from Proposition 3.7.9 that contact flows are hard to find, but actually this proposition enables us to construct many—in fact, all—contact flows. Given a contact plane field π on a three-manifold M, let $N(\pi)$ be the quotient bundle TM/π—this is a way to describe the normal bundle for π without having to choose a Riemannian metric.

Corollary 3.7.11 (contact flow classification). *Let π be a contact plane field on a three-manifold M. Projection from TM to $N(\pi)$ gives a one-to-one correspondence between vector fields preserving the contact structure on M and smooth sections of $N(\pi)$.*

Proof of 3.7.11. Given a smooth section s of $N(\pi)$, let X be a smooth vector field which projects to s. Probably X doesn't preserve π, but by Proposition 3.7.9, whatever it does can be corrected by adding a uniquely determined vector field contained in π, giving a contact vector field.

This shows surjectivity of the map from contact vector fields to sections of $N(\pi)$. Injectivity was established in the remarks before 3.7.9. 3.7.11

Exercise 3.7.12 (area and contact). Explain the relation among Corollary 3.7.11, Example 3.7.2 and Exercise 3.1.13.

The amount of twisting, or non-integrability, of a plane field can be nicely measured, much like the curvature of a connection. To see how, consider any k-plane field τ on an n-manifold M, and let x and y be vectors in τ at some point $p \in M$. Extend x and y locally to smooth vector fields X and Y contained in τ, and form the Lie bracket $[X, Y]$. The *twistedness* of τ is the two-form with values in the normal bundle $N(\tau)$ which, when evaluated on x and y, gives the image of $[X, Y]$ in $N(\tau)$.

We must show that this does not depend on the vector fields X and Y used to extend the given vectors. This can be done by a standard technique: we will check that $[X, Y]/\tau$ is bilinear, not just using real numbers as coefficients, but using functions as coefficients. Thus, we consider any two vector fields U and V tangent to τ defined in a neighborhood of p, and compare $[U, V]/\tau$ with $[fU, gV]/\tau$, where f and g are functions. The new Lie bracket is

$$[fU, gV] = fg[U, V] + fU(g)V - gV(f)U,$$

so

$$[fU, gV]/\tau = fg[U, V]/\tau.$$

Now consider any set of vector fields $\{X_1, X_2, \ldots, X_n\}$ tangent to τ that form a basis for τ in a neighborhood of p. Express U as a linear combination $U = \sum f_i X_i$ of these vector fields. Then $[U, V]/\tau = \sum f_i [X_i, Y]/\tau$. In particular, if $U_p = 0$ then $([U, V]/\tau)_p = 0$. Similarly, if $V_p = 0$ then $([U, V]/\tau)_p = 0$.

It follows that if X' and Y' are two vector fields that agree with X and Y at p, then

$$\begin{aligned} ([X', Y']/\tau)_p &= ([X' - X, Y']/\tau)_p + ([X, Y']/\tau)_p \\ &= ([X, Y']/\tau)_p = ([X, Y]/\tau)_p. \end{aligned}$$

For example, the standard contact structure on $\mathbf{R}^3$ is spanned by the vector fields $X = (1, 0, 0)$ and $Y = (0, 1, x)$. Their Lie bracket is $(0, 0, 1)$. The twistedness form, measured in the basis $(0, 0, 1)$ for the normal bundle, is the area form pulled back by projection to the xy-plane.

Theorem 3.7.13 (Frobenius). *A k-plane field is integrable if and only if its twistedness is identically* 0.

Proof of 3.7.13. If the plane τ is integrable, a vector field contained in τ carries each leaf of the resulting foliation to itself, and conversely, any vector field that preserves leaves of the foliation is contained in τ. It follows that the Lie bracket of two vector fields contained in τ is another vector field contained in τ. Therefore, the twistedness is identically zero.

To show the converse, consider any k-plane field τ whose twistedness is identically zero. Take local coordinates in $\mathbf{R}^n$ near a point p so that τ is transversal to the $\mathbf{R}^{n-k}$ ("vertical") factor. Let $X_1, \ldots, X_k$ be the vector fields contained in τ that project to the coordinate vector fields in the horizontal factor $\mathbf{R}^k$. The flow by X_i projects to translation in the direction of the i-th coordinate axis of $\mathbf{R}^k$, so the Lie bracket $[X_i, X_j]$ has no horizontal component. If the twistedness is zero, we actually have $[X_i, X_j] = 0$. Thus all k fields commute. By integrating them, we obtain local coordinate charts for a foliation. $\boxed{3.7.13}$

If a plane field τ is a connection for a (G, X)-bundle $p : E \to B$, the plane at any point $x \in E$ is identified by the bundle projection p with the tangent space of the base at $p(x)$, and $N_x\tau = T_xE/\tau$ is identified with the tangent space to the fiber E_x at x. The twistedness gives, for each $x \in E$, a two-form on $T_{p(x)}B$ with values in T_xE_x. For a fixed two-vector in T_yB, for $y \in B$, the resulting vectors at all points of $p^{-1}(y)$ form a vector field on that fiber, which comes from the action of some element of the Lie algebra $\mathfrak{g}$ of G. Considering all the fibers together, we get a two-form on B with values in $\mathfrak{g}$—the curvature of τ.

Exercise 3.7.14. Let ω be a nonsingular one-form on an n-manifold (that is, ω is non-zero everywhere). Show that the kernel of ω is the tangent space to a codimension-one foliation if and only if $\omega \wedge d\omega$ is identically 0.

In the case of two-plane fields on three-manifolds, this criterion is particularly nice. The vector space of alternating bilinear forms on a two-dimensional vector space with values in a one-dimensional vector space has dimension one. If the manifold is oriented and the plane field is oriented (so that the plane field is transversely oriented as well), we can assign a sign to the twistedness. For it to be positive means that when you cycle around in a small counterclockwise loop tangent to τ you go upward, as in a right-handed screw or right-handed helix.

In dual language, the plane field τ is expressible, at least locally, as the kernel of a one-form ω. The exterior derivative $d\omega$, restricted to τ, is the dual of the twistedness of τ applied to the chosen basis ω for $N(\tau)$.

Proposition 3.7.15. *A plane field on a three-manifold defines a contact structure if and only if its twistedness is never* 0.

Proof of 3.7.15. Wherever the twistedness of the plane field τ is nonzero, the proof and conclusion of Corollary 3.7.10 apply. Therefore, Corollary 3.7.11 also applies.

Locally, choose a section s of $N(\tau)$ that is never 0. Let X be the unique vector field projecting to s and preserving τ. The twistedness of τ, expressed in units of $s(p)$ at each point p, is an X-invariant, scalar-valued two-form defined on vectors that are contained in τ. It therefore gives a two-form for the local space of leaves of the foliation associated with the flow of X. Take the corresponding measure on this space of leaves, and construct a measure-preserving map to the plane, as in Exercise 3.1.15. Cover this by a map from the neighborhood in the three-manifold into $\mathbf{R}^3$, as in Example 3.7.2, by sending every curve tangent to τ to a Legendrian curve. This gives a chart for a contact structure. $\boxed{3.7.15}$

Corollary 3.7.16 (Riemannian contact structure). *The Levi-Civita connection on the unit tangent bundle of a Riemannian surface S is a contact structure if and only if the Gaussian curvature of S is strictly positive or strictly negative.*

Exercise 3.7.17. Show that if ω is a nonsingular one-form on a three-manifold, the kernel of ω is a contact plane field if and only if $\omega \wedge d\omega$ is nonsingular.

Nonsingular one-forms ω such that $\omega \wedge d\omega$ are also nonsingular are locally modeled on a standard one-form $\alpha = dz - x dy$ that defines the model contact structure in $\mathbf{R}^3$. The pseudogroup preserving α is the *strict contact pseudogroup*.

Exercise 3.7.18 (strict contact structures). A strict contact structure has a canonical flow, which defines a foliation transverse to the contact plane field and locally equivalent to the foliation by lines parallel to the z-axis in the model contact structure in $\mathbf{R}^3$.

Show that in a contact three-manifold, most foliations transverse to the contact plane field are *not* locally equivalent to this foliation. (Compare Example 3.7.2 and Problem 3.7.8.)

3.8. The Eight Model Geometries

What is a geometry? Up till now, we have discussed three kinds of three-dimensional geometry: hyperbolic, Euclidean and spherical. They have in common the property of being as uniform as possible: their isometries can move any point to any other (homogeneity), and can take any orthonormal frame in the tangent space at a point to any other orthonormal frame at that point (isotropy). There are more possibilities if we remove the isotropy condition, allowing the space to have a grain, so to speak, so that certain directions are geometrically distinguished from others.

An enumeration of additional three-dimensional geometries depends on what spaces we wish to consider and what structures we use to define and to distinguish the spaces. For instance, do we think of a geometry as a space equipped with such notions as lines and planes, or as a space equipped with a notion of congruence, or as a space equipped with either a metric or a Riemannian metric? There are deficiencies in all of these approaches.

The problem with using lines and planes is that they aren't general enough. The five new geometries described in this section don't have any good notion of a plane—there are no totally geodesic surfaces passing through certain tangent planes. Besides, even in Euclidean geometry, information about geometric shapes is not determined by incidence properties of lines and planes.

Using congruence as the essence of the definition leads to an excessive proliferation of geometries. For instance, we would get different geometries by considering Euclidean space with the group of congruences consisting of translations, or the one consisting of horizontal translations together with vertical screw motions (where the amount of rotation is proportional to the vertical motion), and so on. Sometimes it is interesting to distinguish these different structures, but for the broad picture these variations should all be considered under the category of Euclidean geometry.

Using distance to determine the geometry is likewise unsatisfactory. First, by simply rescaling something like $\mathbf{H}^3$ or S^3, we get different metric spaces. Even if we consider as equivalent all metric spaces that are isometric up to a constant scaling factor, many of the spaces have a whole family of homogeneous metrics (sometimes with varying degrees of homogeneity) which are not equivalent up to scaling. The three-sphere, for instance, has an interesting family of homogeneous metrics obtained from the usual metric by picking a Hopf fibration and contracting or expanding the lengths of the circles, while keeping the metric constant in the orthogonal directions.

The best way to think of a geometry, really, is to keep in mind these different points of view all at the same time. If we regard changes of the group of congruences that do not change the metric and changes of the metric that do not change the group of congruences as inessential changes in the geometry, and if we also consider two geometries the same when the sets of compact manifolds modeled on them are identical, we end up with a reasonable enumeration of geometries.

For logical purposes, we must pick only one definition. We choose to represent a geometry as a space equipped with a group of congruences, that is, a (G, X)-space.

Definition 3.8.1 (model geometry). A *model geometry* (G, X) is a manifold X together with a Lie group G of diffeomorphisms of X, such that:

(a) X is connected and simply connected;

(b) G acts transitively on X, with compact point stabilizers;

(c) G is not contained in any larger group of diffeomorphisms of X with compact stabilizers of points; and

(d) there exists at least one compact manifold modeled on (G, X).

Condition (a) selects one representative from each class of locally equivalent geometries having different fundamental groups, since locally equivalent geometries are models for identical classes of manifolds (Exercise 3.3.1). Condition (b) means that the space possesses a homogeneous Riemannian metric invariant by G (Lemma 3.4.11), and that it is complete (see Proposition 3.4.15 and the paragraph following its proof). Condition (c) says that no Riemannian metric invariant by G is also invariant by any larger group. In particular, it selects at most one geometry for each isometry class of metric spaces. Another reason for condition (c) is that by enlarging the structure group G, we do not decrease the set of manifolds with that structure. Condition (d) is not phrased in an intrinsic way, but it is useful because it eliminates a whole continuous family of three-dimensional geometries that do not serve as models for any compact manifolds.

Theorem 3.8.2 (two-dimensional model geometries). *There are precisely three two-dimensional model geometries: spherical, Euclidean and hyperbolic.*

Proof of 3.8.2. Since G acts transitively on X, it follows that any G-invariant Riemannian metric on X has constant Gaussian curvature. When a metric is multiplied by a constant k, the Gaussian curvature is multiplied by k^2, so we can find a metric whose curvature is either 0, 1 or -1. It is a standard fact from Riemannian geometry that the only simply connected complete Riemannian n-manifolds with constant sectional curvature 0, 1 and -1 are $\mathbf{E}^n$, S^n and $\mathbf{H}^n$ (see [dC92, p. 163] or [GHL90, p. 135], for example). 3.8.2

For an alternate proof, see Exercise 3.8.11.

Exercise 3.8.3. The similarities of $\mathbf{R}$ form a two-dimensional group $\mathrm{Sim}(1)$, which acts on itself by left multiplication. Why isn't $(\mathrm{Sim}(1), \mathrm{Sim}(1))$ a two-dimensional model geometry? (Hint: see Exercise 2.2.10.)

In enumerating three-dimensional model geometries (G, X), we will first look at the connected component of the identity of G—call it G'. The action of G' is still transitive, and the stabilizers G'_x of points $x \in X$ are connected. This is because the quotients $G'_x/(G'_x)_0$, where $(G'_x)_0$ is the component of the identity of G'_x, form a covering space of X. Since X is simply connected, the covering is trivial.

Therefore G'_x is a connected closed subgroup of $\mathrm{SO}(3)$. Using the fact that a closed subgroup of a Lie group is also a Lie group, and therefore a manifold, it is easy to see that there are only three possibilities: $\mathrm{SO}(3)$, $\mathrm{SO}(2)$ and the trivial group. The stabilizer G_x is a Lie group of the same dimension.

Theorem 3.8.4 (three-dimensional model geometries). *There are eight three-dimensional model geometries (G, X), as follows:*

(a) *If the point stabilizers are three-dimensional, X is S^3, $\mathbf{E}^3$ or $\mathbf{H}^3$.*

(b) *If the point stabilizers are one-dimensional, X fibers over one of the two-dimensional model geometries, in a way that is invariant under G. There is a G-invariant Riemannian metric on X such that the connection orthogonal to the fibers has curvature* 0 *or* 1.

  (b_1) *If the curvature is zero, X is $S^2 \times \mathbf{E}^1$ or $\mathbf{H}^2 \times \mathbf{E}^1$.*

  (b_2) *If the curvature is* 1, *we have nilgeometry (which fibers over $\mathbf{E}^2$) or the geometry of $\widetilde{\mathrm{SL}}(2, \mathbf{R})$ (which fibers over $\mathbf{H}^2$).*

(c) *The only geometry with zero-dimensional stabilizers is solvegeometry, which fibers over the line.*

The geometries in (a) we have already discussed extensively, and those in (b_1) are self-explanatory. The remaining ones will be described in more detail in the course of the proof. We start by giving X a G-invariant Riemannian metric (see the discussion after Definition 3.8.1).

Proof of 3.8.4(a). If G' acts with stabilizer SO(3), any tangent two-plane at any point can be taken by G to any tangent two-plane at any other point, so the metric has constant sectional curvature. As in the two-dimensional case, it follows that the geometry is spherical, Euclidean, or hyperbolic.

The full group of isometries G contains G' with index 2, and can be obtained by adding any orientation-reversing isometry. $\boxed{3.8.4(a)}$

Proof of 3.8.4(b). If G' acts with stabilizer SO(2), there is a non-zero, G'-invariant vector field V on X whose direction at each point gives the axis of rotation of the elements of G' that fix that point. The trajectories of V form a G'-invariant one-dimensional foliation $\mathcal{F}$. Also, the flow of V—call it ϕ_t at time t—commutes with the action of G', so if an element of G' fixes some point on a leaf F of $\mathcal{F}$, it fixes any other point on F: all points on the same leaf have the same stabilizer. This also implies that if an element of G' takes a point $x \in F$ to another point $y \in F$, it commutes with any element of the stabilizer $G'_x = G'_y$.

Now fix a leaf F and a point $x \in F$, and let g_t be an element of g taking $\phi_t(x)$ back to x. Then $g_t \circ \phi_t$ fixes x, and its derivative at x is a linear automorphism of $T_x M$. The derivative is the identity along the axis of the action of G'_x. It commutes with rotations around this axis, that is, with elements of G'_x. Then it must be itself a rotation around this axis, possibly composed with an expansion or contraction. But an expansion or contraction is ruled out, because the assumption that there is a compact manifold modeled on (G, X) implies that V must preserve volume:

Exercise 3.8.5 (divergence). The *divergence* of a vector field V on a manifold X with a volume form ω is a measure of how much V expands or contracts volume. More precisely, $\operatorname{div} V$ is the Lie derivative $L_V\omega$ (Problem 3.1.16), expressed in units of ω.

Now suppose that X is a manifold on which a Lie group G acts transitively, and that V and ω are a vector field and a volume form on X, both invariant under G. Show that $\operatorname{div} V$ is constant over X.

In the situation of the proof, if there is a compact manifold modeled on (G, X), this manifold inherits the vector field and the volume form from X. The vector field must preserve the total volume, and so must preserves volume at every point. Therefore V has divergence zero. Show that this implies that $g_t \circ \phi_t$ acts as a rotation on $T_x M$.

We conclude that the derivative of ϕ_t maps $T_x M$ to $T_{\phi_t(x)} M$ isometrically. Since x was arbitrary, the flow of the vector field V is by isometries.

By considering a neighborhood of a point on a leaf and the fact that the leaf is invariant under a subgroup G'_x isomorphic to $\mathrm{SO}(2)$, we conclude that the leaf does not accumulate on itself, but is an embedded image of either S^1 or $\mathbf{R}$. In fact, it is easy to see that distinct leaves have disjoint neighborhoods. Therefore the quotient space $X/\mathcal{F}$ is a two-dimensional manifold Y. Since V acts by isometries, Y inherits a Riemannian metric from X (just ignore the component of the metric of X in the direction of the leaves), and a transitive action of G' by isometries. Also, Y is connected and simply connected because X is. By the proof of Theorem 3.8.2, Y must be one of the two-dimensional model geometries: $\mathbf{E}^2$, S^2 or $\mathbf{H}^2$. In addition, X is a principal fiber bundle over Y, with fiber and structure group equal to S^1 or $\mathbf{R}$.

The plane field τ orthogonal to $\mathcal{F}$ is a connection for this bundle. Since the group of isometries of X acts transitively, τ has constant curvature.

(b_1) If the curvature is zero, τ defines a foliation. Since Y is simply connected, the bundle is trivial by Proposition 3.6.15. There are three possibilities, depending on Y (an open circle indicates that no new geometry arises from this possibility):

- • If $Y = S^2$, we obtain the model geometry $S^2 \times \mathbf{E}^1$. As a compact manifold modeled on this geometry, we can take $S^2 \times S^1$ (see also Exercise 4.7.1).
- ∘ If $Y = \mathbf{E}^2$, then $X = \mathbf{E}^2 \times \mathbf{E}^1 = \mathbf{E}^3$. Thus G' (and hence G) is contained in a bigger group of isometries, and we don't get a new model geometry.
- • If $Y = \mathbf{H}^2$, we obtain the model geometry $\mathbf{H}^2 \times \mathbf{E}^1$. Any compact hyperbolic surface cross a circle is an example.

In each of these two new geometries, the full group of isometries G contains G' with index 4, since we can reverse the orientation of either factor independently.

(b$_2$) If the curvature of τ is non-zero, τ defines a contact structure. After rescaling our metric in the direction of the fibers and choosing appropriate orientations for the base and the fiber, we can assume that the curvature is 1, expressed in terms of the standard bases for $\bigwedge^2 TY$ and TF. This, together with the condition that X is simply connected, essentially determines the geometry. If Y has non-zero curvature, X can be taken as the tangent circle bundle of Y (or rather, its universal cover) with the Levi-Civita connection (Corollary 3.7.16). The group is made of derivatives of isometries of Y, together with rotations of unit tangent vectors keeping the base point fixed.

- ○ If $Y = S^2$, the tangent circle bundle is SO(3), whose universal cover is S^3 (see Example 3.7.6). For G, we get the group of isometries of S^3 that preserve the Hopf fibration (Exercise 3.7.7(c)). This is not a maximal group acting with compact stabilizers, so it is not a model geometry.
- If $Y = \mathbf{E}^2$, we obtain *nilgeometry*. This can be defined in terms of our model contact structure τ of Section 3.7 (page 168) as the group of contact automorphisms that are lifts of isometries of the xy-plane (see Example 3.7.2). Exercise 3.8.6(e) shows that there exist compact manifolds modeled on this geometry.
- If $Y = \mathbf{H}^2$, the unit tangent bundle is $\mathrm{PSL}(2, \mathbf{R})$, the group of orientation-preserving isometries of $\mathbf{H}^2$ (see Exercise 2.6.6). Passing to the universal cover, we get $X = \widetilde{\mathrm{SL}}(2, \mathbf{R})$. The unit tangent bundle of a compact hyperbolic surface is an example of a three-manifold with this geometry.

For $\widetilde{\mathrm{SL}}(2, \mathbf{R})$ and nilgeometry, the contact structure determines an orientation of the geometry which cannot be reversed. However, the orientation of the base two-dimensional geometry can be reversed simultaneously with the orientation of the fiber, so the index of G' in G is 2. 3.8.4(b)

Exercise 3.8.6 (isometries of nilgeometry). Let G be the group of isometries of nilgeometry.

(a) Compute the difference in elevation between the endpoints of a Legendrian curve for τ whose projection to the xy-plane $\mathbf{E}^2$ is a line segment with endpoints (x_0, y_0) and (x_1, y_1). (Answer: $\frac{1}{2}(x_1 + x_0)(y_1 - y_0)$.)

(b) Using this and the construction in Example 3.7.2, write down the formula for the elements of G that project to a given isometry of $\mathbf{E}^2$. (Remember that all such isometries differ by a vertical translation.)

(c) As a special case, an isometry that projects to a translation of $\mathbf{E}^2$ is of the form

$$(x, y, z) \mapsto (x + x_0, y + y_0, z + x_0 y + z_0),$$

where x_0, y_0, z_0 can be arbitrary real numbers. Show that the group H of such isometries is isomorphic to the *Heisenberg group* of real upper triangular 3×3 matrices with ones on the diagonal. (Hint: the action of H on $\mathbf{R}^3$ is free and transitive, so it makes $\mathbf{R}^3$ into a group isomorphic to H, with multiplication

$$(x_0, y_0, z_0)(x, y, z) \mapsto (x + x_0, y + y_0, z + x_0 y + z_0).$$

Now compare with the multiplication of two Heisenberg matrices.)

(d) If $h_a, h_b \in H$ project to translations by vectors a and b, the commutator $[h_a, h_b]$ is a vertical translation by a distance equal to the signed area of the parallelogram with sides a and b. Conclude that vertical translations form the center of H, of the identity component $G' \subset G$, and of G.

(e) Let h_a and h_b be as above, an assume that the translations a and b are linearly independent. Show that h_a and h_b generate a discrete cocompact subgroup. In particular, if a and b are the unit coordinate vectors, we get the group of upper triangular 3×3 matrices with integer coefficients, also called the *integer Heisenberg group* (Figure 3.26).

The Heisenberg group is †nilpotent—in fact, it is the only three-dimensional nilpotent but non-abelian connected and simply connected Lie group. This explains the term "nilgeometry" (see also Theorem 4.7.12).

Exercise 3.8.7 (geodesics in nilgeometry). Show that the Legendrian curves in Exercise 3.8.6(a) are geodesics of nilgeometry. Apart from these curves and vertical lines, what other geodesics are there?

It is easier to work with the radially symmetric model based on the contact structure σ.

Proof of 3.8.4(c). If G' acts with trivial stabilizers, it can be identified with its single orbit $X = G'/G'_x$, so X is itself a Lie group. Our task, then, is to investigate connected and simply connected three-dimensional Lie groups, asking which ones admit a discrete cocompact subgroup and are not subsumed by one of the preceding seven geometries.

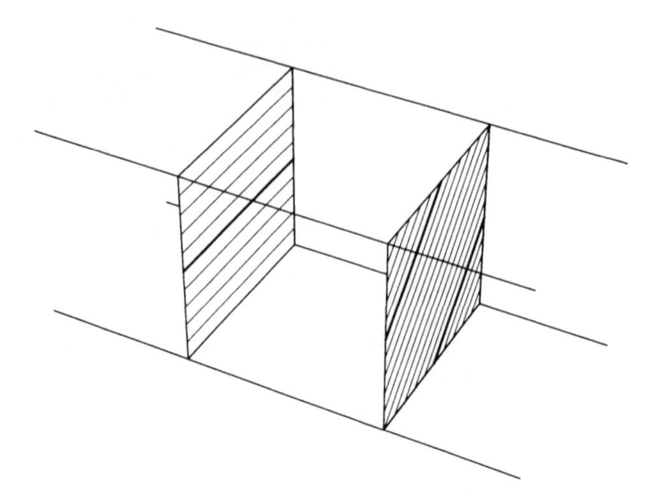

Figure 3.26. The quotient of $\mathbf{R}^3$ by the Heisenberg group. The quotient of $\mathbf{R}^3$ by H is also the quotient of $\mathbf{R} \times T^2$ by the cyclic group generated by the map $(x, y, z) \mapsto (x+1, y, z+y)$. The figure shows $\mathbf{R} \times T^2$ (opposite walls of the cylinder should be thought of as being identified); the shaded squares represent the toruses at $x = 0$ and $x = 1$. We get $\mathbf{R}^3/H$ by taking the region between the two toruses and identifying the toruses so the shadings match.

Two connected Lie groups have isomorphic Lie algebras if and only if they have isomorphic universal covers, so we start by looking at three-dimensional Lie algebras. We will think of a Lie group's Lie algebra sometimes as the tangent space at the identity, and sometimes as the set of left-invariant vector fields on the group.

Suppose G is a Lie group, and Y is a left-invariant vector field on G. Up to a factor, there is a unique volume form ω on G that is invariant under left multiplication. The flow of Y is given by *right* multiplication by a one-parameter subgroup of G, and does not, in general, preserve ω. As in the proof of 3.8.4(b), however, the assumption that there is a compact manifold modeled on (G, X) implies that Y must preserve volume (see Exercise 3.8.5).

Exercise 3.8.8 (adjoint representation). If G is a Lie group with Lie algebra $\mathfrak{g}$, the action of an element $g \in G$ by conjugation is a diffeomorphism of G fixing the identity, and therefore its derivative at the identity is a linear automorphism of $\mathfrak{g}$, which we denote by $\operatorname{ad} g$. The homomorphism $\operatorname{ad} : G \to \operatorname{GL}(\mathfrak{g})$ is called the *adjoint representation* of G. Since ad itself is a smooth map, we can consider its derivative at the identity, also called the *adjoint representation* of the Lie algebra $\mathfrak{g}$, and denoted by $\operatorname{ad} : G \to \operatorname{GL}(\mathfrak{g})$.

(a) Show that $(\operatorname{ad} v)(w) = [v, w]$ for elements $v, w \in \mathfrak{g}$.

(b) Show that the divergence (with respect to the invariant volume form on G) of the left-invariant vector field corresponding to v is constant over G, and equal to the trace of the linear map $\operatorname{ad} v : \mathfrak{g} \to \mathfrak{g}$. (Hint: the derivative at the identity of the determinant map $\det : \operatorname{GL}(V) \to \mathbf{R}$, for V a vector space, is the trace map.)

A Lie group where every left-invariant vector field preserves volume is called *unimodular*. From now on we assume that G is unimodular, so that $\operatorname{tr} \operatorname{ad} V = 0$ for all V in the Lie algebra $\mathfrak{g}$ of G.

Being skew-symmetric, the product (Lie bracket) on $\mathfrak{g}$ can be thought of as a map $\wedge^2 \mathfrak{g} \to \mathfrak{g}$. For G three-dimensional, if we fix a positive definite quadratic form (dot product) and an orientation for $\mathfrak{g}$, we get an identification between $\wedge^2 \mathfrak{g}$ and $\mathfrak{g}$, sending $V \wedge W$ to the cross product $V \times W$. Then the Lie bracket can be seen as a linear map $L : \mathfrak{g} \to \mathfrak{g}$, which turns out to be symmetric with respect to the quadratic form: $L(V) \cdot W = V \cdot L(W)$ for all $V, W \in \mathfrak{g}$.

To see this, fix an orthonormal, positively oriented basis for $\mathfrak{g}$, say $\{e_1, e_2, e_3\}$, and let l_{ij} be the entries of the matrix of L with respect to that basis. The unimodularity condition $\operatorname{tr} \operatorname{ad} V = 0$, applied to $V = e_1$, becomes $\sum_{i=1,2,3} [e_1, e_i] \cdot e_i = 0$, or

$$L(e_3) \cdot e_2 - L(e_2) \cdot e_3 = l_{23} - l_{32} = 0.$$

Repeating for the other basis vectors, we see that unimodularity is equivalent to the symmetry of L.

Every symmetric linear transformation has an orthonormal basis of eigenvectors. Changing $\{e_1, e_2, e_3\}$ to such a basis, the matrix of L becomes diagonal, with entries $c_i = l_{ii}$. In other words,

$$[e_i, e_{i+1}] = c_{i+2} e_{i+2},$$

where the subscripts are taken modulo 3. If we change the quadratic form so as to make the basis $\{a_1 e_1, a_2 e_2, a_3 e_3\}$ orthonormal, where $a_1, a_2, a_3 > 0$, each c_i gets replaced by $c_i \, (a_{i+1} a_{i+2} / a_i)$. Therefore we can arrange for all the c_i's to be equal to ± 1 or 0. To normalize further, we can permute the basis elements, but if it is an odd permutation we must also change the orientation of the vector space, so the net effect is that the c_i's are permuted and change signs. To undo the sign changes, we can replace the basis elements by their negatives. Therefore, using these operations, we can arrange to have $c_1 \geq c_2 \geq c_3$, and to have at least as many positive c_i's as negative ones. Up to isomorphism, then, there are six possibilities for $\mathfrak{g}$, and therefore for the Lie group G. Only one gives rise to a new geometry:

- ∘ $c_1 = c_2 = c_3 = 1$ gives $G = S^3$.
- ∘ $c_1 = c_2 = 1$ and $c_3 = -1$ gives $G = \widetilde{\mathrm{SL}}(2, \mathbf{R})$.
- • $c_1 = c_2 = 1$ and $c_3 = 0$ gives rise to *solvegeometry*. The generators e_1 and e_2 commute, so the Lie group G contains a copy of $\mathbf{R}^2$. This is a normal subgroup, with quotient group $\mathbf{R}$. The Lie group is therefore a semidirect product of $\mathbf{R}^2$ with R, and determined by the action of the one-parameter group generated by e_3 on $\mathbf{R}^2$. The derivative of this action at the identity is $\left(\begin{smallmatrix} 0 & 1 \\ 1 & 0 \end{smallmatrix}\right)$. By rotating e_1 and e_2 by $45°$ we can make the action of e_3 diagonal, with derivative at $t = 0$ equal to $\left(\begin{smallmatrix} 1 & 0 \\ 0 & -1 \end{smallmatrix}\right)$, so the actual transformations in this basis are given by

  $$t \to \begin{pmatrix} e^t & 0 \\ 0 & e^{-t} \end{pmatrix}.$$

  Thus G is an extension $0 \to \mathbf{R}^2 \to G \to \mathbf{R} \to 1$, consisting of maps of the form

  $$(x, y, t) \to (e^{t_0}x + x_0, e^{-t_0}y + y_0, t + t_0),$$

  for arbitrary real x_0, y_0 and t_0.
- ∘ $c_1 = 1$ and $c_2 = c_3 = 0$ gives the Heisenberg group.
- ∘ $c_1 = c_2 = c_3 = 0$ gives $G = \mathbf{R}^3$.
- ∘ $c_1 = 1$, $c_2 = 0$ and $c_3 = -1$ gives the universal cover of the group of isometries of $\mathbf{E}^2$, where e_1 and e_3 act as translations in the x and y directions, while e_2 rotates about the origin.

Example 3.8.9 gives an example of a compact three-manifold modeled on solvegeometry, so this is actually a model geometry.

We still have to compute the full group G of isometries of solvegeometry. There are independent reflections in the two eigenspaces in the $\mathbf{R}^2$ direction. One can also reflect in the $\mathbf{R}$ direction while at the same time interchanging the two eigenspaces. Thus, the index of G' in G is 8. 3.8.4(c)

Example 3.8.9 (solvegeometry manifold). Examples of three-manifolds modeled on solvegeometry can be obtained by forming the mapping torus M_ϕ for a diffeomorphism ϕ of the torus to itself. For example, let $\phi : T^2 \to T^2$ come from the linear automorphism of $\mathbf{R}^2$ with matrix $\left(\begin{smallmatrix} 2 & 1 \\ 1 & 1 \end{smallmatrix}\right)$. Arrange the universal cover of the torus in $\mathbf{R}^2$ so that the two orthogonal eigenspaces of ϕ line up with the x- and y-axes. Since the eigenvalues of ϕ are reciprocal to each other, there is some t_0 such that the transformation

$$\psi : (x, y, t) \mapsto (e^{t_0}x, e^{-t_0}y, t + t_0)$$

of $\mathbf{R}^3$ induces the given automorphism ϕ between $\mathbf{R}^2 \times \{0\}$ and $\mathbf{R}^2 \times \{t_0\}$. Therefore M_ϕ is a solve-manifold: it is the quotient of $\mathbf{R}^3$ by the discrete group of automorphisms of solvegeometry generated by ψ together with unit translations along the x- and y-axes.

Exercise 3.8.10 (mapping tori of the torus). Generalize the example above to the case when ϕ comes from an arbitrary linear automorphism of $\mathbf{R}^2$ that preserves the lattice $\mathbf{Z}^2 \subset \mathbf{R}^2$. If λ_1 and λ_2 are the roots of the characteristic polynomial of ϕ, show that:

(a) If λ_1 and λ_2 are not real, they are roots of unity, and ϕ has finite order. T^2 has a Euclidean structure for which ϕ is an isometry, and M_ϕ has a Euclidean structure.

(b) If λ_1 and λ_2 are real and distinct, ϕ has two eigenspaces. The torus has a Euclidean metric for which these eigenspaces are orthogonal, and M_ϕ has a solvegeometry structure. Find two linear maps that have the same characteristic polynomials with real distinct roots, but give rise to mapping tori that are not homeomorphic.

(c) In the remaining case, $\lambda_1 = \lambda_2 = \pm 1$, either we have $\phi^2 = 1$ and M_ϕ has a Euclidean structure, or ϕ is conjugate to a map with matrix $\pm\left(\begin{smallmatrix} 1 & n \\ 0 & 1 \end{smallmatrix}\right)$, for $n \neq 0$, and M_ϕ has a nilgeometry structure.

In general, show that M_ϕ is a cocompact discrete subgroup of a three-dimensional Lie group. How does this statement generalize for linear maps of the n-torus?

It is a curious fact that each of the eight three-dimensional model geometries is isometric to a three-dimensional Lie group with a left-invariant metric. All the Lie groups except for $\mathbf{H}^3$ are unimodular. The Lie group for $\mathbf{H}^3$ is the group of homotheties of the plane, which acts simply transitively as a group of isometries of upper half space. In every case but $\mathbf{H}^3$, the group of automorphisms of the geometry is the semidirect product of the simply connected Lie group with its group of isometric automorphisms.

Exercise 3.8.11 (two-dimensional geometries from groups). Derive the classification of two-dimensional model geometries from the classification of three-dimensional unimodular Lie algebras, but without using Riemannian geometry.

3.9. Piecewise Linear Manifolds

Example 3.2.11 suggests that triangulated spaces, even when they are manifolds, can be difficult to deal with. If we want manifolds of somewhat better quality, we must impose some additional structure, such as differentiability (Example 3.1.3). In the context of gluings it is natural to consider *piecewise linear* structures, which we introduce now.

A map from a subset of an affine space into another affine space is called *piecewise linear* if it is the restriction of a simplicial map defined on the polyhedron of some simplicial complex. A *piecewise linear manifold* is a manifold based on PL, the pseudogroup generated by piecewise linear homeomorphisms between open subsets of $\mathbf{R}^n$.

Problem 3.9.1. (a) The reason a piecewise linear map is defined in terms of the restriction of a simplicial map is that this definition is clearly satisfied for simple maps, such as the identity. Show that, given a piecewise linear map on an open set, the domain itself can be triangulated so that the map is simplicial.

(b) Show that piecewise linear homeomorphisms between open subsets of $\mathbf{R}^n$ form a pseudogroup. In particular, the "generated by" in the definition of PL is superfluous.

The notion of a piecewise linear map can be immediately extended to maps from a subset of an affine space to a piecewise linear manifold, by looking at the expression of the map in charts.

Convention 3.9.2. When we talk of a triangulation of a piecewise linear manifold X, we will normally have in mind a *piecewise linear triangulation*, that is, one where the homeomorphism $|\Sigma| \to X$ is a piecewise linear map.

Problem 3.9.3 (triangulating piecewise linear manifolds). (a) Show that any piecewise linear manifold has a triangulation. (Hint: start from a finite atlas—see Convention 3.1.4. See also the proof of Theorem 3.10.2.)

(b) Fix a triangulation and a locally finite atlas for the manifold. Prove that one can refine the atlas and subdivide the triangulation in such a way that each chart of the refined atlas is actually linear on each simplex.

Example 3.9.4 (standard piecewise linear sphere). Take an $(n+1)$-dimensional convex polyhedron $K \subset \mathbf{R}^{n+1}$, and consider orthogonal projection π onto a hyperplane P, which we identify with $\mathbf{R}^n$. When restricted to an appropriate subset of ∂K, which is a sphere, π is a homeomorphism onto a subset of P. Such subsets cover ∂K as P ranges over all possible hyperplanes. We take the corresponding homeomorphisms as charts for ∂K, and call the resulting piecewise linear manifold a *standard piecewise linear sphere.*

Exercise 3.9.5. The definition of a standard piecewise linear sphere does not depend on K, in the sense that if K' is another convex polyhedron to which we apply the same procedure, there is a piecewise linear homeomorphism $\partial K \to \partial K'$.

Let Σ be a simplicial complex. Can $|\Sigma|$ be made into a piecewise linear manifold X so that the identity map $|\Sigma| \to X$ is piecewise linear? If so, we say that Σ *is a piecewise linear manifold*, or *has a piecewise linear structure*. The characterization here is much easier than in the topological case (compare Proposition 3.2.5 and the subsequent discussion).

Proposition 3.9.6 (piecewise linear manifolds have spherical links). *A simplicial complex is a piecewise linear manifold if and only if the link of each simplex is a piecewise linear manifold equivalent to the standard piecewise linear sphere.*

Proof of 3.9.6. If the link of each vertex is a standard piecewise linear sphere, the star of each vertex is piecewise linear homeomorphic to a piecewise linear ball (compare Exercise 3.2.4). The collection of these homeomorphisms is a piecewise linear atlas. The converse depends on the following fact:

Exercise 3.9.7. Let Σ' be a subdivision of a simplicial complex Σ. Show that, if $\sigma \in \Sigma$ and $\sigma' \in \Sigma'$ have the same dimension and $|\sigma'| \subset |\sigma|$, the links of σ and σ' are piecewise linear equivalent.

Now given a vertex v of the simplicial complex and a piecewise linear chart ϕ whose domain contains v, we can subdivide the triangulation in the neighborhood of v so that ϕ is affine on each simplex that contains v. It is easy to see that the link of any simplex containing

v in the subdivided triangulation is piecewise linear homeomorphic to a piecewise linear sphere of the appropriate dimension. [3.9.6]

It follows that the triangulated space $\Sigma^2 P$ of Example 3.2.11, although homeomorphic to S^5, is not a piecewise linear manifold. The homeomorphism cannot be piecewise linear—in fact, it maps the suspension circle to a wild circle in S^5.

This answers in the negative the famous *Hauptvermutung*, or fundamental conjecture, which asks whether any two triangulations of a space admit isomorphic subdivisions. (The first counterexample to the Hauptvermutung was found by Milnor [Mil61], but it did not involve a manifold.) The Hauptvermutung is true for many classes of spaces, notably two-manifolds [Rad25], two-dimensional complexes [Pap43], and three-manifold [Bin54, Moi52].

A slightly weaker form of the Hauptvermutung asks whether any two piecewise linear structures (stiffenings) of a topological manifold are equivalent. This and the related question of whether every topological manifold admits a piecewise linear structure have also been answered in the negative: Kirby and Siebenmann [KS69] showed there exist six-manifolds that do not admit any piecewise linear structure, and five-manifolds with inequivalent piecewise linear structures. But in dimension three or lower all topological manifolds have a piecewise linear structure (see [Bin59] and the references in the previous paragraph), and any two such structures must be equivalent by the Hauptvermutung.

We conclude this section with some further examples of $\mathcal{G}$-structures. Although related to piecewise linear structures, these notions will not be used later.

Example 3.9.8 (piecewise projective manifolds). A map between open subsets of $\mathbf{R}^n$ is called *piecewise projective* if it can be extended to a map on the polyhedron of some simplicial complex, whose restriction to each simplex is a projective transformation. A PP-manifold, where PP is the pseudogroup of piecewise projective homeomorphisms between open subsets of $\mathbf{R}^n$, is called a *piecewise projective manifold.*

Let K be a convex polyhedron as in Example 3.9.4. The boundary ∂K has a piecewise projective structure whose charts are defined by projection from a point inside the convex hull to a hyperplane. When the point changes and the hyperplane changes, the transition functions are piecewise projective. This defines the *standard piecewise projective sphere.*

Problem 3.9.9 (piecewise linear and piecewise projective structures). Any piecewise linear map is piecewise projective, so a piecewise linear manifold automatically has a piecewise projective relaxation.

(a) Show that every piecewise projective structure admits a piecewise linear stiffening, unique up to piecewise projective equivalence.

(b) Show that the standard piecewise linear sphere is a stiffening of the standard piecewise projective sphere. Does the link of a vertex in a piecewise linear-manifold have a canonical piecewise linear-structure, or only a canonical piecewise projective structure?

Not all interesting pseudogroups act transitively. Here is one of the most intriguing examples:

Example 3.9.10 (piecewise integral projective manifolds). The group $\mathrm{PGL}(n+1, \mathbf{R})$ of projective transformations of $\mathbf{RP}^n$ contains the discrete subgroup $\mathrm{PGL}(n+1, \mathbf{Z})$. The pseudogroup PIP of *piecewise integral projective* transformations consists of homeomorphisms between open subsets of $\mathbf{R}^n$ for which there is a subdivision of the domain into n-simplices, such that on each one, the homeomorphism is induced by the action of $\mathrm{PGL}(n+1, \mathbf{Z})$ on $\mathbf{R}^n \subset \mathbf{RP}^n$.

Since PIP takes points in $\mathbf{R}^n$ with rational coordinates to points with rational coordinates, it is not transitive.

Problem 3.9.11 (piecewise linear and piecewise integral projective structures). Show that every piecewise linear manifold admits a piecewise integral projective-stiffening unique up to piecewise linear equivalence.

The unusual aspect to this structure is that the group of piecewise integral projective homeomorphisms of a manifold is only countable. Richard Thompson has shown that the group of orientation-preserving piecewise integral projective transformations of the circle is finitely presented and simple. It is not known whether the piecewise integral projective homeomorphism group of a surface is finitely presented.

It turns out that the Teichmüller space of a surface (Section 4.6) can be given a sphere at infinity, analogous to that for hyperbolic space, such that the transformations induced by homeomorphisms of the surface are piecewise integral projective.

3.10. Smoothings

In Section 3.9 we got a pretty good theoretical understanding of which gluings of polyhedra give piecewise linear three-manifolds, but there remains a significant issue. Can a space obtained by gluing polyhedra be smoothed, that is, given a differentiable structure? (We'll use the words "smooth" and "differentiable" interchangeably in this section, since the same results apply to all differentiability classes: compare Example 3.1.6.)

It turns out that for dimensions up to three, the answer is yes: one can go back and forth between piecewise linear and differentiable structures at will. It is a great convenience to be able to do so. The standard ideas of three-dimensional topology can be developed in either differentiable or piecewise linear form, but some work much better in one form, others in the other.

To begin with, how can we reasonably compare a piecewise linear structure on a manifold to a smooth structure? The concepts of stiffening and relaxation do not directly apply, because diffeomorphisms are not generally piecewise linear maps, or vice versa. In fact, the intersection of the PL and $\mathcal{C}^r$ pseudogroups, for $r \geq 1$, is the pseudogroup arising from affine transformations (Example 3.3.4).

The key is the idea of a *smooth triangulation* of a smooth manifold. A smooth triangulation is one where the domain is a triangulated piecewise linear manifold (Convention 3.9.2) and the map is *simplicially smooth*, that is, its restriction to each simplex is a diffeomorphic embedding. A homeomorphism from a piecewise linear manifold to a smooth manifold is *piecewise smooth* if some triangulation of the domain induces a smooth triangulation of the range. A *smoothing* of a piecewise linear manifold M is a smooth manifold N, together with a piecewise smooth map $M \to N$.

Problem 3.10.1. Since $\mathbf{R}^n$ has both a piecewise linear structure and a smooth structure, it makes sense to talk about piecewise smooth maps from $\mathbf{R}^n$ to itself.

(a) Show that such maps don't form a group. Thus, piecewise smooth maps serve as bridges between piecewise linear and smooth structures; they don't work well alone.

(b) Let $\mathcal{G}$ be the pseudogroup generated by piecewise smooth homeomorphisms of $\mathbf{R}^n$ to itself. How well can you characterize elements of $\mathcal{G}$?

Theorem 3.10.2 (triangulating smooth manifolds). *Every smooth manifold admits a smooth triangulation. Any two smooth triangulations have the same underlying piecewise linear structure, that is, the domains are isomorphic piecewise linear manifolds.*

Proof of 3.10.2. What follows is an outline. Munkres [Mun66, Chapter II] gives a very clear and detailed proof, loosely based on the original proof by J. H. C. Whitehead [Whi40].

First, consider any smooth triangulation of a compact subset of $\mathbf{R}^n$. Make a very fine subdivision using a pattern similar to Problem 1.3.4. Replace each diffeomorphic embedding of a simplex in the subdivision by the affine map having the same vertices, to obtain a piecewise linear triangulation. The map remains piecewise linear wherever it was already piecewise linear.

This can be used to show uniqueness of the piecewise linear structure underlying a smooth triangulation. For, given two smooth triangulations, one can subdivide and approximate them until the map from one to the other is piecewise linear. This can be done coordinate patch by coordinate patch, treating each patch as $\mathbf{R}^n$.

Existence can be proven similarly, along the lines of Problem 3.9.3. Start with a finite atlas for the manifold M. For each chart (U_i, ϕ_i), take a relatively compact open subset U_i' so that the U_i' still cover M, and choose a finite triangulation for a compact subset of U_i containing U_i'. The idea now is to modify these partial triangulations, chart by chart, until they are PL-compatible whenever they overlap. To extend over a new chart, work in $V = \phi_i(U_i') \subset \mathbf{R}^n$. One has a smooth triangulation already defined on some subset of V; the object is to extend it over V. Subdivide the existing triangulation if necessary, and approximate it by a piecewise linear map. This makes the previously defined piecewise linear structure compatible with the one in V, and gives a way to extend the smooth triangulation over V. 3.10.2

Thus, every smooth manifold is a smoothing of some (essentially unique) piecewise linear manifold. The opposite problem, finding a smoothing for a given piecewise linear manifold, is harder. The hard part is to get a clear conceptual idea of what needs to be done—the actual process of smoothing then becomes relatively easy.

For any point x in a piecewise linear manifold M we can define a tangent space T_xM, as (for example) the set of piecewise linear curves $[0, t) \to M$ with starting point x, modulo the equivalence given by agreement between right derivatives at 0. This space is homeomorphic to $\mathbf{R}^n$, where n is the dimension of M—the pullback by a chart provides a homeomorphism. Unlike the differentiable case, the homeomorphism is not unique up to a linear map, so T_xM has no natural vector space structure.

An even bigger problem is that there is no natural topology on the total space $TM = \bigcup_x T_x M$ that makes it into a bundle over M. In the differentiable case, the topology of TM is such that the derivative map $df : TM \to TN$ is continuous, for any smooth map $f : M \to N$. For a piecewise linear manifold, this doesn't work: the derivative of a piecewise linear map $f : M \to N$, defined in an obvious way (Figure 3.27), can change discontinuously from point to point. Therefore, if we insist that the derivative of an arbitrary piecewise linear map $TM \to \mathbf{R}^n$ be continuous, we end up with a topology on TM where each fiber is essentially independent (though all fibers are joined at 0 to form a subspace homeomorphic to M).

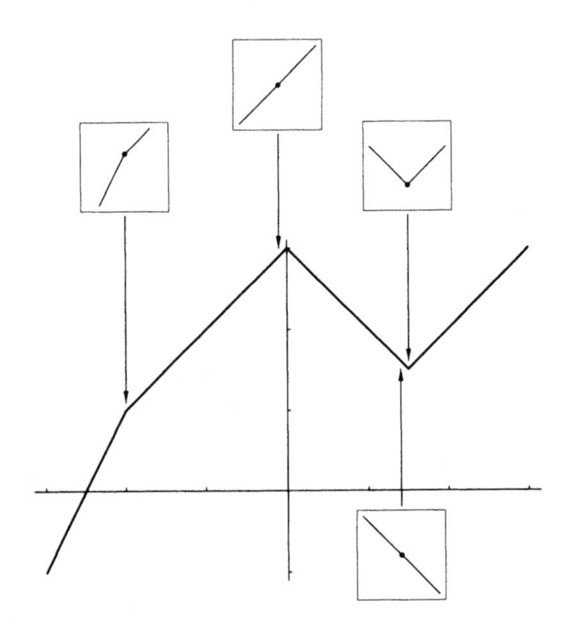

Figure 3.27. The derivative of a piecewise linear map. A piecewise linear map $f : \mathbf{R} \to \mathbf{R}$ has a derivative $df : T\mathbf{R} \to T\mathbf{R}$, taking $T_x\mathbf{R}$ to $T_{f(x)}\mathbf{R}$ for each x. At points x where the graph of f bends, df_x is not linear, and in particular it is very different from df_y for values of y close to x on either side.

The situation is better if we fix a triangulation for M and stick to maps $M \to \mathbf{R}^n$ that are affine on each simplex. The derivative of such a map is constant along any open simplex of any dimension, so the tangent space over each open simplex $\mathring{\sigma}$ containing a point x is simply $\mathring{\sigma} \times T_x M$. Moreover $T_x M$ is subdivided into polyhedral *tangent cones*, one for each simplex containing x; the derivative df_x is linear when restricted to each tangent cone.

How do the pieces of the total tangent space TM meet? Let's look first at a triangulation of $\mathbf{R}$. The tangent space of the differentiable real line is $\mathbf{R}^2$, where we think of the horizontal coordinate as the position of the basepoint and the vertical coordinate as the tangent vector proper. Once we triangulate the line we are in the

Figure 3.28. The tangent space for a triangulated line. At a vertex, the fiber does not have a linear structure: it is the union of two *tangent cones* (rays). A neighborhood basis for a tangent vector there consists of closed half-disks pointing to the right if the vector is positive, and to the left if it is negative.

situation of Figure 3.27: at any vertex, directional derivatives in positive directions agree with the limit from the right but not in general with the limit from the left. Therefore, along the tangent space to any vertex, we make a slit just to the left of the positive vectors and just to the right of the negative vectors, as in Figure 3.28. (It's like pulling apart two segments of a grapefruit, with half the membrane sticking to each segment.)

The picture is similar for a triangulation of $\mathbf{R}^n$. Begin with the disjoint union of the tangent spaces of $\mathbf{R}^n$ along every open simplex. For a tangent vector v based at a point p, define C_v as the union of all simplices σ that contain some initial portion of the ray with origin p and direction v. The directional derivative along v varies continuously when its base point is moved into C_v; glue there. (Formally, a neighborhood basis for (p, v) consists of the products of neighborhoods of v in the tangent space of $\mathbf{R}^n$ at p with neighborhoods of p in C_v.)

We now have a topology for the tangent space of a triangulated piecewise linear manifold M. If M maps to a smooth manifold N by a simplicially smooth homeomorphism, the derivative map is a continuous bijection between TM and TN. But it is not a homeomorphism because the topology in the range is coarser. So one set of data needed to turn a triangulated piecewise linear manifold into a smoothly triangulated smooth manifold is a coarsening of the topology of the tangent space TM—a recipe for gluing the flaps in Figure 3.28, and turning TM into a bundle. Another set of data is a *linearization* for each fiber T_xM—a linear structure compatible with the partial linear structures of the tangent cones that make up the fiber. (In fact, either set of data determines the other: see Exercise 3.10.5.)

Let's analyze linearizations of the polyhedral cone structure on T_xM. A linearization can be seen as a homeomorphism $L : T_xM \to \mathbf{R}^n$ that is linear on each polyhedral cone, but we regard as equivalent homeomorphisms that differ only by composition with a linear map of $\mathbf{R}^n$. If we normalize by prescribing the images in $\mathbf{R}^n$ of n linearly independent vectors inside a particular cone, we can record a linearization by noting where it takes a finite number of additional points.

For instance, in one dimension, the tangent space to a vertex is $\mathbf{R}$, divided into the positive and negative real axis. Normalize the homeomorphism $L : \mathbf{R} \to \mathbf{R}$ so the positive real axis is mapped to itself by the identity; then the linearization is determined by $L(-1)$, which can be any negative real number. (Any point on the negative real axis will do instead of -1.) The space of linearizations is the open half-line.

In two dimensions, consider first a point along an edge. We can choose coordinates so the tangent space is $\mathbf{R}^2$, divided into the lower and upper half-planes. Normalize $L : \mathbf{R}^2 \to \mathbf{R}^2$ so it is the identity on the upper half-plane, and record where it takes $(0, -1)$ as a result. It can be anywhere in the lower half-plane. Therefore, the space of linearizations is a half-plane.

The tangent space to a vertex in a two-dimensional triangulation is $\mathbf{R}^2$, subdivided into $k \geq 3$ angles coming together. Normalize L so one of these angles goes to the first quadrant of $\mathbf{R}^2$. Pick a point p_i on each of the remaining $k-2$ rays, in order around the vertex; the linearization is determined by the images $L(p_i)$. The polar coordinates (r_i, θ_i) of these images are subject to the constraints

$$r_i > 0 \quad \text{and} \quad 0 < \theta_1 < \ldots < \theta_{k-2} \leq \tfrac{3}{2}\pi,$$

so the space of linearizations is an open orthant in $\mathbf{R}^{k-2}$ (for the radii) cross an open $(k-2)$-simplex (for the angles). In particular, it is contractible.

Exercise 3.10.3 (linearizations form manifold). Extend this reasoning to show that the space of linearizations for T_xM is an open subset of

$$\left(\mathbf{R}_+ \times S^{n-m-1} \times \mathbf{R}^m\right)^{j-(n-m)},$$

where n is the dimension of the triangulation, m is the dimension of the simplex σ in whose interior x lies, and j is the number of vertices in $\operatorname{lk}\sigma$. In particular, this space is always a manifold. Compute it in the cases $m = n - 1$ and $m = n - 2$.

Definition 3.10.4 (welding). A *welding* of a triangulated piecewise linear manifold M is an assignment of a linearization to the tangent space of every point of M, continuous in the following sense:

If x is a point in M and $y \in M$ is in the interior of a simplex σ containing x, a certain open subset of T_xM is canonically identified, by translation, with a subset of T_yM. (If x and y belong to the same open simplex, the tangent spaces are identified in their entirety.) Continuity means that, as y approaches x within $\mathring{\sigma}$, the chosen linearization for T_yM converges to the one for T_xM, where they can be compared (Figure 3.29).

Clearly, if M is a triangulated manifold and $f : M \to N$ is a simplicially smooth homeomorphism into a smooth manifold, the linearizations defined by f on the tangent spaces of points of M form a welding. Not every welding of M comes from a piecewise differentiable map, however. Those that do are called *smooth*.

Exercise 3.10.5 (welding and coarsening). Show that the welding induced by f is determined by the topology on TM induced by f, and vice versa.

Problem 3.10.6. Investigate when a welding can be smooth. In particular, consider a neighborhood of a vertex in a triangulation of the plane. Can you find nonvacuous necessary conditions that the second derivatives of a smooth welding must satisfy along edges at that point?

Proposition 3.10.7 (weldings can be smoothed). *Any welding of a triangulated piecewise linear manifold M can be approximated by a smooth welding.*

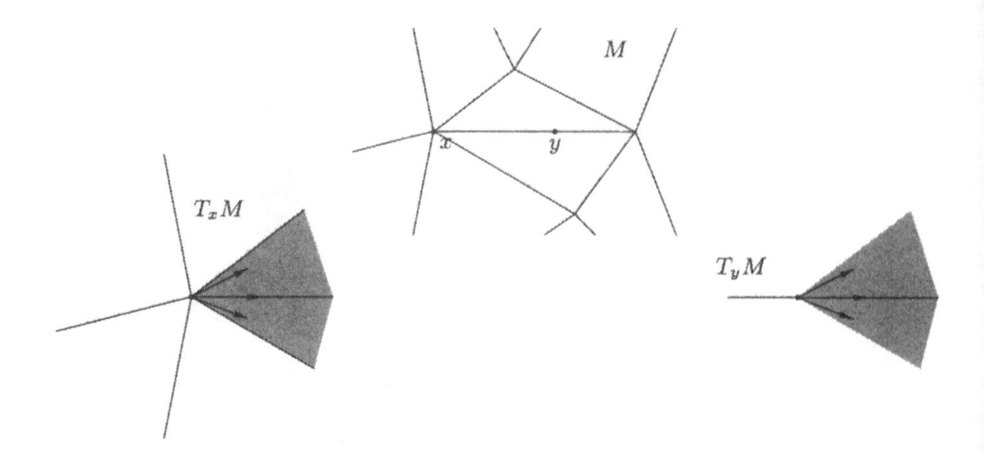

Figure 3.29. The welding continuity condition. The shaded regions indicate portions of T_xM and T_yM that are canonically identified. A linearization on T_yM is determined by the images of the three indicated vectors at y (for example). As y approaches x along the edge, the images should approach the corresponding images at x.

Proof of 3.10.7. We assume, by induction, that we have found smooth charts covering an open neighborhood of the k-skeleton of M, so that the (suitably modified) welding coincides in this neighborhood with the welding induced by the identity map.

To start the induction, consider a vertex v. Some neighborhood of the origin in T_vM is canonically identified with a neighborhood U of v in M. Use the linearization of T_vM given by the welding to define a coordinate chart in U. Homotope the welding within U so that it agrees with the welding defined by the chart on a smaller neighborhood of v. Repeat for each vertex.

Now let the induction assumption be satisfied for some k, and take a $(k+1)$-simplex σ. The welding defines a continuous map from $\mathring{\sigma}$ to the appropriate space of linearizations, which is a manifold (Exercise 3.10.3). The map is smooth in a neighborhood of the boundary. We can approximate such a map by one that is smooth everywhere in $\mathring{\sigma}$, without changing it in a smaller neighborhood of the boundary.

This gives the tangent bundle to M along the open simplex the structure of a smooth vector bundle. The tangent space to the simplex determines a smooth subbundle of dimension $k+1$. We choose a smooth complementary subbundle, and identify it with $\mathring{\sigma} \times \mathbf{R}^{n-k-1}$. The exponential map identifies a neighborhood of the zero section of this subbundle with a neighborhood U of the open simplex. Composing the two identifications, we get a coordinate chart on U, compatible with the preexisting charts on neighborhoods of lower-dimensional simplices.

Doing this for every $(k+1)$-simplex, we extend the smooth structure to a neighborhood of the $(k+1)$-skeleton. As in the case of the vertices, we can homotope the welding so that it agrees with that defined by the smooth structure in this neighborhood. This completes the induction step. 3.10.7

Theorem 3.10.8 (existence of smoothing). *Every piecewise linear manifold of dimension up to three can be smoothed.*

Proof of 3.10.8. Fix a triangulation for the manifold (Problem 3.9.3). By Proposition 3.10.7, all we have to do is find a welding for the triangulated manifold.

In one dimension, this is trivial. Choose any linearization for the tangent space at each vertex; the tangent space for a point on an edge is already linear. The continuity condition or Definition 3.10.4 is automatically satisfied.

In two dimensions, things are almost as easy. Again, choose the linearization at each vertex at will. As we have seen, the space of linearizations of the tangent space along an edge is parametrized by a half-plane. Extending the welding along an edge therefore amounts to finding a path between two points in a half-plane—easily done. The extension to two-simplices is automatic.

In three dimensions, the same reasoning applies. By Exercise 3.10.3, the space of linearizations for the tangent space along an edge is connected, so we can extend a welding from vertices to edges. Then, for each triangle, we must extend the welding from the triangle's boundary to the interior. The space of linearizations in this case is a half-space in $\mathbf{R}^3$, and hence contractible, so the extension can be done. 3.10.8

We have not addressed the question of uniqueness of smoothings. Could it happen, for instance, that a piecewise linear manifold M has smoothings $M \to N_0$ and $M \to N_1$, with N_0 and N_1 not diffeomorphic? Again, things are fairly simple in low dimension:

Theorem 3.10.9 (uniqueness of smoothing). *If $M \to N_0$ and $M \to N_1$ are two smoothings of a piecewise linear manifold of dimension at most three, the composite homeomorphism $N_0 \to N_1$ can be approximated by a diffeomorphism.*

Proof of 3.10.9. The two smoothings give two weldings of M (with a suitable triangulation). If M has dimension one or two, the space of weldings of M is contractible, so we can homotope one welding to the other. Now consider $M \times [0,1]$, subdivided into polyhedra of the form simplex $\times[0,1]$. These cells can be further subdivided into simplices. The homotopy between weldings gives a welding of $M \times [0,1]$ that agrees with the given weldings on the two ends. Use this to construct a smoothing of $M \times [0,1]$ (Proposition 3.10.7), inducing the smoothing N_0 at $M \times \{0\}$ and the smoothing N_1 at $M \times \{1\}$. By taking a little care with the construction, we can ensure that each section $M \times \{t\}$ is a smooth submanifold. The unit vector field in the $[0,1]$ direction may not be smooth under this structure, but we can convolute with a C^∞ bump function to approximate it by a smooth vector field V having unit speed in the $[0,1]$ direction and whose projection to M is very small. Then the integral of the vector field from 0 to 1 is a diffeomorphism $N_0 \to N_1$ approximating the composite map.

The space of weldings of a triangulated three-manifold is harder to analyze: the trouble comes at the vertices.

Challenge 3.10.10 (space of triangulations of S^2). Is the space of linearizations for the tangent space at a vertex in a triangulation of $\mathbf{R}^3$ contractible?

This can be reformulated in terms of the intersections of polyhedral cones with a unit sphere. Given a triangulation τ of S^2, let T_τ be the space of isometry classes of triangulations with geodesic sides, equipped with a combinatorial isomorphism to τ. Is T_τ contractible?

This question was resolved positively by Bloch, Connelly and Henderson in [EDBH84]. Igor Rivin [Riv] has found an elegant solution using three-dimensional hyperbolic polyhedra.

We will not attempt to analyze further the space of weldings of a triangulated three-manifold. Instead, we will use another method to compare smoothings, making use of the following theorem of Smale, interesting in its own right:

Theorem 3.10.11 (extending diffeomorphism). *Every diffeomorphism of S^2 to itself extends to a diffeomorphism of the three-ball to itself.*

We will prove this theorem and a generalization later, but first let's apply it to the proof of Theorem 3.10.9. We have two weldings induced on M by the smoothings $f_0 : M \to N_0$ and $f_1 : M \to N_1$. Since the space of weldings along an edge or a two-simplex is contractible, we can homotope the weldings so they agree except in small neighborhoods of the vertices. Let P_0 and P_1 be obtained from N_0 and N_1 by removing small open balls around the vertices. As before, the homotopy between weldings enables us to construct a diffeomorphism from P_0 to P_1 which is close to $f_1 \circ f_0^{-1}$. By 3.10.11, we can extend to a homeomorphism $N_0 \to N_1$ which is a diffeomorphism on P_0 and on each of the closed balls around vertices. This piecewise-differentiable map $N_0 \to N_1$ can easily be smoothed to be a diffeomorphism. 3.10.9

Proof of 3.10.11. We will show that any diffeomorphism f of the unit sphere $S^2 \in \mathbf{R}^3$ is smoothly isotopic to an orthogonal map $g \in O(3)$. Given such an isotopy F, which we can assume to be constant near the endpoints of the interval, we extend f to a diffeomorphism of the unit ball by mapping each sphere S_r of radius r to itself under the diffeomorphism F_r, where $0 \le r \le 1$.

Thus we have to show that any diffeomorphism of S^2 can be isotoped to an orthogonal map. In fact more is true: for $n \le 3$, the inclusion $O(n+1) \subset \operatorname{Diff} S^n$ is a †deformation retract, where $\operatorname{Diff} S^n$ is the group of smooth diffeomorphisms of the n-sphere, topologized with the C^1 topology (that is, a diffeomorphism is near the identity if both its values and its derivatives are uniformly near the identity). This result is easy for $n = 1$, and due to Smale [Sma59] for $n = 2$. The case $n = 3$ was conjectured by Smale and solved by Hatcher [Hat83b]; the proof is much more arduous, and is beyond our scope.

So let f be a diffeomorphism of S^2. We will perform various operations on f to change it to an orthogonal transformation. (Exercise 3.10.13 and Problem 3.10.14 ask you to check that, if each step is carried out simultaneously on all diffeomorphisms, the effect is continuous as a function of f. Thus, this process actually gives a deformation retraction from $\operatorname{Diff} S^2$ to $O(3)$.)

First we will deform the diffeomorphism so it coincides with an orthogonal map inside a disk. This part of the proof works equally well for a diffeomorphism of S^n, so we do it in this setting. We let $(e_0, \dots, e_n)$ be the standard orthonormal basis for $\mathbf{R}^{n+1}$.

Lemma 3.10.12 (nice patch). *Let D be a hemisphere of S^n with center $e_0 \in S^n$. Any diffeomorphism f of S^n is smoothly isotopic to* *a diffeomorphism taking e_0 and $-e_0$ to antipodal points and whose restriction to D is equal to the restriction of some element of $O(n+1)$.*

Proof of 3.10.12. We define the isotopy in four phases, making it constant near the endpoints of the interval corresponding to each phase in order to maintain smoothness.

First we make the derivative orthogonal rather than linear at e_0. Let $x_0 = f(e_0)$ and let $x_1, \dots, x_n$ be the image of $e_1, \dots, e_n$ under the derivative of f at e_0. Let X_0 be the $(n+1) \times (n+1)$ matrix with columns the x_i, and let X_1 be the orthogonal matrix obtained by applying the Gram–Schmidt process to the columns of X_0, from left to right (note that the first column of X_t is still x_0). Homotope X_0 to X_1 by applying the straight-line homotopy to one column at a time, from left to right (again easing in an out of each one to maintain smoothness). Each instance X_t of the homotopy is still in $\mathrm{GL}(n, \mathbf{R},)$.

Now isotope f by composing it with $X_t X_0^{-1}$ and projecting back to the sphere, using central projection. The isotopy keeps the image of e_0 at x_0 and causes the derivative at e_0 to change from X_0 to X_1. Let the diffeomorphism at the end of this first phase be f_1.

Next, compose with a parabolic transformation fixing x_0, so that the image of $-e_0$ migrates from $f_1(-e_0)$ to $-f_1(e_0) = -x_0$ along a geodesic arc. At each moment in this process, the desired parabolic transformation is well defined (see Exercise 2.5.25). The derivative at e_0 remains the same, since the derivative of a parabolic transformation at the fixed point is the identity. Let the map at the end of this phase be f_2.

Now let $\phi : \mathbf{R}^{n+1} \to [0,1]$ be a smooth bump function that equals 0 outside the unit ball and 1 inside the ball of radius $\frac{1}{2}$. Translate ϕ to e_0 and scale it so it has support in a small ball around e_0. Take a weighted average, with weights $1 - t\phi$ and $t\phi$, of f_2 and the orthogonal map that approximates it at e_0. If the support of ϕ is small enough, each intermediate map is still a diffeomorphism (show this), so we get the third phase of the isotopy. The end map of this phase coincides with an orthogonal map near e_0 and with f_2 outside the support of ϕ.

Finally, by conjugating with a hyperbolic transformation with fixed points x_0 and $-x_0$, expand the disk where the diffeomorphism is orthogonal until it encompasses the hemisphere D from the statement of the lemma. 3.10.12

Exercise 3.10.13 (continuous dependence 1). Show that the isotopy defined in the proof of Lemma 3.10.12 depends continuously on the diffeomorphism f. Among other things, you will need to specify a rule for choosing the scaling factor for the bump function ϕ, and show that the choice depends continuously on f. (Recall that Diff S^n has the C^1 topology.)

Because of lemma 3.10.12, we can assume from now on that our diffeomorphism f is an orthogonal map in a hemisphere D with center e_0. We work with the composition of f with the inverse of this orthogonal map. By stereographic projection from e_0, we can consider this composition as a diffeomorphism $h : \mathbf{R}^n \to \mathbf{R}^n$ that is the identity outside the unit ball. We want to isotope h to the identity.

Here we return to the assumption that $n = 2$ (the case $n = 1$ being easily solved by averaging). Let V be the image of the constant vector field $(0, 1)$ under h. We normalize V to retain only the information about direction, thus defining a map $\mathbf{R}^2 \to S^1$ that is constant outside the unit circle, and therefore outside the square $Q = [-1, 1] \times [-1, 1]$. We can homotope this map smoothly to the constant map, without changing it outside Q, as we can see by lifting it to a map $\mathbf{R}^2 \to \mathbf{R}$ into the universal cover of S^1. Let V_t be the corresponding unit vector field at time t; we will recover from V_t a diffeomorphism of $\mathbf{R}^2$.

We claim that the trajectory under V_t of any point in the square starts from the bottom edge of the square and ends on the top edge, as in Figure 3.30. For otherwise the orbit accumulates in the square at some point p. By adjusting the vector field in a neighborhood of p, we then obtain a closed orbit, which is impossible (see problem 3.10.15).

We now multiply V_t by a scalar function, constant on each trajectory, so as to make the time needed to go from the bottom to the top edge along any trajectory equal to 2. We define h_t as the homeomorphism of the square Q into itself that takes the point (x, y) to the point with parameter y on the trajectory that starts at $(x, -1)$, the trajectory being parametrized by $[-1, 1]$. The entrance and exit points of a trajectory under V_t generally don't have the same x-coordinate (Figure 3.30); they define a diffeomorphism ϕ_t between the bottom and top edges of the square. We compose h_t with a diffeomorphism that maps each horizontal line to itself, the map varying from the identity at the bottom to ϕ_t^{-1} at the top. After this correction, h_t is equal to the identity on the boundary of the

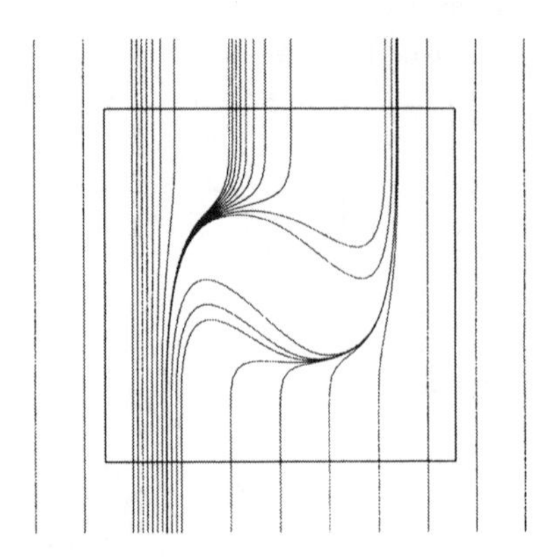

Figure 3.30. Vector field on the square. Trajectories of a smooth vector field that is constant and vertical outside a square.

square and can be extended to a diffeomorphism of $\mathbf{R}^2$. As t ranges from 0 to 1, these diffeomorphisms give the required isotopy between f and the identity. $\boxed{3.10.11}$

Problem 3.10.14 (continuous dependence 2). Check that the construction just given actually shows that the space of diffeomorphisms of $\mathbf{R}^2$ that equal the identity outside the square Q is contractible. (With Exercise 3.10.13, this implies that $O(3)$ is a deformation retract of Diff S^2.)

(a) Prove that, with an appropriate choice of homotopy, V_t depends continuously on h, where the space of smooth vector fields on the square is given the C^1 topology.

(b) Prove the scalar function used to multiply V_t in the last paragraph of the proof varies continuously with the vector field. (Hint: use the inverse function theorem on a single vector field. Use facts about approximate solutions to differential equations and continuous dependence of inverse functions on parameters for the continuity.)

(c) Show that the correction needed to compensate for ϕ_t also varies continuously.

Problem 3.10.15 (closed orbits). Prove that a never zero vector field in the plane has no closed orbits. (Hints: Use Problem 1.3.12 and Proposition 1.3.10. Move the vector field to make it point inwards along the closed orbit.) The answer to this question is a part of the famous Poincaré-Bendixson theorem.

Exercise 3.10.16. Why doesn't the second part of the proof of 3.10.11 go through in higher dimensions?

Challenge 3.10.17 (smoothing four-manifolds). Show that every piecewise linear four-manifold has a smooth structure. If you follow the outline as in the proof of Theorem 3.10.8, the difficulty is in trying to extend a welding along the edges of a triangulated four-manifold. One suggestion is that, rather than sticking with a fixed triangulation, you can subdivide:

(a) Show that for any piecewise projective map f of S^2 to itself, there is some fixed triangulation such that f is isotopic to the identity through homeomorphisms which are projective on each triangle. (Hint: use Smale's theorem).

(b) Show that any piecewise linear four-manifold has a triangulation that admits a welding.

We emphasize again that smooth and piecewise linear structures diverge in higher dimensions. There are piecewise linear manifolds that cannot be made smooth, and piecewise linear manifolds that have essentially distinct smoothings. See also Example 3.1.6 and the discussions following the proof of Proposition 3.9.6.

Problem 3.10.18. (a) Suppose a smooth manifold M of dimension m is embedded smoothly as a closed subset of a smooth manifold N of dimension $m+1$ and that M and N are orientable. Suppose this is done twice, with the same M, but with two different manifolds N, namely N_1 and N_2. Prove that there is a neighborhood of M in N_1 which is diffeomorphic to a neighborhood of M in N_2, such that the diffeomorphism is the identity on M. For the purpose of this chapter, this only needs to be proved for $m = 0$, 1 and 2.

(b) Prove that any diffeomorphic embedding of the unit ball B^n in $\mathbf{R}^n$ is isotopic among diffeomorphic embeddings to the linear map having the same derivative at the origin.

(c) Prove that any diffeomorphism of the unit disk D^2 to itself that fixes three points on its boundary is isotopic among such diffeomorphisms to the identity. [Hint: use the Riemann mapping theorem creatively.]

Problem 3.10.19 (canonical smoothing of a two-manifold). From Theorems 3.10.8 and 3.10.9 we know that a two-manifold has a differentiable structure canonical up to diffeomorphism.

Find a canonical definition for a differentiable structure on a triangulated two-manifold M which depends only on the local combinatorics of the triangulation, so that no arbitrary choices need to be made in the course of the construction.

More explicitly, construct a finite differentiable atlas canonically associated with a triangulated two-manifold M refining the open star cover of

M, such that for any isomorphism $f : U \to V$ between open star neighborhoods in triangulated two-manifolds, f carries the smooth coordinate charts contained in U to the smooth coordinate charts contained in V.

Challenge 3.10.20 (canonical smoothing of a three-manifold). Generalize Problem 3.10.19 to three-manifolds. To do this, you probably need some heavy machinery such as the uniformization theorem for Riemannian metrics on S^2, used with ingenuity.

Chapter 4

The Structure of Discrete Groups

There are often strong consequences for the topology of a manifold M which can be expressed as the quotient space of a homogeneous space (G, X) by a discrete group $\Gamma \subset G$. These consequences arise from geometric and algebraic restrictions on discrete groups. In some cases, the information is strong enough to enable a complete classification of closed (G, X)-manifolds.

In this chapter, we will investigate the structure of discrete subgroups of automorphisms of a homogeneous space, with an emphasis on the three-dimensional model geometries.

4.1. Groups Generated by Small Elements

Let G be a Lie group with a left invariant metric d (Lemma 3.4.11). We denote the identity element of G by 1, and call an element of G *small* if it is near the identity. To get an overall image of discrete groups Γ of G, we need to understand not only the well-behaved examples but the extreme examples as well, when Γ is almost indiscrete, in the sense that it has many very small elements.

For this, the *Hausdorff topology* is quite helpful. If X is a compact metric space and $A, A' \subset X$ are closed subsets, the *Hausdorff distance* $d_H(A, A')$ between A and A' is the greatest distance separating a point in either set from the other set. In symbols, $d_H(A, A')$ is the infimum of the values of r such that $A \subset \bigcup_{x \in A'} B_r(x)$ and $A' \subset \bigcup_{x \in A} B_r(x)$, where $B_r(x)$ is the ball of radius r around x.

Exercise 4.1.1. The Hausdorff metric makes the set of closed subsets of X into a compact metric space. The underlying topology depends only on the topology of X, not on the metric on X.

For subsets of a noncompact metric space, the appropriate definition is not as clear, and in fact there are two good definitions. Most obvious is to use exactly the same definition as for compact spaces. Unfortunately, with this topology, the space of closed subsets is not compact, or even paracompact. For our purposes, the best is to define the Hausdorff topology in terms of the intersections with compact sets. That is, if X is a locally compact, complete metric space and $A \subset X$ is a closed subset, we define neighborhoods $\mathcal{N}_{K,\varepsilon}(A)$, for $K \subset X$ compact and $\varepsilon > 0$, to consist of sets B such that

$$d_H(A \cap K, B \cap K) < \varepsilon.$$

Exercise 4.1.2. Show that, with this topology, the resulting space is compact.

Proposition 4.1.3 (closed subgroups form Hausdorff closed set). *Among the closed subsets of a Lie group G, closed subgroups form a closed, hence compact, set.*

Proof of 4.1.3. Let $A \subset G$ be a closed subset which is a limit point of closed subgroups of G. Let $a, b \in A$. We need to verify that $a^{-1} \in A$ and that $ab \in A$. Choose a compact set $K \subset G$ whose interior contains a, b, ab and a^{-1}. Then $\mathcal{N}_{K,\varepsilon}(A)$ contains subgroups with elements within ε of a and b, hence elements near to a^{-1} and to ab. Taking the limit as $\varepsilon \to 0$, the result follows. $\boxed{4.1.3}$

A property that closed subgroups of a Lie group may or may not have is a *closed property* if the set of closed subgroups that satisfy the property is closed in the Hausdorff topology.

Exercise 4.1.4. Show that the following properties of closed subgroups H are closed:

(a) H is connected.

(b) H is abelian.

(c) A group H is *nilpotent of class k* if all k-times iterated commutators $[h_0, [h_1, \dots, [h_{k-1}, h_k] \dots]]$ equal the identity. Show that the being nilpotent of class k is a closed property. (Hint: show first that the closure of a nilpotent subgroup is nilpotent of the same class.)

Clearly, discreteness is not a closed property: it fails, for example, for the sequence of subgroups $n^{-1}\mathbf{Z} \subset \mathbf{R}$, for $n = 1, 2, \dots$ One important goal of this section is to understand indiscrete limits of discrete groups: see Theorem 4.1.7.

Any closed subgroup $H \subset G$ is a Lie subgroup. The connected component of the identity in H, denoted H_0, is also a Lie subgroup, and, being connected, it is determined by its tangent space at 1. The entire subgroup H is a union of a discrete set of cosets of H_0; in the special case that H_0 is just the identity, H is a discrete group.

In Section 3.6 we saw that the Lie algebra $\mathfrak{g}$ of a Lie group G can be thought of either as the tangent space to the identity in G, or as the space of left-invariant vector fields in G. The flow of a left-invariant vector field that has the value v at the identity is given by the action *on the right* of a one-parameter subgroup of G, whose elements we denote by $\exp tv$, where $t \in \mathbf{R}$ is the time parameter of the flow. The correspondence $v \mapsto \exp v$ is called the *exponential map* $\mathfrak{g} \to G$. Since its derivative at 1 is the identity, exp is bijective from a neighborhood of 0 in $\mathfrak{g}$ to a neighborhood of 1 in G. In such a neighborhood, we can define the *logarithm map* log to be the inverse function of exp. The exponential and logarithm maps are important tools in the study of Lie groups.

Example 4.1.5. The usual exponential function is the exponential map for the multiplicative group of positive real numbers.

The tangent space at 1 to $\mathrm{GL}(n, \mathbf{R})$ is the vector space $\mathrm{Mat}(n, \mathbf{R})$ of linear maps of $\mathbf{R}^n$ into itself, and the exponential map

$$\exp\colon \mathrm{Mat}(n, \mathbf{R}) \to \mathrm{GL}(n, \mathbf{R})$$

is defined by the usual power series for the exponential function. The logarithm function is also defined by the usual power series, which converges provided one is in a suitable neighborhood of 0 in $\mathrm{Mat}(n, \mathbf{R})$.

It is easy to check that $\exp(nv) = (\exp v)^n$ for any integer n and any $v \in \mathfrak{g}$, just as with the usual exponential function. However, $\exp(v + w) = \exp v \exp w$ for all v and w if and only the identity component of G is abelian.

Theorem 4.1.6 (discrete and small-generated is nilpotent). *In any Lie group G, there exists a neighborhood U of the identity such that any discrete subgroup Γ of G generated by $\Gamma \cap U$ is nilpotent.*

Proof of 4.1.6. The basic idea is that, for very small elements, multiplication in G is approximated by vector addition in $\mathfrak{g}$. Since addition of vectors is commutative, multiplication of small elements is commutative up to second order. In symbols, there is an $\varepsilon > 0$

and a constant C such that, for any $a, b \in G$ with $d(1,a) < \varepsilon$ and $d(1,b) < \varepsilon$, we have

$$d(1,[a,b]) < Cd(1,a)d(1,b).$$

In particular, $[a,b]$ is considerably closer to 1 than either a or b. This inequality follows from Taylor's theorem applied to the map $[\,,\,] : G \times G \to G$, taking into account that $[1,b] = [a,1] = 1$.

Take ε small enough so that $d(1,[a,b]) < \frac{1}{2}d(1,a)$ for any $a, b \in U = B_\varepsilon(1)$, and let Γ be a discrete subgroup of G generated by $\Gamma \cap U$. Then any nested commutator

$$[g_1, [g_2, [g_3, \dots, [g_{k-1}, g_k] \cdots]]]$$

of elements $g_i \in \Gamma \cap U$ is within distance $2^{-k}\varepsilon$ of 1. Since there is a minimum distance of nontrivial elements of Γ from 1, there is some k for which all such k-th nested commutators are trivial. This implies that Γ is nilpotent, because we have a central series

$$\Gamma = \Gamma_0 \supset \Gamma_1 \supset \cdots \supset \Gamma_k = 1,$$

where Γ_i is generated by i-th nested commutators of elements in $\Gamma \cap U$. 4.1.6

The key result describing how discreteness fails to be closed is a generalization of Theorem 4.1.6:

Theorem 4.1.7 (nilpotent limits). *Let G be a Lie group.*

(a) *There exists a neighborhood U of 1 in G such that any discrete subgroup Γ of G generated by $\Gamma \cap U$ is a cocompact subgroup of a connected, closed, nilpotent subgroup N of G whose intersection with U is connected.*

(b) *The closure of the set of discrete subgroups of G consists of closed subgroups (but not necessarily all closed subgroups) whose identity component is nilpotent.*

Proof of 4.1.7. There is no loss of generality in assuming that G is connected.

We will take for U the embedded image under the exponential map of a ball $B_\varepsilon(1)$ in $\mathfrak{g}$, where ε is small enough to ensure that U satisfies certain conditions stated below. We also assume that U satisfies the statement of Theorem 4.1.6. Assuming that Γ is generated by $\Gamma \cap U$, we will construct a sequence of closed nilpotent subgroups

$$\Gamma = N_0 \subset N_1 \subset \cdots \subset N_k = N$$

of increasing dimension, the last of which will be connected.

The first step is the easiest. We can assume that Γ is not the trivial group. Let a be a nontrivial element of Γ as close to the identity as possible, and let v be its logarithm. We know that $a \in U$ (since otherwise $\Gamma \cap U = \{1\}$ cannot generate Γ), so the commutator of a with any element of $\Gamma \cap U$, being smaller than a, must be trivial. Thus a is central in U. We show that the one-parameter subgroup $M_1 = \exp \mathbf{R}v$ is centralized by Γ.

To do this, we assume that U is so small that any element in U has a unique square root in U, and that the set of such square roots is mapped into U under conjugation by elements of U. Then, if $\gamma \in U$ is a generator of Γ, the square root $a^{1/2} = \exp \frac{1}{2}v$ commutes with γ, because $\gamma a^{1/2} \gamma^{-1}$ is also a square root of a contained in U. By induction, all iterated square roots $\exp 2^{-p}v$ commute with γ. Powers of these elements are dense in M_1, so γ is in the centralizer of M_1.

M_1 is also closed, because the powers $a^n = \exp nv$, for $n \in \mathbf{Z}$, are equally spaced in M_1, and the group $\exp \mathbf{Z}v$ is closed. More formally, $M_1 / \exp \mathbf{Z}v$ is compact, hence closed, in the coset space $G/\exp \mathbf{Z}v$, since it consists of a circle's worth of translates of $\exp \mathbf{Z}v$. Together with the closedness of $\exp \mathbf{Z}v$, this implies that M_1 itself is closed.

It follows that $N_1 = M_1\Gamma$ is a closed nilpotent subgroup of G on which Γ acts cocompactly.

The induction step is similar. We assume (Figure 4.1) that we have a sequence $N_0 \subset \cdots \subset N_i$ of closed nilpotent groups containing Γ, each of the form $N_j = M_j\Gamma$, where M_j is the connected component of the identity in N_j, and $\dim M_j = j$. Each M_j and N_j, for $j \leq i$, is normal in N_i, and in fact N_i acts trivially by conjugation on M_j/M_{j-1}, that is, $[N_i, M_j] \subset M_{j-1}$. We also assume that Γ is a cocompact subgroup of N_i.

In order to extend M_i one dimension higher, we will adjoin to it the one-parameter subgroup of G generated by some element $a \in \Gamma \cap U$ whose associated one-parameter subgroup $\exp(t \log a)$ is not contained in M_i; if no such element exists, the induction is complete. The result will be an $(i+1)$-dimensional connected closed Lie group if the one-parameter subgroup that we adjoin normalizes M_i.

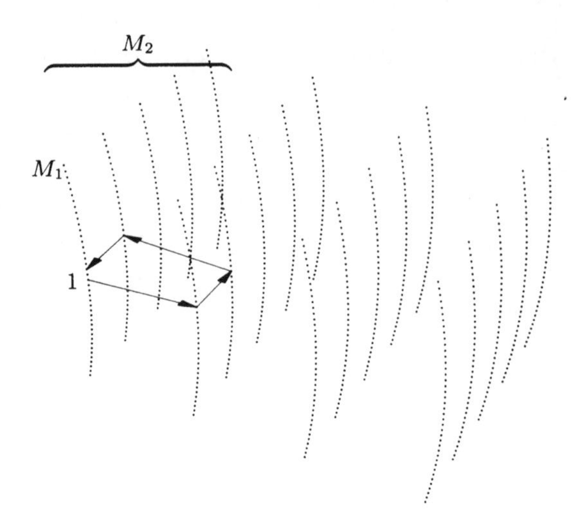

Figure 4.1. A discrete Lie subgroup generated by small elements. Such a subgroup Γ must look something like a lattice in $\mathbf{R}^n$, but multiplication might not be quite commutative: the commutators of elements in spaced out directions might be nontrivial in directions where the dots are spaced much closer together. In the notation of the proof, the M_i are i-dimensional manifolds containing the indicated subgroups of Γ.

Let L_i be the component of the identity in $M_i \cap U$. Choose for a a smallest element of $\Gamma \setminus L_i$ (so that $a \in U$), and set $v = \log a$. Then a is also the smallest element of U whose associated one-parameter group is not contained in M_i. We show that, for all $t \in \mathbf{R}$, the inner automorphism $\phi(t)$ coming from the action of $\exp tv$ by conjugation leaves M_j invariant, for $j \leq i$. This is true for $t = 1$ by the induction assumption, since $a \in \Gamma$. Since M_j is a connected subgroup, it is determined by its tangent space $\mathfrak{m}_j$ at the identity, so we just need to show that the derivative of $\phi(t)$ at the identity leaves $\mathfrak{m}_j$ invariant. (This derivative is none other than $\operatorname{ad}(\exp tv)$: see Exercise 3.8.8.)

We now use the fact that a map that is sufficiently near the identity fixes a certain point if and only if its square fixes the same point. More precisely, we have:

Lemma 4.1.8 (faithful square). *Let V be a spherical neighborhood of $0 \in \mathbf{R}^n$ and let $f\colon V \to \mathbf{R}^n$ be a diffeomorphism such that the derivative of $g = f - \mathrm{Id}$ is bounded in norm by $\eta > 0$. Then $|f(f(u)) - u - 2g(u)| \leq \eta |g(u)|$ for any point $u \in V$ where $f(f(u))$ is defined. In particular, if $\eta < 2$, a point fixed by $f^2 = f \circ f$ must also be fixed by f.*

Proof of 4.1.8. We have

$$f^2(u) - u - 2g(u) = g(f(u)) + g(u) - 2g(u) = g(u + g(u)) - g(u).$$

Now Taylor's Theorem, applied to g, gives the required result. 4.1.8

To apply the lemma, we think of $\mathfrak{m}_j$ as a point in the Grassmannian manifold of j-dimensional subspaces of $\mathfrak{g}$. If $\mathfrak{m}_j$ is fixed under $\phi(t)$, for $0 < t \leq 1$, it is also fixed by $\phi(\frac{1}{2}t)$, provided that the neighborhood U that we started with is small enough (where "small enough" can be chosen independently of the group G). It follows that $\mathfrak{m}_j$ is invariant under $\phi(2^{-p}q)$ for all integers p and q, and therefore under all $\phi(t)$, with t real.

Exercise 4.1.9. Without using Lemma 4.1.8, prove that $\phi(\mathbf{R})$ normalizes M_j, by mapping into $\mathrm{GL}(\mathfrak{g})$ and using the exponential and logarithm functions there.

We can now set $M_{i+1} = M_i \exp \mathbf{R}v$, as explained above. Then M_{i+1} is a connected subgroup of G, and M_j is normal in M_{i+1} for $j \leq i$.

The same reasoning used above to show that M_1 is closed works to show that M_{i+1} is closed: we work in the coset space $G/(M_i \exp \mathbf{Z}v)$, and notice that $M_i \exp \mathbf{Z}v$ is closed because it is a union of connected components of N_i, which is a closed subgroup by the induction assumption.

Next we prove that conjugation by $\exp tv$, for $t \in \mathbf{R}$, acts trivially on M_j/M_{j-1} for $1 \leq j \leq i+1$. This is the same as saying that the action is trivial on the quotient of tangent spaces $\mathfrak{m}_j/\mathfrak{m}_{j-1}$, which is a one-dimensional vector space. This is true for $t = 1$, so also for $t = 2^{-p}q$, and finally for $t \in \mathbf{R}$. Also, since M_i acts trivially on M_j/M_{j-1} by the induction assumption, so does $M_{i+1} = M_i \exp \mathbf{R}v$.

Now we claim that each $\gamma \in \Gamma$ normalizes M_{i+1} and acts trivially on M_{i+1}/M_i. We can restrict ourselves to $\gamma \in \Gamma \cap U$. Then the commutator $[a, \gamma]$ is an element of Γ that is smaller than a, and therefore, by our choice of a, it lies in M_i. This means that the map $\phi(1)$ of conjugation by a fixes the coset of γ in the coset space G/M_i; we want to show that the same is true of $\phi(t)$, for $t \in \mathbf{R}$, and this will imply the claim, since then $\gamma m \exp tv\, \gamma^{-1} \in M_i \exp tv$ if $m \in M_i$.

We would like to apply Lemma 4.1.8 to G/M_i to conclude that if $\phi(t)$ fixes the coset of γ, so does $\phi(\frac{1}{2}t)$. We must be careful, however, because the constant in the lemma would need to be chosen uniformly for all closed subgroups, and M_i may very nearly fill out G. So instead we look at the foliation $\mathcal{F}_U$ of U coming from the coset decomposition of G, and work in the quotient space of the foliation, which is cleaner (Figure 4.2). For $0 \leq t \leq 1$ the map $\phi(t)$ maps each

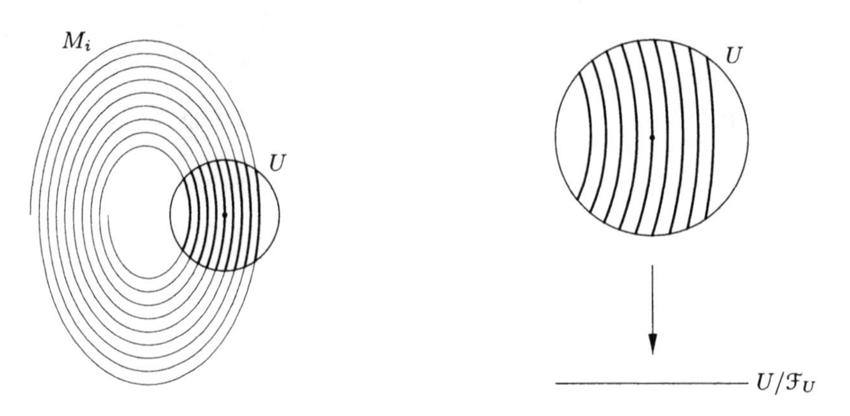

Figure 4.2. Foliating G by cosets of M_i. Even a small neighborhood U of the identity may be penetrated by the subgroup M_i many times (left), so instead of working with the identification space "U/M_i", which depends heavily on M_i, we work with "$U/\mathfrak{F}_U$" (right), where we can apply Lemma 4.1.8.

leaf of $\mathfrak{F}_U$ into a leaf of $\mathfrak{F}_{U'}$, where U' is slightly larger than U. We know that $\phi(1)$ fixes the leaf containing γ, and now we can apply Lemma 4.1.8 to the space of leaves of $\mathfrak{F}_U$ to show that each $\phi(t)$ does the same.

We can now set $N_{i+1} = M_{i+1}\Gamma$. It follows from the results of the previous paragraphs that N_{i+1} is a subgroup of G, and that its action by conjugation leaves M_j invariant and is trivial on M_j/M_{j-1}. Moreover, N_{i+1}/Γ equals a circle's worth of translates of N_i/Γ, and is therefore compact because N_i/Γ is. It follows that N_{i+1} is closed.

To show that N_{i+1} is nilpotent, let $\{1\} = \Gamma_0 \subset \cdots \subset \Gamma_k = \Gamma$ be a central series for Γ. Then

$$M_0 \subset \cdots \subset M_{i+1} \subset M_{i+1}\Gamma_1 \subset \cdots \subset M_{i+1}\Gamma_k$$

is a central series for N_{i+1}.

This concludes the induction step, and therefore the proof of part (a), since G is finite-dimensional. Part (b) is just a rephrasing of part (a). 4.1.7

Exercise 4.1.10 (characterization of N). Show that the connected Lie group N constructed in the proof of Theorem 4.1.7 is the closure of the group generated by iterated square roots of all elements of $\Gamma \cap U$.

Deduce that N grows with Γ in the following sense: Fix a neighborhood U of the identity as in the proof. For an arbitrary (but fixed) discrete subgroup Γ of G, denote by $N(V)$ the connected Lie group obtained by applying the proof's construction to the discrete group generated by $\Gamma \cap V$, where $V \subset U$ is a neighborhood of the identity. Then $U'' \subset U'$ implies $N(U'') \subset N(U')$, and in fact $g^{-1}U''g \subset U'$ for $g \in \Gamma$ implies $g^{-1}N(U'')g \subset N(U')$.

In particular cases, Theorem 4.1.7 can be made much more specific.

Corollary 4.1.11 (discrete orthogonal almost abelian). *For any dimension n, there is an integer m such that any discrete subgroup of $O(n)$ contains an abelian subgroup with index at most m.*

Proof of 4.1.11. Let U be a neighborhood of the identity as specified in Theorem 4.1.7. For any discrete (or equivalently, finite) subgroup Γ of $O(n)$, the subgroup Γ_U generated by its intersection with U has bounded index. Specifically, we can find a symmetric neighborhood $W = W^{-1}$ of 1 such that $W^2 \subset U$. If c_i are coset representatives for Γ_U, then $c_i W$ and $c_j W$ are disjoint for $i \neq j$. Therefore the index of Γ_U in Γ is at most the volume of $O(n)$ divided by the volume of W.

Since $O(n)$ is compact, it has a *bi-invariant metric*, that is, a Riemannian metric that is invariant by both left and right multiplication; this is shown by an averaging process essentially identical to that in Lemma 3.4.11 (existence of invariant metric).

According to Theorem 4.1.7, Γ_U is contained in a connected nilpotent subgroup N of $O(n)$, which inherits from $O(n)$, by restriction, a bi-invariant metric. But a connected nilpotent Lie group with a bi-invariant metric is abelian. To see this, let Z be the center of N. The action ad of N on its Lie algebra $\mathfrak{n}$ (Exercise 3.8.8) preserves the metric on $\mathfrak{n}$ coming from the bi-invariant metric on N, and it fixes pointwise the tangent space $\mathfrak{z}$ to Z. Conversely, any tangent vector fixed by ad generates a central one-parameter subgroup, and therefore lies in $\mathfrak{z}$.

In particular, the perpendicular subspace $\mathfrak{z}^{\perp}$ of $\mathfrak{z}$ in $\mathfrak{n}$ contains no (nonzero) invariant vectors. But, because ad preserves the metric on $\mathfrak{n}$, the whole space $\mathfrak{z}^{\perp}$ is invariant by ad, and it projects isomorphically to the Lie algebra of N/Z. Therefore the Lie algebra of N/Z cannot have invariant vectors under the adjoint representation, that is, the center of N/Z is trivial. Since N is nilpotent, this can only happen if N/Z is itself trivial, that is, if N is abelian. 4.1.11

We will derive a similar corollary for Euclidean isometry groups, but first let's recall some basic facts about Euclidean isometries.

Exercise 4.1.12 (isometries of $\mathbf{E}^n$). (a) Every isometry ϕ of $\mathbf{E}^n$ can be written as $v \mapsto A(v) + v_0$, where $A \in O(n)$ of ϕ and v_0 is some vector. A does not change if we move the origin of $\mathbf{E}^n$ (that is, A is well-defined if we consider $\mathbf{E}^n$ as an affine space). We call A the *rotational part* of ϕ.

(b) If 1 is not an eigenvalue of A, then ϕ has a unique fixed point.

(c) A discrete group of translations of $\mathbf{E}^n$ is isomorphic to $\mathbf{Z}^m$, where $m \leq n$. The group is cocompact if and only if $n = m$.

(d) The commutator of an arbitrary isometry $\phi \in \mathbf{E}^n$ with a translation by v is the translation by $v - A^{-1}v$, where A is the rotational part of ϕ. In particular, if ϕ commutes with n linearly independent translations, it is a translation.

(e) If G is some group of isometries of $\mathbf{E}^n$, the *translation subgroup* of G, that is, the subgroup T_G consisting of all the elements of G that are pure translations, is normal in G.

Corollary 4.1.13 (discrete Euclidean almost abelian). *For each dimension n, there is an integer m such that any discrete subgroup of isometries of $\mathbf{E}^n$ has an abelian subgroup of index at most m.*

Proof of 4.1.13. Let U be a neighborhood of the identity in $\operatorname{Isom} \mathbf{E}^n$ as specified in Theorem 4.1.7. We may assume that $U = RT$, where R is a neigborhood of 1 in $O(n-1)$ and T is a neighborhood of 1 in the group of translations. In fact T can be replaced by $\mathbf{R}^n$, the whole group of translations!

To see this, let Γ be a subgroup generated by its intersection with $R\mathbf{R}^n$. For each $\lambda = 1, 2, \ldots$, the group Γ_λ generated by $\Gamma \cap R(\lambda T)$ is conjugate to the group generated by $(\lambda \Gamma \lambda^{-1}) \cap RT$, and therefore we can apply Theorem 4.1.7 to it. We get a sequence of connected, closed, nilpotent subgroups $N_\lambda \supset \Gamma_\lambda$, and by Exercise 4.1.10 the N_λ increase with λ. But the only way they can increase is by going up in dimension, so eventually the N_λ stabilize and we conclude that Γ is contained in a connected, closed, nilpotent subgroup of $\operatorname{Isom} \mathbf{E}^n$.

Now if Γ is any discrete subgroup of $\mathbf{E}^n$, it follows as in the proof of Corollary 4.1.11 that the index in Γ of the subgroup generated by $\Gamma \cap R\mathbf{R}^n$ is bounded.

To complete the proof, we must show that any connected, closed, nilpotent subgroup $N \subset \operatorname{Isom} \mathbf{E}^n$ is abelian. Let T_N be the translation subgroup of N, as in Exercise 4.1.12(e). We think of T_N as a subset of $\mathbf{E}^n$, and let V be the subspace spanned by T_N. Then V and $V^\perp$ are each preserved by N, and N maps homomorphically into $O(V^\perp) \times \operatorname{Isom} V$. This map is injective, because an element of its kernel acts on $V^\perp$ as a translation and on V as the identity, and by assumption all pure translations are in V.

Thus N is isomorphic to a subgroup of $O(n-k) \times \operatorname{Isom} \mathbf{E}^k$, where k is the dimension of V. Since N is nilpotent, its projections to the two factors are also nilpotent. If $k < n$, it follows by induction and by Corollary 4.1.11 that the projections are abelian, and so is N itself. There remains the case $k = n$, when N contains translations in "all directions".

We claim that in this case N contains only translations. For if there were an element ϕ with nontrivial rotational part A, the commutator $[\phi^{-1}, T_N]$ would be a group of translations spanning the image of $A - \mathrm{Id}$, by Exercise 4.1.12(d). But $\operatorname{im}(A - \mathrm{Id})$ is mapped to itself isomorphically by $A - \mathrm{Id}$, so repeated commutators with ϕ^{-1} would forever span the same subspace $\operatorname{im}(A - \mathrm{Id})$, contradicting the nilpotency of N unless A is the identity. 4.1.13

Problem 4.1.14 (abelian subgroups translate). For every abelian subgroup A of $\operatorname{Isom} \mathbf{E}^n$, show that there is a unique maximal Euclidean subspace $\mathbf{E}_A$ on which A acts by translations.

In order to apply Theorem 4.1.7 to groups of hyperbolic isometries, we need to extend our analysis to the case of a group action on a manifold X, where the group is generated by elements that move a fixed point in X a short distance, but are not necessarily close to the identity. We start with a simple lemma. Recall that a *word* of length n over a set A of elements of a group G is simply an n-tuple $(a_1, \ldots, a_n)$ of elements of $A \cup A^{-1}$; we say the word $(a_1, \ldots, a_n)$ *represents* the element $a_1 \ldots a_n \in G$, and we often write the word itself as $a_1 \ldots a_n$ when no confusion can arise. The *empty word*, or word of length 0, represents the identity element of G. The *norm* of an element of G (with respect to A) is the length of the shortest word over A that represents G.

Lemma 4.1.15 (small group, short words). *If G is a group with more than m elements, and A is a set of generators for G, there exist at least $m + 1$ distinct elements of G with norm no greater than m.*

More generally, if G acts transitively on a space S having more than m elements and $s \in S$, there exist at least $m + 1$ elements of G of norm no greater than m and taking s to distinct points.

Proof of 4.1.15. To prove the first statement, let $r(n)$, for $n \geq 0$, be the number of distinct elements of G with norm at most n. We call r the *growth function* of G (with respect to A). We want to show that $r(m) > m$. Otherwise there would exist $i < m$ such that $r(i+1) = r(i) \leq m$; but this would imply $r(n) = r(i)$ for all $n \geq i$, that is, the group would have order $r(i) \leq m$.

The second statement is proved similarly. 4.1.15

Proposition 4.1.16 (short motion almost nilpotent). *Let G be a Lie group acting on a manifold X so that stabilizers of points are compact. Let $x \in X$ be any point. There exists an integer m and an $\varepsilon > 0$ such that any discrete subgroup Γ of G generated by elements that move x a distance less than ε has a normal nilpotent subgroup of finite index no more than m. Furthermore, Γ is contained in a closed subgroup of G with no more than m components and with a nilpotent identity component.*

Proof of 4.1.16. Let U be a neighborhood of the identity as stipulated in Theorem 4.1.7. Let V be a neighborhood of G_x having compact closure, and let $m > 0$ be such that V can be covered with $m-1$ translates of U. Let $W = W^{-1}$ be a symmetric neighborhood of G_x satisfying $W^m \subset V$. We claim that, if Γ is a discrete group generated by $\Gamma \cap W$, the index in Γ of the subgroup Γ_U generated by $\Gamma \cap U$ is at most m. To see this, consider the action of G on the coset space Γ/Γ_U. If Γ/Γ_U has more than m elements, we can by Lemma 4.1.15 find elements of Γ representing $m+1$ distinct cosets of Γ_U, each having norm at most m with respect to $\Gamma \cap W$. Since $W^m \subset V$, all these elements lie in V, so by our choice of m there are two of them, say g and g', in the same translate of U. We therefore have $g'g^{-1} \in \Gamma \cap U$, contradicting the assumption that g and g' represent different cosets of Γ_U.

We can now take $\varepsilon > 0$ such that W contains all elements of G that move x a distance less than ε. This essentially proves the proposition, except for the statement that the subgroup of finite index can be taken to be normal.

To deal with this complication, we inductively define sets $U_0 \supset U_1 \supset \cdots$ by the formula

$$U_{i+1} = \bigcap_{g \in V} g U_i g^{-1}, \qquad U_0 = U.$$

Each U_i is still a neighborhood of the identity, because V has compact closure. We redefine m, W and ε by replacing U in the first paragraph of the proof by U_n, where n is the dimension of G.

Now if Γ is a discrete group generated by its intersection with W, we form the subgroup Γ_i generated by $\Gamma \cap U_i$. By Theorem 4.1.7, each Γ_i is nilpotent, and is a cocompact subgroup of a connected nilpotent subgroup $N_i \subset G$; moreover, $N_i \subset N_{i-1}$ for each i, by Exercise 4.1.10. The N_i can decrease only by dropping in dimension, so we must have $N_i = N_{i-1}$ for some $1 \leq i \leq n$. Then Γ normalizes N_i, because, for any $g \in \Gamma \cap W \subset V$, we have $g^{-1}U_i g \subset U_{i-1}$ by the definition of the U_i, and this implies $g^{-1}N_i g \subset N_{i-1} = N_i$, again by Exercise 4.1.10. So $H = \Gamma \cap N_i$ is a normal, nilpotent subgroup of Γ, having bounded index (since it is contained in Γ_n, which has index at most m).

ΓN_i is a closed subgroup of G with no more than m components and with a nilpotent identity component N_i. 4.1.16

Corollary 4.1.17 (small discrete hyperbolic almost abelian). *For every dimension n there is an integer m and a distance $\varepsilon > 0$ such that any discrete subgroup Γ of* $\operatorname{Isom} \mathbf{H}^n$ *generated by elements that move some point x by less than ε contains a normal abelian subgroup with index at most m.*

Proof of 4.1.17. In light of the preceding result, it is enough to show that every connected closed nilpotent subgroup of $\operatorname{Isom} \mathbf{H}^n$ is abelian. Let ϕ be a non-trivial element in the center of N.

If ϕ is hyperbolic, N preserves the axis of ϕ, and so is a subgroup of $O(n-1) \times \mathbf{R}$. The conclusion follows from Corollary 4.1.11 (discrete orthogonal almost abelian).

If ϕ is parabolic, N preserves the fixed point of ϕ on S^{n-1}_∞, so it is a subgroup of the group of Euclidean similarities. However, it must actually be a subgroup of the group of Euclidean isometries, since ϕ does not commute with a similarity that actually expands or contracts. The result follows from Corollary 4.1.13 (discrete Euclidean almost abelian).

If ϕ is elliptic, N preserves the hyperbolic subspace of dimension $d < n$ fixed by ϕ, and it is a subgroup of $\operatorname{Isom} \mathbf{H}^d \times O(n-d)$. The result follows by induction, using again Corollary 4.1.11. 4.1.17

4.2. Euclidean Manifolds and Crystallographic Groups

Corollary 4.1.13 says that every discrete group of isometries of $\mathbf{E}^n$ contains an abelian subgroup with finite index. There is a beautiful classical theory, primarily due to Bieberbach [Bie11, Bie12], that generalizes this result (see also [Zas48]). This theory has important consequences for the study of closed Euclidean manifolds and of *crystallographic groups*, which we now define.

Definition 4.2.1 (crystallographic group). An *n-dimensional crystallographic group* is a cocompact discrete groups of isometries of $\mathbf{E}^n$. Crystallographic groups are also sometimes called *Bieberbach groups*.

Theorem 4.2.2 (Bieberbach). (a) *A group Γ is isomorphic to a discrete group of isometries of $\mathbf{E}^m$, for some m, if and only if Γ contains a subgroup of finite index that is free abelian of finite rank.*

(b) *An m-dimensional crystallographic group Γ contains a normal subgroup of finite index that is free abelian of rank m and equals its own centralizer. This subgroup is characterized as the unique maximal abelian subgroup of finite index in Γ, or as the translation subgroup of Γ.*

Conversely, if Γ is a group having a normal subgroup A of finite index that is free abelian of rank m and equals its own centralizer, Γ is isomorphic to an m-dimensional crystallographic group. If Γ is torsion-free, we need not require that A be normal or equal to its centralizer.

(c) *If Γ and Γ' are crystallographic groups of dimension m and m' that are isomorphic as groups, we have $m = m'$ and there is an affine isomorphism $a : \mathbf{E}^m \to \mathbf{E}^m$ conjugating Γ to Γ'.*

(d) *For any given m, there are only a finite number of m-dimensional crystallographic groups, up to affine equivalence.*

Corollary 4.2.3 (classification of Euclidean manifolds). *Diffeomorphism classes of closed Euclidean m-dimensional manifolds are in one-to-one correspondence, via their fundamental groups, with torsion-free groups containing a subgroup of finite index isomorphic to $\mathbf{Z}^m$.*

Proof of 4.2.3. Theorem 4.2.2(c), together with Corollary 3.5.12 (classification of complete (G, X)-structures on a manifold), implies that two closed Euclidean m-manifolds are diffeomorphic if and only if they have isomorphic fundamental groups. Therefore what we must show is that a group Γ is the fundamental group of a closed Euclidean m-manifold if and only if it is torsion-free and has a free abelian subgroup of rank m and finite index. But a discrete group Γ

of Euclidean isometries acts freely if and only if it is torsion-free (you should be able to prove this by now, or else see Corollary 2.5.19). Therefore the corollary follows from Theorem 4.2.2(b). 4.2.3

Proof of 4.2.2(a). The "only if" part follows from Corollary 4.1.13.

There is a straightforward construction for the "if" part, which produces an action of Γ on a (generally high-dimensional) Euclidean space. Let A be a free abelian subgroup of rank n and index p, and choose some effective action of A on $\mathbf{E}^n$ by translations. Let X be the quotient of $\Gamma \times \mathbf{E}^n$ by the equivalence relation $(ga, x) \sim (g, ax)$ for $a \in A$. Then X consists of one copy of $\mathbf{E}^n$ for each coset in Γ/A; in the language of Section 3.6, X is the $\mathbf{E}^n$-bundle associated to the principal A-bundle Γ over the coset space Γ/A (where A acts on Γ by right multiplication). Γ acts naturally on X, and therefore also on the space of sections of X, which is $\mathbf{E}^{np}$ (the product of the fibers, since the base is discrete). An element $\gamma \in \Gamma$ acts on $\mathbf{E}^{np}$ by permuting the fibers $\mathbf{E}^n$ and translating the model fiber, so the action of Γ is by isometries. It is also effective and discrete. (The result of this construction, which given an action of a subgroup $H \subset G$ on a space X produces an action of G on $X^{G/H}$, is known as the *induced representation*.) 4.2.2(a)

To prove part (b), we need two results, interesting in their own right, about invariant subspaces. We start with a definition: if G is a group of isometries of $\mathbf{E}^n$, a *translation subspace* of G is a (nonempty) Euclidean subspace of $\mathbf{E}^n$ that is preserved by G and on which G acts by translations.

Proposition 4.2.4 (maximal translation subspace). *An abelian subgroup A of* $\operatorname{Isom} \mathbf{E}^n$ *has a unique maximal translation subspace* $\mathbf{E}_A$. *If A is discrete and cocompact,* $\mathbf{E}_A = \mathbf{E}^n$.

Proof of 4.2.4. Suppose, by induction, that every abelian subgroup of $\operatorname{Isom} \mathbf{E}^k$, for $k < n$, has a unique maximal translation subspace, and let A be a subgroup of $\operatorname{Isom} \mathbf{E}^n$. If all elements of A are translations, there is nothing to prove. Otherwise, let ϕ be any element of A that is not a translation. Let U be the eigenspace of the rotational part of ϕ corresponding to the eigenvalue 1 (U may be trivial). Because ϕ preserves the foliation of $\mathbf{E}^n$ by translates of U, it induces an isometry in the quotient Euclidean space $\mathbf{E}^n/U$. In $\mathbf{E}^n/U$ there are no eigenvectors with eigenvalue 1, so there is a unique fixed point, by Exercise 4.1.12(b). This fixed point corresponds in $\mathbf{E}^n$ to a translation subspace for ϕ, which we write $\mathbf{E}_\phi$. Any other translation subspace would be contained in a translate of U, hence in $\mathbf{E}_\phi$, so $\mathbf{E}_\phi$ is unique maximal.

Since A is abelian and $\mathbf{E}_\phi$ is canonically associated with ϕ, every element of A must map $\mathbf{E}_\phi$ to itself. By induction, there is a unique maximal translation subspace for the action of A on $\mathbf{E}_\phi$; this is also the unique maximal translation subspace $\mathbf{E}_A$ for the action of A on $\mathbf{E}^n$.

If A is discrete and cocompact, we can find a bounded fundamental domain F intersecting $\mathbf{E}_A$. Every point in $\mathbf{E}^n$ can be mapped into F by some element of A. But A preserves distances measured orthogonally to $\mathbf{E}_A$, so a point far form $\mathbf{E}_A$ cannot be taken to F. We conclude that $\mathbf{E}_A = \mathbf{E}^n$. 4.2.4

Proposition 4.2.5 (invariant translation subspace). *Let Γ be any discrete group of isometries of $\mathbf{E}^n$, and suppose Γ has a free abelian subgroup A of finite index and rank m. Then there is an m-dimensional Euclidean subspace of $\mathbf{E}^n$ invariant by Γ on which A acts effectively by translations.*

Proof of 4.2.5. Let $\mathbf{E}_A$ be the maximal subspace on which A acts by translations (Proposition 4.2.4). The action of A on $\mathbf{E}_A$ is effective, since any element of Γ that has a fixed point has finite order, and A is torsion-free. Then m, the rank of A, is also the dimension of the space T spanned by the image of A in the group of translations of $\mathbf{E}_A$ (we cannot have $m > \dim T$ by Exercise 4.1.12(c)). We foliate $\mathbf{E}_A$ by parallel copies of T.

Assume for the moment that A is normal in Γ. Then the leaves of this foliation are preserved by Γ, and we get an action of Γ on the space of leaves, which is a Euclidean space of dimension $\dim \mathbf{E}_A - m$. The action factors through Γ/A, which is a finite group, so there is a fixed point, that is, an invariant leaf. This proves the proposition if A is normal.

In the general case, since A has finite index in Γ, the intersection of its conjugates is a normal subgroup B of finite index, which is also free abelian of rank m. Applying the proposition to B, we find an m-dimensional subspace $\mathbf{E}^m \subset \mathbf{E}^n$ invariant by Γ on which B acts effectively by translations. Then A also acts by translations on $\mathbf{E}^m$, since any $a \in A$ commutes with translations that span $\mathbf{E}^m$, so the rotational part of a acts trivially on $\mathbf{E}^m$. Also, A acts effectively on $\mathbf{E}^m$: for any $a \in A$ we have $a^k \in B$, where k is the index of B in A, and if a is nontrivial so is a^k, because A is torsion-free. 4.2.5

Proof of 4.2.2(b). Let Γ be a discrete cocompact group of isometries of $\mathbf{E}^m$. By Corollary 4.1.13, Γ has an abelian subgroup $B \subset \Gamma$ of finite index (hence still cocompact). By Proposition 4.2.4, B acts by translations on $\mathbf{E}^m$. Therefore B is a subgroup of $A = T_\Gamma$, the translation subgroup of Γ, which is normal by Exercise 4.1.12(e). A is isomorphic to $\mathbf{Z}^m$ by Exercise 4.1.12(c). Finally, any element of Γ that commutes with everything in A must be a translation, by Exercise 4.1.12(d); since A contains all translations in Γ, the centralizer of A equals A, and there is no abelian subgroup of Γ strictly bigger than A.

To prove the converse, suppose Γ is a group having a free abelian subgroup A of rank m and finite index. By Theorem 4.2.2(a), Γ acts effectively as a discrete group on some Euclidean space $\mathbf{E}^n$, for $n \geq m$. We now use Proposition 4.2.5 to find an m-dimensional subspace $\mathbf{E}^m \subset \mathbf{E}^n$ that is invariant under Γ and on which A acts effectively by translations. Let $\Gamma_m \subset \Gamma$ be the subgroup that fixes $\mathbf{E}^m$. If we can show that Γ_m is trivial, it will follow that Γ acts as a discrete cocompact group of isometries of $\mathbf{E}^m$, as desired.

Since Γ_m is a subgroup of $O(n-m)$, it is finite. Therefore if Γ is torsion-free, Γ_m is trivial.

Otherwise, assume that A is normal. Then the commutator subgroup $[A, \Gamma_m]$ is contained in both A and Γ_m. But $A \cap \Gamma_m$ is trivial, so, if A equals its own centralizer, Γ_m is again trivial.

4.2.2(b)

Proof of 4.2.2(c). Let $\rho : \Gamma \to \operatorname{Isom} \mathbf{E}^m$ and $\rho' : \Gamma \to \operatorname{Isom} \mathbf{E}^{m'}$ be injective homomorphisms. We know that $m = m'$ because, by part (b), the dimension is the rank of the maximal abelian group of finite index in Γ. Consider the diagonal action (ρ, ρ') of Γ on the product $\mathbf{E}^m \times \mathbf{E}^m$. By Proposition 4.2.5 (invariant translation subspace), there is a copy of $\mathbf{E}^m \subset \mathbf{E}^m \times \mathbf{E}^m$ invariant by the action. This subspace must project surjectively to each factor, since it is invariant, so it is the graph of an affine isomorphism $\mathbf{E}^m \to \mathbf{E}^m$ conjugating the two actions.

4.2.2(c)

Proof of 4.2.2(d). This part may seem plausible now, because part (b) says that, if Γ is an m-dimensional crystallographic group, we have a group extension (exact sequence)

$$0 \to A \to \Gamma \to F \to 1,$$

where A is isomorphic to $\mathbf{Z}^m$ and the order of F is bounded. (The bound follows from the use of Corollary 4.1.13 (discrete Euclidean almost abelian) in the proof.)

However, there are many ways in which Γ can be built up from A and F. Notice that $F = \Gamma/A$ acts on A by conjugation. We therefore break the rest of the proof into two steps. In Theorem 4.2.6, we show that there are only so many ways in which a finite group can act on A. Then we show that there are only a finite number of possibilities for Γ after the action of F on A has been fixed. (Actually we give two proofs of this fact, one of them in the form of Exercise 4.2.12.)

Theorem 4.2.6 (finite subgroups of $\mathrm{GL}(n, \mathbf{Z})$). *There are only finitely many finite subgroups of* $\mathrm{GL}(n, \mathbf{Z})$, *up to conjugacy in* $\mathrm{GL}(n, \mathbf{Z})$. *Thus, if F is a finite group acting on a rank-n free abelian group A, there are only a bounded number of possibilites for the action.*

Proof of 4.2.6. The second statement follows from the first because the action of F is given by a homomorphism from F to a finite subgroup of $\mathrm{Aut}(A)$, and $\mathrm{Aut}(A)$ is isomorphic to $\mathrm{GL}(n, \mathbf{Z})$. We now prove the first statement.

Given a finite subgroup $G \subset \mathrm{GL}(n, \mathbf{Z})$, we will find a basis for the lattice $\mathbf{Z}^n$ so that elements of G are represented by matrices with entries whose absolute values are bounded by a constant depending only on n.

Consider the standard embedding $\mathbf{Z}^n \subset \mathbf{R}^n$, so that G can be seen as a subset of $\mathrm{GL}(n, \mathbf{R})$. By averaging over G any positive definite quadratic form on $\mathbf{R}^n$, find one such quadratic form that is invariant under G. Scale so that the lattice points nearest the origin have length 1. Let $V \subset \mathbf{R}^n$ be the vector subspace spanned by these lattice points. Scale the orthogonal complement $V^\perp$ until the lattice points in $\mathbf{R}^n \setminus V$ nearest the origin have length 1. The scaled quadratic form is still G-invariant because G preserves V and $V^\perp$. Continue in this way until the vector space spanned by lattice points of minimum (unit) norm is all of $\mathbf{R}^n$.

There may not be a basis for $\mathbf{Z}^n$ consisting of elements of norm 1 (see Exercise 4.2.9), but we can find a basis consisting of elements of norm at most $\frac{1}{2}(n+1)$. To do this, choose linearly independent elements $a_1, \ldots, a_n \in \mathbf{Z}^n$ of norm 1. Suppose inductively that, for some $k < n$, we have a basis $b_1, \ldots, b_k$ for $\mathbf{Z}^n \cap W_k$, where W_k is the vector space spanned by $a_1, \ldots, a_k$. Let $W_k' \subset W_{k+1}$ be a translate of W_k as close as possible to W_k, and take for b_{k+1} a point of W_k' nearest the origin. The component of b_{k+1} in the direction orthogonal to W_k has length at most 1, and its component in the direction of W_k has length at most $\frac{1}{2}k$, because the parallelepiped P spanned by $a_1, \ldots, a_k$ has diameter less than k, and the projection of b_{k+1} to W_k lies in a copy of P centered at the origin (which is contained in a ball of radius $\frac{1}{2}k$). Therefore, b_{k+1} has norm at most $\sqrt{(\frac{1}{2}k)^2 + 1} \leq \frac{1}{2}(k+2)$. This completes the inductive step.

We conclude by showing that matrices representing elements of G with respect to our basis $b_1, \ldots, b_n$ have bounded entries. Since G preserves length, it suffices to show that the coordinates of any vector v of length at most $\frac{1}{2}(n+1)$ are bounded. The i-th coordinate of v is the ratio of the volume of the parallelepiped P spanned by $b_1, \ldots, b_{i-1}, v, b_{i+1}, \ldots, b_n$ to the volume of the parallelepiped Q spanned by $b_1, \ldots, b_n$. The volume of P is at most $(\frac{1}{2}(n+1))^n$, while the volume of Q is greater than the volume of a ball in $\mathbf{E}^n$ of radius $\frac{1}{2}$, since the translate of Q with center at the origin contains such a ball. $\boxed{4.2.6}$

In the preceding proof, we fixed a lattice $\mathbf{Z}^n \subset \mathbf{R}^n$ and changed the inner product on $\mathbf{R}^n$ so that the lattice acquired certain properties. It is sometimes more intuitive to visualize this process from the dual point of view, where we alter the lattice but maintain the same inner product. Recast in these terms, the proof of Theorem 4.2.6 implies that, if F is a finite subgroup of $\mathrm{GL}(n, \mathbf{Z})$, there exists an embedding $\rho : \mathbf{Z}^n \to \mathbf{R}^n$ such that (with respect to the standard metric in $\mathbf{R}^n$) the origin's nearest neighbors in the lattice $\rho(\mathbf{Z}^n)$ span $\mathbf{R}^n$, and $\rho f \rho^{-1}$ is an isometry for every $f \in F \subset \mathrm{GL}(n, \mathbf{R})$.

More generally, it is interesting to consider *all* embeddings $\rho : \mathbf{Z}^n \to \mathbf{R}^n$ such that F acts by isometries on the image. Considering F up to conjugacy in $\mathrm{GL}(n, \mathbf{Z})$ is the same as forgetting about the particular homomorphism ρ and retaining only its image, a lattice in $\mathbf{R}^n$. In this way, for each conjugacy class of finite subgroups of $\mathrm{GL}(n, \mathbf{Z})$, we get a family of lattices with the corresponding symmetries. Of course, some of these lattices can have additional symmetries as well. If some lattice with symmetry F has no other symmetries, we say that F is an *exact lattice group*. An example of an F that is not exact is the trivial group—every lattice has this symmetry, but also the symmetry $-\mathrm{Id}$. If every lattice with symmetry F has no other symmetries, F is a maximal finite subgroup of $\mathrm{GL}(n, \mathbf{Z})$, and the corresponding lattices are *maximally symmetric*.

Exercise 4.2.7 (finite subgroups of $\mathrm{SL}(2,\mathbf{Z})$**).** Show that any finite subgroup of $\mathrm{SL}(2,\mathbf{Z})$ is conjugate to a subgroup of the rotational symmetries of one of the two lattices in Figure 4.3. State the corresponding result for subgroups of $\mathrm{GL}(2,\mathbf{Z})$. What are the exact lattice subgroups of $\mathrm{GL}(2,\mathbf{Z})$?

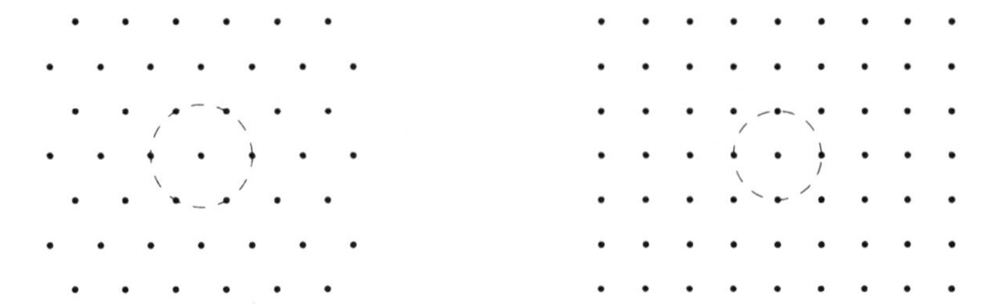

Figure 4.3. The equilateral triangular lattice and the square lattice. These are the maximally symmetric lattices in dimension two; their symmetry groups in $O(2)$ are the dihedral groups D_{12} and D_8 with 12 and 8 elements, respectively.

Problem 4.2.8 (isometric lattices). (a) Show that the group of symmetries of an octahedron, which has order 48, can be embedded in $\mathrm{GL}(3,\mathbf{Z})$ in exactly three ways, up to conjugacy in $\mathrm{GL}(3,\mathbf{Z})$. Lattices having each of these three types of symmetry are illustrated in Figure 4.4.

(Hint: To show that these are the only possibilities, use the observation that if L is a lattice invariant by a finite subgroup $F \subset O(n)$ and f is an element of F, any vector $v \in L$ of minimal length and not fixed by f maps to a vector $f(v)$ making an angle of at least 60° with v.)

(b) A body-centered cubic lattice becomes a face-centered cubic lattice if you stretch the vertical direction by an appropriate factor. Likewise, all three isometric lattices can be transformed into one another by scaling in the direction of the cube diagonal in Figure 4.4. This means the three embeddings of part (a), although distinct in $\mathrm{GL}(3,\mathbf{Z})$, are conjugate in $\mathrm{GL}(3,\mathbf{R})$.

Exercise 4.2.9 (thin cubic lattice). Find an n-dimensional lattice where the origin's nearest neighbors span $\mathbf{R}^n$, but don't form a basis for the lattice. (Hint: start with the standard cubic lattice in dimension $n \geq 5$ and add more points.)

Problem 4.2.10 (estimating lattice automorphisms). Let C_m be the cyclic group of order m. For any m, we consider the canonical homomorphism $\mathrm{SL}(n,\mathbf{Z}) \to \mathrm{SL}(n,C_m)$.

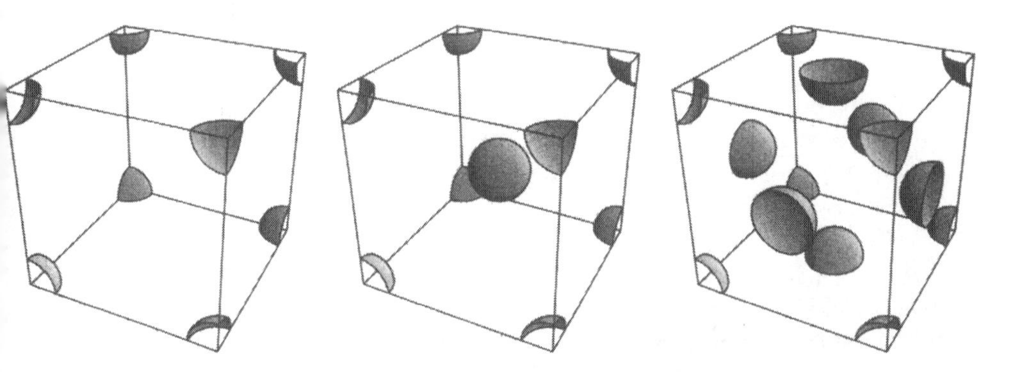

Figure 4.4. Lattices with octahedral symmetry. These three lattices have the largest possible symmetry group, that of the octahedron. (The regions shown represent 1, 2 and 3 fundamental domains, respectively.) In crystallography, they are called generically the *isometric lattices*, and specifically the *cubic lattice*, the *body-centered cubic lattice* and the *face-centered cubic lattice*. See also Problem 4.3.12 and Problem 4.3.13.

(a) If $m \geq 3$, any finite subgroup F of $\mathrm{SL}(n, \mathbf{Z})$ injects into $\mathrm{SL}(n, C_m)$. (Hint: In the metric used for the proof of Theorem 4.2.6, the unit ball in the lattice injects into C_m^n.)

(b) The kernel of $F \to \mathrm{SL}(n, C_2)$ has order a power of 2.

(c) The order of $\mathrm{SL}(n, C_p)$ is $(p-1)^{-1}(p^n - 1)(p^n - p) \cdots (p^n - p^{n-1})$, for p a prime.

(d) Using this information, what restrictions can you find for finite subgroups of $\mathrm{SL}(n, \mathbf{Z})$, for small n? Do you get any more information by considering $\mathrm{SL}(n, C_m)$ for general m?

We continue with the proof of Theorem 4.2.2(d). We're working with a group extension $0 \to A \to \Gamma \to F \to 1$, where Γ is an m-dimensional crystallographic group, F a finite group and A a free abelian group of rank m. The action of F in A is given. We want to know how many non-isomorphic possibilities there are for Γ.

We identify Γ (non-canonically) with $A \times F$ by choosing representatives $c(g) \in \Gamma$ for each coset $g \in F = \Gamma/A$, and by mapping (a, g) to $c(g)a$, for $a \in A$ and $g \in F$. Of course, under this identification, the group law on $A \times F$ is not the usual product law—the discrepancy is given partly by the action of F on A, and partly by a "correction function" $\alpha : F \times F \to A$, defined by

$$c(g)c(h) = c(gh)\alpha(g, h).$$

(If α is identically zero, Γ is the *semidirect product* $A \ltimes F$.)

There is a necessary and sufficient compatibility condition for a function $\alpha : F \times F \to A$ to be the correction function for a group extension with a given action, but we don't need to write it down here (see Problem 4.2.11). What's important is that in α we have enough data to reconstruct the group law on $A \times F$, and that, as we will see, all inequivalent group laws arise from a bounded number of choices for α. "Equivalent" here means this: if we change the choice of coset representatives $c(g)$ in the construction above, we change the group law on $A \times F$ and the correction function α; but the isomorphism type is obviously the same.

To see that one can always choose the coset representatives $c(g)$ in such a way that the $\alpha(g, h)$ lie in a bounded subset of A, work as follows. By scaling as in the proof of Theorem 4.2.6, change the Euclidean structure in $\mathbf{R}^n$ to one in which the origin's nearest neighbors in A have norm 1 and span $\mathbf{R}^n$. The new Euclidean structure is still invariant under Γ. Let $a_1, \ldots, a_m$ be a set of linearly independent elements of A of norm 1, and let P be the parallelepiped they span. P is a fundamental domain for the subgroup of A generated by $a_1, \ldots, a_m$, and has diameter less than m. Therefore, if $c(g)$ is chosen as a representative of $g \in F = \Gamma/A$ that moves the center of P a minimal distance, this distance is at most $\frac{1}{2}m$. Then $\alpha(g, h) = c(gh)^{-1}c(g)c(h)$ moves the origin a distance less than $\frac{3}{2}m$. The number of elements of A moving a point a bounded distance is bounded, so the number of possibilities for α is bounded. All these bounds depend on F and on its action on A, but not on Γ. 4.2.2(d)

Problem 4.2.11 (group cohomology). Write down the compatibility condition needed for a correction function $\alpha : F \times F \to A$ to come from a group extension $0 \to A \to \Gamma \to F \to 1$ (with a given action $\rho : F \to \mathrm{Aut}(A)$).

The correction function is also called a *cocycle*, and the compatibility condition a *cocycle condition*, like the condition for bundle transition maps in Section 3.6. The group cohomology point of view in the present context is developed from scratch, and in a fair amount of detail, in [Cha86].

Exercise 4.2.12 (semidirect crystallographic supergroup). Prove that every crystallographic group Γ whose maximal abelian subgroup of finite index is A is a subgroup of the semidirect product $A \ltimes \Gamma/A$, with bounded finite index.

(a) Γ acts as a group of affine maps of A^n, where n is the order of Γ/A. (Hint: construct the induced representation, as in the proof of 4.2.2(a)).

(b) Let $B \subset A^n$ be the orbit of the action of A which contains the origin $(0, \ldots, 0) \in A^n$. There is an induced affine action of Γ on A^n/B, which factors through Γ/A.

(c) Embed A^n in a larger free abelian group C of the same rank by adjoining n-th roots to all elements. Γ acts on C. Let D be the maximal subgroup containing B that has the same rank as B. Then the action of Γ on C/D has a fixed point, so there is a coset E of D which is invariant by Γ.

(d) The action of Γ on E is effective. The group ΓE consisting of transformations in Γ followed by translations in E is isomorphic to the semidirect product $A \ltimes \Gamma/A$.

(e) The above shows that Γ with n-th roots adjoined for all translations in Γ is a semidirect product. The index is n^m, where m is the rank of A. Show that the minimal index with which Γ can be embedded in a semidirect product is at most n. (Note: the action of Γ/A on the lattice, in this minimal embedding, might not be isomorphic to the action of Γ/A on A).

4.3. Three-Dimensional Euclidean Manifolds

The geometric theory of three-dimensional crystallography was mainly developed during the nineteenth century.

There are 32 subgroups of $O(3)$ that occur as images of three-dimensional crystallographic groups. They are known in crystallography as the *point groups*, and were first enumerated by Hessel [Hes30].

There are fourteen crystallographic groups that are the full groups of symmetries of a lattice (as an array of points) in $\mathbf{R}^3$, up to affine equivalence. They were enumerated by Bravais [Bra49], and the classes of lattices, according to their symmetry, are known as Bravais lattices (see Problem 4.3.13).

There are 65 orientation-preserving crystallographic groups, and they were classified by Sohncke [Soh79]. Finally, three people independently derived the full list of 230 crystallographic groups [Fed49, Sch91, Bar94].

From a list of crystallographic groups, one can of course derive a list of closed Euclidean three-manifolds by crossing out the groups that do not act freely [Now34]. This involves much more work than necessary, though, since there turn out to be only ten such manifolds. A direct classification was given in [HW35].

Convention 4.3.1. In this section and the next, we understand manifolds to be closed.

The rest of this section is devoted to the classification of Euclidean three-manifolds. The first step is to understand better how the finite group $F \subset O(3)$ of rotational parts of elements of Γ relates to the lattice of translations in Γ. If F is the trivial group, Γ is of course $\mathbf{Z}^3$, and its quotient is T^3. The next more complicated case is when F is cyclic and non-trivial.

Lemma 4.3.2 (Euclidean splitting). *Let Γ be a torsion-free three-dimensional crystallographic group whose projection in $O(3)$ is the cyclic group C_k, for $k > 1$.*

(a) *There is an orthogonal splitting $\mathbf{E}^3 = \mathbf{E}^1 \oplus \mathbf{E}^2$ preserved by Γ, in the sense that Γ preserves the foliation by lines parallel to $\mathbf{E}^1$ and the foliation by planes parallel to $\mathbf{E}^2$.*

(b) *If Γ preserves orientation, Γ is a subgroup of $T_1 \times \operatorname{Isom} \mathbf{E}^2$, and its translation subgroup T_Γ equals $(\Gamma \cap T_1) + (\Gamma \cap T_2)$, where T_i is the group of translations of $\mathbf{E}^i$.*

(c) *If Γ contains orientation-reversing elements, then $k = 2$ and Γ is a subgroup of $\operatorname{Isom} \mathbf{E}^1 \times T_2$. In addition, T_Γ contains $(\Gamma \cap T_1) + (\Gamma \cap T_2)$ with index at most 2.*

Proof of 4.3.2. To prove (a), choose $a \in \Gamma$ such that the rotational part α of a generates C_k. By Exercise 4.1.12(b), 1 must be an eigenvalue of α. Either the 1-eigenspace is one-dimensional and α is a rotation of order k about a line, or it is two-dimensional and α is reflection through a plane. In either case we have a splitting of $\mathbf{R}^3$ into one- and two-dimensional subspaces invariant under α. Since the rotational part of any element of Γ is a power of α, the corresponding splitting of $\mathbf{E}^3$ is preserved by Γ. In particular, Γ is a subgroup of $\operatorname{Isom} \mathbf{E}^1 \times \operatorname{Isom} \mathbf{E}^2$. We denote the projections on the factors by p_1 and p_2.

If Γ preserves orientation, α is a rotation and a is a screw motion in the direction of $\mathbf{E}^1$. For any $g \in \Gamma$, the projection $p_1(g)$ is a translation, because α acts as the identity on the one-dimensional factor. As shown in Figure 4.5, $p_1(g)$ may not be in Γ, but its k-th power is. Thus the subgroup $kp_1(\Gamma)$ is contained in $\Gamma \cap T_1$, and therefore $p_1(\Gamma)$ contains $\Gamma \cap T_1$ with index at most k. On the other hand, the kernel of p_1, a subgroup of $\operatorname{Isom} \mathbf{E}^2$, consists entirely of translations, for anything else would have fixed points. Thus,

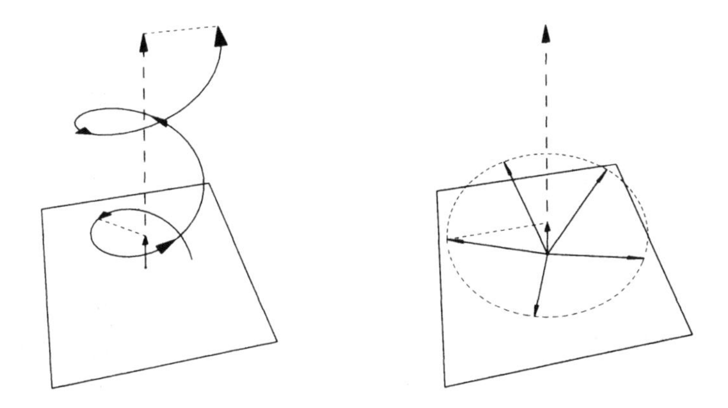

Figure 4.5. The k-th power of $p_1(g)$ is in Γ. Either g is a screw motion in the direction of $\mathbf{E}^1$, in which case $g^k = p_1(g)^k$, or g is a translation, in which case k of its conjugates add up to $p_1(g)^k$. (In the two diagrams, the short vertical arrow represents $p(g)$ and the long, dashed arrow represents its k-th power.)

combining the translations of Γ along $\mathbf{E}^1$ with those along $\mathbf{E}^2$ we get a subgroup of Γ of index at most k. But T_Γ also has index k, so these two subgroups must be identical. This proves part (b).

If Γ contains orientation-reversing elements, $k = 2$ and α is a reflection in the two-dimensional factor. To prove (c), we can almost get away with simply interchanging the indices 1 and 2 in the preceding argument, except that the subgroup $2p_2(\Gamma)$ has index 4, rather than 2, in $p_2(\Gamma)$. Therefore the image of p_2 contains $\Gamma \cap T_2$ with index up to 4, and that's also the maximum index of $(\Gamma \cap T_1) + (\Gamma \cap T_2)$ in Γ. 4.3.2

Problem 4.3.3 (splitting crystallographic groups). Is the conclusion of the lemma false if Γ is not torsion-free? If so, what weaker conclusion is valid? What can you say for n-dimensional crystallographic groups?

Theorem 4.3.4 (Euclidean covered by torus cross $\mathbf{E}^1$). *If Γ is a torsion-free three-dimensional crystallographic group, the translation subgroup T_Γ has a rank-two subgroup Z normal in Γ. In geometric language, every closed Euclidean three-manifold M is the quotient of $T^2 \times \mathbf{E}^1$, where T^2 is a Euclidean torus, by the action of a discrete group $\hat{\Gamma}$.*

Proof of 4.3.4. To get the geometric version from the algebraic one, we write $M = \mathbf{E}^3/\Gamma$ and $\hat{\Gamma} = \Gamma/Z$. The torus is the quotient modulo Z of any Z-invariant plane $\mathbf{E}^2$, and $\mathbf{E}^1$ is the plane's orthogonal complement. Any isometry of $\mathbf{E}^3$ that normalizes Z preserves the splitting $\mathbf{E}^3 = \mathbf{E}^2 \oplus \mathbf{E}^1$, so $\hat{\Gamma} \subset \operatorname{Isom} T^2 \times \operatorname{Isom} \mathbf{E}^1$.

To prove the theorem we will consider the various possibilities for the group $F \subset O(3)$ of rotational parts of elements of Γ. In each case we go on to find out what the corresponding manifolds are. Two observations will be useful throughout:

(1) If we take Z to be maximal, the kernel of the projection $\hat{\Gamma} \to \operatorname{Isom} \mathbf{E}^1$ is a torsion-free group of symmetries of the torus not containing translations, so it is either the trivial group or C_2 acting by a glide-reflection; in the second case F must include a reflection in a plane containing the $\mathbf{E}^1$ factor, which only happens in case (e) below.

(2) The image of $\hat{\Gamma}$ in $\operatorname{Isom} \mathbf{E}^1$ is a discrete cocompact subgroup of $\operatorname{Isom} \mathbf{E}^1$, and so equals either $\mathbf{Z}$ or $C_2 * C_2$.

Case (a): If F is trivial, $T_\Gamma = \Gamma$ and we can take for Z any maximal rank-two subgroup of Γ. The generator of $\hat{\Gamma} = \mathbf{Z}$ acts as a translation both in the $\mathbf{E}^1$ direction and in the T^2 direction. The quotient manifold is, geometrically, the mapping torus (Example 3.6.5) of T^2 by a translation. Since the isometry is isotopic to the identity, the manifold is topologically $T^2 \times S^1 = T^3$, also known as G_1 in the notation of [Wol67] and as $(^{\oplus}|)$ in the notation of Chapter 5.

Case (b): If F is orientation-preserving and cyclic of order greater than 1, with generator $\alpha \in \mathrm{SO}(2)$, it follows from Lemma 4.3.2(b) that

$$T_\Gamma = (T_\Gamma \cap T_\alpha) + (T_\Gamma \cap T_\alpha^\perp),$$

where T_α is the axis of α (more precisely, the space of translations along that axis) and $T_\alpha^\perp$ is its orthogonal complement. We take $Z = T_\Gamma \cap T_\alpha^\perp$.

We know that Γ acts in the T_α direction by translations; we take any element of Γ such that the amount of this translation is minimal (but nonzero). The corresponding action on $T_\alpha^\perp$ is a rotation preserving Z, that is, it factors to an isometry of $T^2 = T_\alpha^\perp / Z$. In other words, each of the resulting manifolds is a mapping torus of T^2 by an isometry that lifts to a rotation of $\mathbf{E}^2$. By Exercise 4.2.7, we have only the following possibilities:

F	C_2	C_3	C_4	C_6
Angle of rotation	180°	120°	90°	60°
Type of two-torus	arbitrary	hexagonal	square	hexagonal
Name in [Wol67]	G_2	G_3	G_4	G_5
Name in Chapter 5	$(2_1 2_1 2_1 2_1)$, $(\|^{\ominus\ominus})$	$(3_1 3_1 3_1{}')$	$(4_1 4_1 2_1{}')$	$(6_1 3_1 2_1{}')$

The last three of these manifolds have two different oriented forms, distinguished by the handedness of the screw motion with shortest translation distance.

Case (c): F is orientation-preserving, but not cyclic. For each nontrivial $\alpha \in F$ we apply Lemma 4.3.2(b) to the subgroup of Γ consisting of maps whose rotational part is a power of α, to conclude that T_Γ is generated by its intersections with T_α and $T_\alpha^\perp$ (see the previous case). If $\alpha, \beta \in F$ are rotations about different axes, the axes must be orthogonal, because orthogonal projection from T_α onto T_β and back to T_α takes the first nonzero lattice point to a lattice point, which must necessarily be zero. Therefore all axes of rotation are mutually orthogonal. By Exercise 4.3.5 there are exactly three of them, and $F = C_2 \oplus C_2$ acts by 180° rotations.

Exercise 4.3.5 (composing rotations). Let f and g be rotations of $\mathbf{E}^2$, $\mathbf{H}^2$ or the elliptic plane $\mathbf{RP}^2$, with given fixed points. Give a description of $f \circ g$, including a construction for its fixed point (see Figure 4.6; the fixed point may be at ∞ or beyond in the projective embeddings of $\mathbf{E}^2$ or $\mathbf{H}^2$).

Use this to show that an orientation-preserving group of isometries of S^2 having all axes of rotation mutually orthogonal either is cyclic or equals $C_2 \oplus C_2$ acting by 180° rotations about three mutually orthogonal axes.

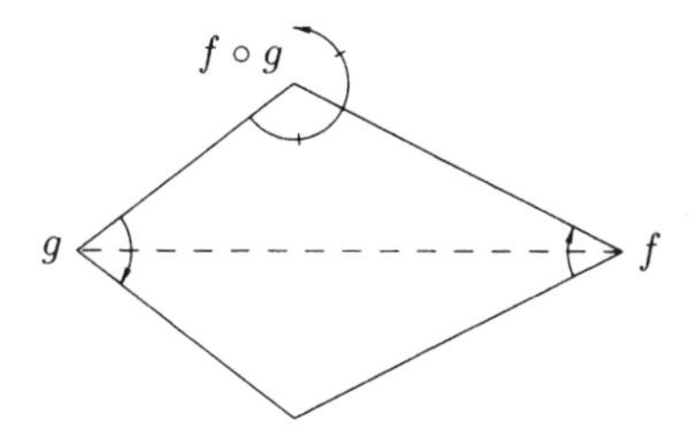

Figure 4.6. Finding the fixed point of a composition of rotations

We take Z be the group generated by translations along the directions of two of these axes, say T_α and T_β. We must have $\hat{\Gamma} = C_2 * C_2$, because elements of Γ with rotational part α or β act on the $\mathbf{E}^1$ factor as reflections. Two elements of this type that generate $\hat{\Gamma}$ will perforce have different axes, one T_α and one T_β (Figure 4.7). They are screw motions, whose translational components cannot be in Z, but their squares will be in Z. The resulting manifold is unique, and is called G_6 in [Wol67] and $(2_1 2_1{}'^{\ominus})$ in Chapter 5.

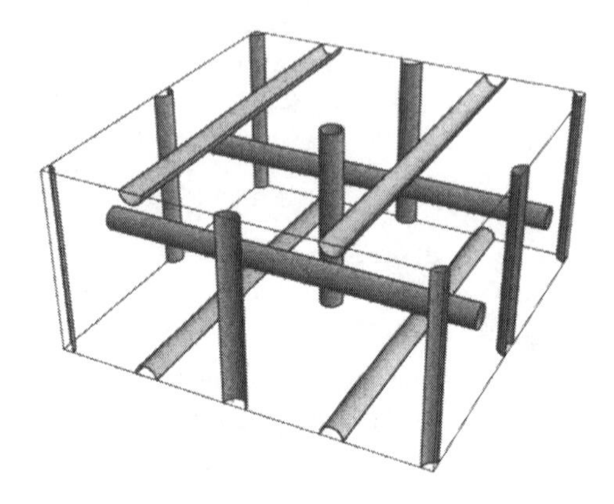

Figure 4.7. Two fundamental domains for the Euclidean manifold $(2_1 2_1{}'^{\ominus})$. The cylinders represent axis of screw motions, with rotation angle $180°$ and translation component equal to the height of the box (and half its width or depth). The manifold is the quotient of $T^2 \times \mathbf{E}^1$ by $\hat{\Gamma}$, where T^2 is the Euclidean torus obtained by identifying opposite sides of the top square, $\mathbf{E}^1$ runs vertically, and $\hat{\Gamma} = C_2 * C_2$ is generated by screw motions around horizontal axes in the top and middle layers.

Exercise 4.3.6. Show that this is the manifold of Figure 3.2.

This completes the classification of the orientable Euclidean three-manifold. We now consider the non-orientable cases.

Case (d): $F = C_2$, acting by reflection in a plane V. Take $Z = T_\Gamma \cap T_V$, where T_V is the group of translations of $\mathbf{E}^3$ parallel to V. By Lemma 4.3.2(c) and its proof, Γ preserves the splitting $\mathbf{E}^3 = V \oplus V^\perp$, the V-component of any element of Γ is a translation, and the group Z' of such translations contains Z with index two or four.

Orientation-reversing elements of Γ are glide-reflections in planes parallel to V. They act as reflections in the $V^\perp$ factor, so $\hat{\Gamma} = C_2 * C_2$.

Let a and b be generators of Z. Among glide-reflections in a given plane, there is exactly one that translates by one of the vectors

$\frac{1}{2}a$, $\frac{1}{2}b$ and $\frac{1}{2}(a+b)$, for if there were glide-reflections translating by any two of these vectors, the third would be in Z, contradicting our choice of a and b. If the set of translational components of glide-reflections is the same for all planes, Z has index two in Z' (Figure 4.8, left); if it alternates between consecutive planes, the

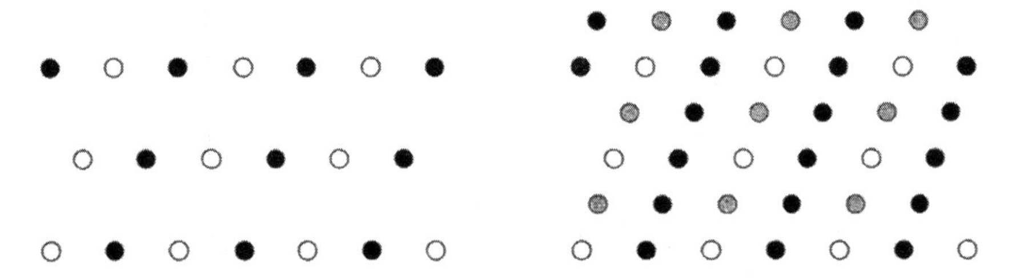

Figure 4.8. Translational components of glide-reflections. The white dots represent the lattice Z. Glide-reflections in the plane of the page have translational parts indicated by dark gray dots. Glide-reflections in the plane immediately above or below may have the same translational parts (left) or different ones (light gray dots on the right). In the latter case the composition of two glide-reflections in adjacent planes is a translation not in Z (black dots).

index is four, and Γ contains translations whose V-component is not in Z (Figure 4.8, right). The resulting manifold in the first case is denoted B_1 in [Wol67], and $(^{\ominus}|)$ or $(^{+}{:}^{+}{:})$ or $(|^{\oplus\oplus})$ in Chapter 5. In the second case it is denoted B_2 and $(^{+}{:}^{\oplus}) = (^{\ominus\prime})$.

Exercise 4.3.7. In addition to their description as quotients of $T^2 \times \mathbf{E}^1$ by $C_2 * C_2$, show that these two manifolds can be described as mapping tori of a Euclidean Klein bottle K^2 by an isometry ϕ that lifts to a translation of the plane. State the difference between the two in terms of the action of ϕ on the two "special" simple closed geodesics on K^2 (those that come from the axes of the covering transformations of K^2 that are glide-reflections).

Case (e): F is not orientation-preserving and not cyclic. Then the fixed plane of any reflection must contain every axis, since the composition of a reflection with a rotation whose axis is not contained in the plane of reflection does not have 1 as an eigenvalue, by Exercise 4.3.8; an isometry with this rotational part would have a fixed point.

Exercise 4.3.8 (composing rotation and reflection). Describe the composition of the reflection in a line in $\mathbf{E}^2$ with a rotation about a point not on that line, and show that it has no fixed points. Generalize to S^2 and to $\mathbf{H}^2$; use the spherical case to describe the composition of a reflection in a plane in $\mathbf{E}^3$ with a rotation about an axis intersecting that plane in a point.

It follows that there can be only one axis T_α, since the composition of rotations about two different axes has an axis not in the same plane, by Exercise 4.3.5. Γ preserves the splitting $\mathbf{E}^3 = T_\alpha \oplus T_\alpha^\perp$, and it acts on $T_\alpha = \mathbf{E}^1$ by translations. We take for Z the set of translations in $T_\alpha^\perp = \mathbf{E}^2$. Here, unlike in the previous cases, the map $\hat{\Gamma} \to \operatorname{Isom} \mathbf{E}^1$ is not one-to-one, for if $\hat{\Gamma}$ were isomorphic to its image (which consists of translations only), it would be cyclic, and therefore so would F, contrary to assumption. So the subgroup of Γ that acts trivially on the T_α factor is the fundamental group of the Klein bottle, obtained by adjoining to Z some glide-reflection in a (uniquely determined) plane V containing T_α. Since α normalizes this subgroup, it fixes V. It follows that α must be a 180° rotation, and F is $C_2 \oplus C_2$ generated by reflection in two orthogonal planes containing T_α.

The rotational parts of elements of Γ that act trivially on T_α are the identity and reflection in V, denoted γ. The rotational parts of elements of Γ that translate by a minimal nonzero distance along T_α must be α and $\alpha\gamma$. There are two possibilities: the axes of rotation of elements with rotational part α can lie on the glide-reflection planes parallel to V (Figure 4.9, left), or they can lie equidistantly in between these planes (right). The resulting manifolds are called B_3 and B_4 in [Wol67], and $({}^-{:}{}^-{:}) = (|^{\oplus\ominus}) = (2_1 2_1{}^+{:})$ and $({}^+{:}^{\ominus}) = ({}'^{\oplus\ominus}) = (2_1 2_1{}^{\oplus})$ in Chapter 5. $\boxed{4.3.4}$

Exercise 4.3.9. Express these two manifolds as mapping tori of the Klein bottle (compare Exercise 4.3.7).

Problem 4.3.10. Generalize Exercise 4.3.5 to the case that f and g are arbitrary orientation-preserving isometries of $\mathbf{E}^3$, $\mathbf{H}^3$ or $\mathbf{RP}^3$. (Compare Proposition 2.5.4.)

Problem 4.3.11 (other product coverings). In what other ways can the ten Euclidean three-manifolds be described as $(M^2 \times \mathbf{E}^1)/\hat{\Gamma}$, where M^2 is a Euclidean two-manifold and $\hat{\Gamma}$ is $\mathbf{Z}$ or $C_2 * C_2$?

Problem 4.3.12 (finite subgroups of $\mathrm{SL}(3, \mathbf{Z})$). There are four maximal finite subgroups F of $\mathrm{GL}(3, \mathbf{Z})$ up to conjugacy, corresponding to four types of maximally symmetric lattices (compare Exercise 4.2.7). There are another ten exact lattice subgroups of $\mathrm{GL}(3, \mathbf{Z})$. Here is an outline of the classification. (Some of the logic here overlaps with the classification of finite subgroups of $\mathrm{SO}(3)$, which we tackle in Problem 4.4.6. You may want to look at that problem now.)

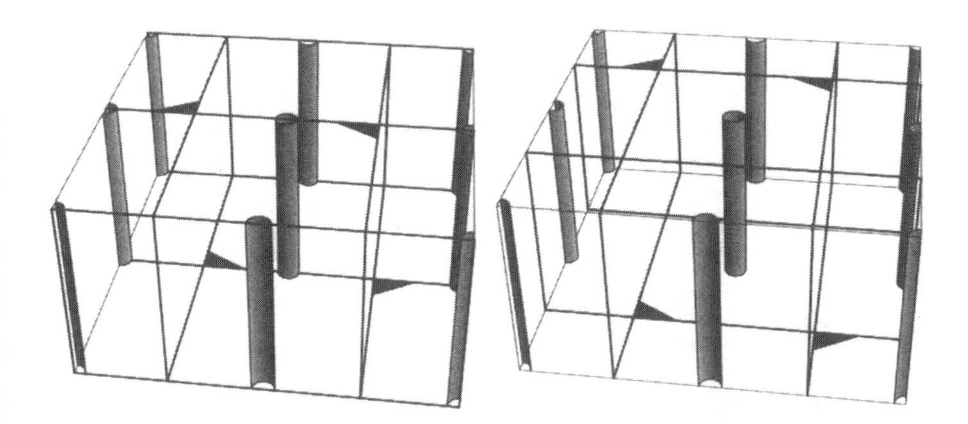

Figure 4.9. Two fundamental domains for the Euclidean manifolds $(2_1 2_1{}^{+}{:})$ **and** $(2_1 2_1{}^{\oplus})$**.** Cylinders represent axes of screw motions that rotate by $180°$ and translate by odd multiples of the box height. Vertical rectangles drawn with heavy lines represent planes of glide reflections, whose action is indicated by the black half-arrowheads. In each case, the manifold is of the form $(T^2 \times \mathbf{E}^1)/\hat{\Gamma}$, where T^2 is the torus coming from the top square, $\mathbf{E}^1$ runs vertically, and $\hat{\Gamma} = C_2 * C_2$ is generated by one screw motion and one glide reflection.

(a) Since $-\mathrm{Id}$ is central in $\mathrm{GL}(3, \mathbf{Z})$ and preserves every lattice, we can restrict our attention to finite subgroups of $\mathrm{SL}(3, \mathbf{Z})$.

(b) If A is a lattice invariant by a rotation of order m about an axis V, show that $mA \subset (A \cap V) \oplus (A \cap V^{\perp}) \subset A$. (Hint: see Figure 4.5, right.) Deduce that $m = 2$, 3, 4 or 6.

(c) If $m = 6$, then A is the product of a triangular lattice in $\mathbf{E}^2$ with a lattice in $\mathbf{E}^1$. (Hint: apply (a) to the square and to the cube of the order-six rotation.) The group of orientation-preserving symmetries for such a lattice is D_{12}, the dihedral group of order 12, generated by $180°$ rotations about two lines in $V^{\perp}$. This group is maximal.

(d) If $m = 3$ and we're not in case (b), the orthogonal image of A in $V^{\perp}$ contains $A \cap V^{\perp}$ with index 3. When seen from the direction of V, the three cosets must look as in Figure 4.10. A automatically has the symmetry of D_6, the dihedral group of order six. This gives another exact lattice group, but not a maximal one—it can be enlarged by adjusting the scale in the direction of V, so as to make A into any of the lattices of Problem 4.2.8.

(e) If $m = 4$, then A either equals $(A \cap V) \oplus (A \cap V^{\perp})$ or contains this with index 2. In either case, the exact lattice group is isomorphic to D_8. These are not maximal, because, by adjusting the scale in the direction

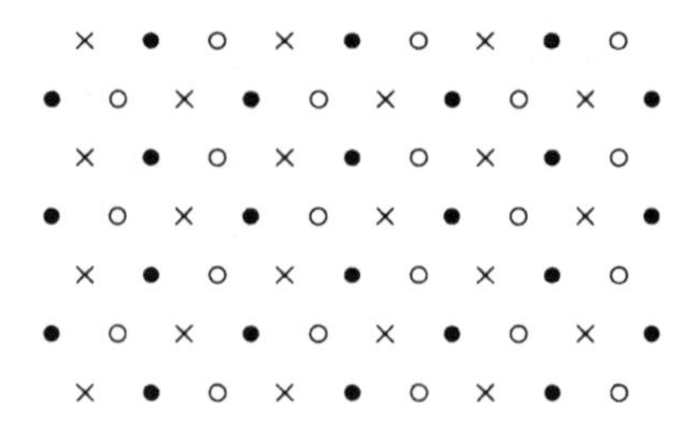

Figure 4.10. Three layers of points in a cubic lattice, seen along a diagonal

of V, one can make A into the cubic lattice (in the first case) or the face-centered or body-centered cubic lattice (in the second case).

(f) C_2 has two embeddings in $\mathrm{SL}(3, \mathbf{Z})$ up to conjugacy, and $C_2 \oplus C_2$ has four. This gives six more exact lattice groups. None of them is maximal; they are each subsumed in the embeddings of D_8.

(g) If A is invariant by a rotation about an axis V and has other symmetries that do not preserve V, then A has the symmetries of an octahedron, and so is either the cubic, the face-centered cubic or the body-centered cubic lattice.

(h) There are lattices with no orientation-preserving symmetry apart from the identity.

A crystallographic group in dimension n that is the full group of symmetries of some affine (that is, originless) lattice in $\mathbf{E}^n$ is called a *Bravais group*, after the person who first enumerated such groups in dimension three [Bra49]. If two lattices have isomorphic Bravais groups, the groups must be conjugate under an affine map by Theorem 4.2.2(c), and this map necessarily takes one lattice to the other. The classes of lattices according to this equivalence are known as *Bravais classes*, or, somewhat improperly, *Bravais lattices*.

Problem 4.3.13 (exact lattice groups and Bravais groups). Show that two n-dimensional lattices are in the same Bravais class if and only if they have their corresponding symmetry groups in $\mathrm{GL}(n, \mathbf{Z})$ are the same. In other words, exact lattice groups in $\mathrm{GL}(n, \mathbf{Z})$, up to conjugacy, are in one-to-one correspondence with n-dimensional Bravais groups, up to affine equivalence.

Using this and Problem 4.3.12, describe the Bravais lattices in three dimensions. Check your results against Table 4.1.

Problem 4.3.14. Show that there are two three-dimensional crystallographic groups G_1 and G_2 (not uniquely determined) such that every other

System	Symmetries in SO(3)	Number of classes	Case in 4.3.12
Triclinic	$\{1\}$	1	(h)
Monoclinic	C_2	2	(f)
Orthorhombic	$C_2 \oplus C_2$	4	(f)
Tetragonal	C_8	2	(e)
Trigonal	C_6	1	(d)
Hexagonal	C_{12}	1	(c)
Isometric	octahedral group	3	(g)

Table 4.1. Bravais lattice classes. In three-dimensional crystallography, Bravais lattices (Problem 4.3.13) are grouped into *systems*, according to their rotational symmetries. (The trigonal system is often subsumed under the hexagonal.)

three-dimensional crystallographic group is isomorphic to a subgroup of G_1 or G_2.

Problem 4.3.15 (two-dimensional groups, three-dimensional manifolds). Show that every closed orientable three-manifold arises as the tangent circle bundle of a two-dimensional Euclidean orbifold. See Chapter 5, and especially Section 5.11, for the appropriate definitions. Which Euclidean three-manifold occurs twice?

Challenge 4.3.16 (intransigent groups). There are interesting discrete subgroups of $\operatorname{Isom} \mathbf{E}^2 \times \operatorname{Isom} \mathbf{E}^2$ whose projections to the factors are not discrete.

Consider the group G with presentation

$$\langle \alpha, \beta, \gamma \mid \alpha^2 = \beta^2 = \gamma^2 = (\alpha\beta)^5 = (\beta\gamma)^5 = (\gamma\alpha)^5 = 1 \rangle,$$

and the golden isosceles triangles T_1 and T_2 with angles $(\pi/5, 2\pi/5, 2\pi/5)$ and $(3\pi/5, \pi/5, \pi/5)$, respectively.

(a) The group generated by reflections in the sides of T_1 is a homomorphic image of G, and likewise for T_2. Let ρ_1 and ρ_2 be the two homomorphisms.

(b) Show that the image of the diagonal homomorphism $\rho_1 \times \rho_2 : G \to \operatorname{Isom} \mathbf{E}^2 \times \operatorname{Isom} \mathbf{E}^2$ is a four-dimensional crystallographic group, yet the images of ρ_1 and ρ_2 are not discrete. (Hint: pass to a subgroup of index 2 of G, and consider how $\rho_1 \times \rho_2$ acts on the ring $\mathbf{Z}[\omega]$, where ω is a primitive fifth root of unity; this ring maps into $\mathbf{C}^2$ as a lattice.)

(c) Show that $\rho_1 \times \rho_2$ is not faithful by comparing it with an action ρ on the hyperbolic plane. Construct a picture of the crystallographic group of (b) by showing that $\mathbf{H}^2/\rho(G_0)$ is a surface M of genus two, where G_0 is the kernel of a map from G to D_{10}. Then G acts on the maximal abelian cover $\tilde{M}$, and its image there has a subgroup of index 10 isomorphic to $\mathbf{Z}^4$, the group of deck transformations of $\tilde{M}$ over M. Map $\tilde{M}$ equivariantly into $\mathbf{E}^4$ (with polyhedral image), using maps of its triangles to copies of the two golden triangles. Can you decide whether this is an embedding?

(d) Generalize this construction for any prime p to produce crystallographic groups in $\mathbf{E}^{p-1}$ which are generated by three elements of order two and contain elements of order p. Show that this is the minimum possible dimension for a crystallographic group having elements of order p.

Problem 4.3.17. Translate Challenge 4.3.16 into a complex analytic form. In other words, make use of the conformal equivalence between any two triangles rather than the affine equivalence. The surface M of genus two is a five-fold regular branched covering of the Riemann sphere $\mathbf{CP}^1$, so it can be put in the form $\{(x, y) \in \mathbf{CP}^1 \times \mathbf{CP}^1 : x^2 + y^5 = 1\}$. The two maps of the universal abelian cover $\tilde{M}$ of M to $\mathbf{C}$ have differentials that can be taken as $y^{-3}dx$ and $y^{-4}dx$. (Derive this from the qualitative form of the branching.) What happens when 5 is replaced by an arbitrary prime p?

4.4. Elliptic Three-Manifolds

There are an infinite number of elliptic three-manifolds (that is, closed ones: see Convention 4.3.1). But from Corollary 4.1.11 we know that there is some integer m such that every elliptic three-manifold has a covering space M, with at most m sheets, whose fundamental group is abelian and is contained in a connected abelian subgroup of SO(4).

An abelian subgroup of SO(4) can be diagonalized over $\mathbf{C}$. Translated into real terms, this means that it preserves a splitting of $\mathbf{R}^4$ into $\mathbf{R}^2 \times \mathbf{R}^2$; these invariant planes give two invariant circles in S^3, by intersection. If the group is connected, it acts by rotations on each circle, and so is a subgroup of the torus $S^1 \times S^1$.

It follows that the fundamental group Γ of the covering manifold M is a subgroup of $S^1 \times S^1$. Its projection to either factor is injective, because a non-trivial element in the kernel would fix a point in the corresponding invariant circle. Injectivity of the projections, together with discreteness, implies that Γ is cyclic. (As usual, this term includes the trivial group.) Thus Γ is generated by the map

4.4.1. $$(z_1, z_2) \mapsto (e^{2\pi i/p} z_1, e^{2\pi i q/p} z_2),$$

for p and q relatively prime, in terms of the coordinates determined by the invariant planes. You should check that the quotient $M = S^3/\Gamma$ is a lens space (Example 1.4.6).

In a sense, then, all elliptic three-manifolds are "almost" lens spaces—they are lens spaces up to bounded-sheeted covers. Nonetheless, the most interesting examples are not lens spaces.

We will work out here the classification of elliptic three-manifolds, which was first given in outline in [Hop26] and in detail in [ST30]. Our analysis does not depend on Corollary 4.1.11. It uses only elementary arguments, starting from two fundamental facts: every elliptic three-manifold is orientable (Exercise 4.4.2), and the group of orientation-preserving isometries of S^3 is almost a product (Equation 2.7.8).

Exercise 4.4.2 (elliptic three-manifolds are orientable). Any isometry of S^3 that has no fixed points is orientation-preserving. In particular, every elliptic three-manifold is orientable.

(Hint: given an element of $O(4)$, add up the dimensions of its 1-eigenspace, its -1-eigenspace, and its two-dimensional irreducible subspaces, if any. Or, if you are familiar with the Lefschetz fixed-point theorem, deduce from it that every orientation-reversing continuous map of an odd-dimensional sphere to itself has a fixed point.)

An elliptic three-manifold M is a quotient of elliptic three-space $\mathbf{RP}^3$ if and only if its holonomy group contains the antipodal map $-\mathrm{Id}$.

Consider any M that is not such a quotient, that is, whose holonomy does not contain the antipodal map. The action of $-\mathrm{Id}$ on S^3 factors down to M, since $-\mathrm{Id}$ is central. The quotient action of $-\mathrm{Id}$ on M has no fixed point; otherwise, there would be $\gamma \in \pi_1(M)$ and $x \in S^3$ with $\gamma(x) = -x$, so we'd have $\gamma^2(x) = x$, then $\gamma^2 = \mathrm{Id}$ and finally $\gamma = -\mathrm{Id}$ by Exercise 4.4.3. We conclude that $M/\{\pm\mathrm{Id}\}$ is itself a manifold, and is covered by $\mathbf{RP}^3$.

Exercise 4.4.3 (free order-two). The only order-two element of $O(n+1)$ that acts freely on S^n is the antipodal map.

In view of this, the classification of elliptic three-manifolds can be broken into two stages. First we find all quotients of $\mathbf{RP}^3$ by groups of orientation-preserving isometries acting freely, and second, we adjoin certain double covers of such quotients, namely those whose holonomy does not contain $-\mathrm{Id}$.

Since second stage is quite simple, we'll analyze it first:

Proposition 4.4.4 (existence of double cover). *A three-manifold $M = \mathbf{RP}^3/\Gamma$ has a double cover by a three-manifold $\tilde{M}$ whose holonomy does not contain* $-\mathrm{Id}$ *if and only if* Γ *has odd order.*

When it exists, the double covering $\tilde{M}$ is unique, with $\pi_1(\tilde{M}) = \Gamma$ and $\pi_1(M) = \Gamma \times C_2$.

Proof of 4.4.4. If Γ has even order, it contains an element $\gamma \in \operatorname{Isom} \mathbf{RP}^3$ of order two. No isometry $\tilde{\gamma}$ of S^3 covering γ cannot satisfy $\tilde{\gamma}^2 = \mathrm{Id}$, for that would imply $\tilde{\gamma} = \pm\mathrm{Id}$, and γ would act trivially on $\mathbf{RP}^3$. Therefore, any lift must satisfy $\tilde{\gamma}^2 = -\mathrm{Id}$. It follows that M does not have a covering of the desired form.

If Γ has odd order, consider the holonomy group $\tilde{\Gamma}$ for M, and the two-to-one projection homomorphism $p : \tilde{\Gamma} \to \Gamma$. Think of left multiplication by an element in $\tilde{\Gamma}$ as a permutation of $\tilde{\Gamma}$. Multiplication by $-\mathrm{Id}$ transposes an odd number of pairs of elements of $\tilde{\Gamma}$, so it is an odd permutation. Therefore the subgroup $\Gamma_0 \subset \tilde{\Gamma}$ consisting of even permutations projects isomorphically to Γ. It follows that Γ_0 is one of the factors in a product decomposition for $\tilde{\Gamma}$, where the other factor is $C_2 = \{\pm\mathrm{Id}\}$.

The desired double cover is S^3/Γ_0.

To see the uniqueness of the double cover when Γ has odd order, consider the fundamental group of such a double cover; this is some subgroup Γ_1 of index 2 in $\tilde{\Gamma}$. The projection $\Gamma_1 \to \Gamma$ is an isomorphism, since its kernel is trivial. Then Γ_1 is the graph of a homomorphism from Γ to C_2. The only homomorphism is trivial, so $\Gamma_1 = \Gamma_0$. 4.4.4

The advantage of working with quotients of $\mathbf{RP}^3$ is that the orientation-preserving isometry group of $\mathbf{RP}^3$ is a product, rather than almost a product as for S^3. In fact, with the usual identification of $\mathbf{RP}^3 = \mathrm{SO}(3)$, the group of orientation-preserving isometries is $\mathrm{SO}(3) \times \mathrm{SO}(3)$, acting on itself by left $\times$ right multiplication $((g, h) : x \mapsto gxh^{-1})$. This action is effective because $\mathrm{SO}(3)$ has a trivial center (unlike S^3). It is easy to check that $\mathrm{SO}(3) \times \mathrm{SO}(3)$ is the full group of isometries, for instance by observing that it acts transitively on the bundle of oriented orthonormal frames on $\mathbf{RP}^3$.

To classify the finite subgroups of $\mathrm{SO}(3) \times \mathrm{SO}(3)$ that act freely on $\mathbf{RP}^3$, we need three ingredients: the classification of finite subgroups of $\mathrm{SO}(3)$, a construction for subgroups of a product given subgroups of the factors, and a criterion for a group to act freely on $\mathbf{RP}^3$. The last item is the easiest.

Exercise 4.4.5 (free action on $\mathbf{RP}^3$). (a) A subgroup $H \subset \mathrm{SO}(3) \times \mathrm{SO}(3)$ acts freely on $\mathbf{RP}^3$ if and only if there is no nontrivial element $(h_1, h_2) \in H$ such that h_1 and h_2 are conjugate in $\mathrm{SO}(3)$.

(b) Every element of $\mathrm{SO}(3)$ is a rotation by some angle about some axis. Write down a necessary and sufficient condition for two elements of $\mathrm{SO}(3)$ to be conjugate.

Problem 4.4.6 (grungy classification of subgroups of $\mathrm{SO}(3)$**).** The finite subgroups of $\mathrm{SO}(3)$ will be classified conceptually in Section 5.5, but it is worth working them out by a straightforward, albeit grungy, approach:

(a) For any spherical triangle with angles α, β and γ there are elements A, B and C of $\mathrm{SO}(3)$ that rotate by angles 2α, 2β and 2γ, and that satisfy $ABC = 1$ (compare Exercise 4.3.5).

(b) Conversely, if $A, B, C \in \mathrm{SO}(3)$ satisfy $ABC = 1$, and if the three axes are not coplanar, they can be described as above using any of eight spherical triangles.

(c) Three real numbers α, β and γ between 0 and π are the angles of a nondegenerate spherical triangle if and only if their sum exceeds π and they satisfy the triangle inequalities $\alpha + \beta > \gamma$, $\beta + \gamma > \alpha$, and $\gamma + \alpha > \beta$.

(d) In a spherical triangle, the shortest side is opposite the smallest angle.

(e) If A and B generate a finite group $F \subset \mathrm{SO}(3)$, and if they have axes closer than any other pair of elements of F, then at least one has order 2 and the other has order 2 or 3.

(f) If A and B as above have order 2 and generate a finite subgroup, the group is a dihedral group D_{2n} of order $2n$.

(g) If A and B as above have order 2 and 3, their product has order 3, 4 or 5, and they generate the group of the orientation-preserving isometries of a regular polyhedron: the tetrahedral group (order 12), the octahedral group (order 24) or the icosahedral group (order 60). (See also Problem 4.4.7.)

(h) Every finite subgroup of $\mathrm{SO}(3)$ is generated by one or two elements.

(i) The finite subgroups of $\mathrm{SO}(3)$ are the cyclic groups, the dihedral groups, the tetrahedral group, the octahedral group and the icosahedral group.

Problem 4.4.7 (polyhedra and permutations). (a) Give geometric descriptions for the following isomorphisms: between the tetrahedral group and the alternating group A_4; between the octahedral group and the symmetric group S_4; and between the icosahedral group and A_5. (Hint: consider the action on various geometrical figures inside a regular polyhedron.)

(b) Describe the structure of the full group of isometries of each regular polyhedron, allowing reversals of orientation.

(c) Show there is an automorphism of the icosahedral group that does not come from an isometry. Give a description of this automorphism by imagining a dodecahedron made of pentagrams (pentagons that form five-pointed stars).

One type of subgroup of a product $G_1 \times G_2$, is the graph of any homomorphism from $G_1 \to G_2$. In fact, all subgroups that project isomorphically to G_1 are obtained in this way. Symmetrically, all subgroups that project isomorphically to G_2 are graphs of homomorphisms $G_2 \to G_1$. These are particular cases of the following construction, which gives all subgroups of a product.

Given groups G_1, G_2 and B, and homomorphisms $h_1 : G_1 \to B$ and $h_2 : G_2 \to B$, the *fiber product* $G(h_1, h_2)$ of h_1 and h_2 is the subgroup of $G_1 \times G_2$ given by

$$G(h_1, h_2) = \{(g_1, g_2) \in G_1 \times G_2 : h_1(g_1) = h_2(g_2)\}.$$

The special case when h_1 or h_2 is an isomorphism reduces to the graph of a homomorphism.

Proposition 4.4.8 (diagonal groups are fiber product). *Every subgroup $H \subset G_1 \times G_2$ that projects surjectively to both factors is the fiber product of a pair of surjective homomorphisms of G_1 and G_2 to some group B.*

Proof of 4.4.8. Let $H_1 = H \cap G_1 \times \{1\}$ and $H_2 = H \cap \{1\} \times G_2$. Using the surjectivity of $H \to G_2$, it is easy to check that H_1 is normal in $G_1 \times \{1\}$. Likewise, H_2 is normal in $\{1\} \times G_2$. Taking the quotient of H and $G_1 \times G_2$ by the normal subgroup $H_1 \times H_2$, we obtain a subgroup $B = H/(H_1 \times H_2)$ of the product $(G_1/H_1) \times (G_2/H_2)$. Since B intersects each factor trivially, it projects isomorphically to each factor. Composing these isomorphisms with the quotient maps $G_1 \to G_1/H_1$ and $G_2 \to G_2/H_2$, we get homomorphisms $h_1 : G_1 \to B$ and $h_2 : G_2 \to B$. You should check that H is the fiber product $G(h_1, h_2)$. $\boxed{4.4.8}$

We start putting the pieces together.

Proposition 4.4.9 (order-two commitment). *Let $H \subset \mathrm{SO}(3) \times \mathrm{SO}(3)$ be a group that acts freely on $\mathbf{RP}^3$, let Γ_1 and Γ_2 be its two projections, and let H_1 and H_2 be its intersections with the factors $\mathrm{SO}(3) \times \{1\}$ and $\{1\} \times \mathrm{SO}(3)$. Then, possibly after interchanging the factors in $\mathrm{SO}(3) \times \mathrm{SO}(3)$, we have the following situation:*

(a) *all order-two elements in* Γ_1 *are in* H_1, *and*

(b) Γ_2 *is cyclic of odd order.*

Proof of 4.4.9. All elements of order two are conjugate in SO(3), so our criterion for the action to be free (Exercise 4.4.5) shows that H_1 and H_2 cannot both have elements of order two. By interchanging the factors, if necessary, we assume that H_2 has no elements of order two.

Given $\alpha \in \Gamma_1$ of order two, take $\beta \in \Gamma_2$ such that $(\alpha, \beta) \in H$, so that $\beta^2 \in H_2$. The order m of β^2 must be odd since H_2 has no elements of order two. Now $(\alpha^m, \beta^m) \in H$ has order two, and this implies $\beta^m = 1$, otherwise β^m would be conjugate to α^m, both having order two. We conclude that $(\alpha^m, 1) \in H$, which shows that $\alpha \in H_1$ since $\alpha = \alpha^m$.

Now compare Γ_2 with the list of finite subgroups of SO(3) in Problem 4.4.6(i). The only entries in that list having no element of order two are the cyclic groups of odd order. Therefore, if Γ_2 has no element of order two, it is cyclic of odd order, and we're done.

Otherwise, since H_2 has no elements of order two, we apply the same reasoning to conclude that H_2 is cyclic of odd order. The proof of 4.4.8 shows that $\Gamma_1/H_1 = \Gamma_2/H_2$ and that H_1 is normal in Γ_1. If $\Gamma_1 = H_1$, then, $\Gamma_2 = H_2$ and again the proposition follows. Otherwise, Γ_1 is a finite subgroup of SO(3) all of whose order-two elements are contained in a certain proper normal subgroup. By checking against Problem 4.4.6(i), we see that Γ_1 is the tetrahedral group, and H_1 is a subgroup of index three (see also Problem 4.4.7). Therefore the index of H_2 in Γ_2 is also three. Since H_2 contains no elements of order two, neither does Γ_2, so it is cyclic. 4.4.9

Exercise 4.4.10. Let the notation be as in Proposition 4.4.9.

(a) Show that, possibly after switching indices, all order-three elements in Γ_1 are in H_1, and H_2 has no elements of order three.

(b) Construct H of order 5 and such that H_1 and H_2 are both trivial.

(c) Construct H with order a power of 2 and such that both Γ_1 and Γ_2 are non-trivial. Ditto with order a power of 3.

Corollary 4.4.11 (Seifert). *If* $H \subset \mathrm{SO}(3) \times \mathrm{SO}(3)$ *acts freely on* $\mathbf{RP}^3$, *there is some one-parameter subgroup* SO(2) *of one of the two factors that commutes with the action of* H.

Proof of 4.4.11. We apply Proposition 4.4.9 to conclude that Γ_2 is cyclic, and therefore generated by a rotation around some axis.

We take for SO(2) the group of all rotations around this axis. If $g \in \mathrm{SO}(2)$ is such a rotation, $(1, g)$ centralizes $\mathrm{SO}(3) \times \Gamma_2$, which contains H. 4.4.11

Exercise 4.4.12. Elliptic three-manifolds are in one-to-one correspondence with subgroups of $U(2)$ that act freely on $\mathbf{C}^2 \setminus \{0\}$.

The significance of Corollary 4.4.11 is that any elliptic three-manifold $M = \mathbf{RP}^3/H$ (or its double cover) is foliated by circles, coming from the orbits under the action of the one-parameter subgroup given by the statement of the corollary. (In S^3, these orbits form a Hopf fibration.) This foliation is not quite a fiber bundle, because there can be special circles such that neighboring circles wind around several times before closing. A foliation like that is called a *Seifert fiber space*; we will discuss the general theory of Seifert fiber spaces in Section 5.11.

We can use these fiberings to better understand the topology and geometry of elliptic three-manifolds, finding interpretations for them in terms of the geometry of S^2. The key is the following identification, already used in Example 3.7.6, between SO(3) and the tangent circle bundle of $S^2 \subset \mathbf{R}^3$. The derivative of an isometry of S^2 takes the tangent circle bundle UTS^2 to itself. The resulting action of SO(3) is simply transitive, so after arbitrarily picking a base vector $v_0 \in UTS^2$, we can identify each element of SO(3) with the image of v_0 under it.

Under this identification, the fibers of the tangent circle bundle are fibers of the Hopf fibration, pushed down to SO(3). Left multiplication by an element $g \in \mathrm{SO}(3)$ corresponds to the action of the derivative dg on UTS^2. Right multiplication by g corresponds to a "recipe" for navigating from v_0 to $gv_0 = v_0 g$ (such as "turn 12.7° to the left, go forward 1219 kilometers, then turn 48.3° to the right"), to be applied to all other unit tangent vectors.

Exercise 4.4.13. By this definition, the derivatives of orientation-preserving isometries of S^2 account for the factor $\mathrm{SO}(3) \times \{1\} \subset \mathrm{SO}(3) \times \mathrm{SO}(3)$. Derivatives of orientation-reversing isometries of S^2, too, are orientation-preserving isometries of $UTS^2 = \mathbf{RP}^3$. Describe such derivatives in terms of $\mathrm{SO}(3) \times \mathrm{SO}(3)$. In particular:

(a) What element of $\mathrm{SO}(3) \times \mathrm{SO}(3)$ corresponds to the antipodal map on S^2? What is the quotient three-manifold?

(b) What is the description of the full group of isometries of an icosahedron as a subgroup of $\mathrm{SO}(3) \times \mathrm{SO}(3)$?

Let $\mathrm{SO}(2) \in \{1\} \times \mathrm{SO}(3)$ consist of those isometries whose recipes don't call for "going forward"—that is, which preserve each fiber of UTS^2 (each Hopf circle). By Corollary 4.4.11, for any finite group Γ acting freely on $\mathbf{RP}^3$, we can choose the base vector v_0 so that $\mathbf{RP}^3$ is contained in $\mathrm{SO}(3) \times \mathrm{SO}(2)$. We can then think of the quotient space reasonably well in terms of a two-dimensional picture. For instance, the quotient space of $\mathbf{RP}^3$ by C_5 acting on the right may be thought of as the set of all unit tangent vectors to S^2, up to rotation by multiples of 72°. If we take a figure P with fivefold symmetry, such as a regular spherical pentagon, the quotient is the set of all figures on S^2 congruent to P. Its fundamental group is C_{10}, generated by the loop described by rotating the figure in place one click. When you rotate it ten clicks, or two full turns, the resulting path is homotopic to the identity. The quotient is, in fact, the lens space $L_{10,1}$: see Exercise 4.4.15. (See also Problem 4.4.16.)

The quotient space of a finite subgroup of SO(3) acting on the left can be thought of as the tangent circle bundle of the quotient orbifold, a concept we will develop in Chapter 5. For instance, the quotient space of the group of orientation-preserving isometries of an icosahedron is topologically a sphere, with three singular points coming from the vertices, the midpoints of edges, and the centers of faces—the points where the group does not act freely. The quotient of the tangent circle bundle of the sphere can be thought of as the tangent circle bundle of the quotient, but with some weirdness at the three singular points; it turns out to be the Poincaré dodecahedral space (Problem 4.4.17). We can also think of it as the space of all arrangements of 60 tangent vectors to the sphere that are symmetrical by the icosahedral group (which has order 60).

These two constructions can be combined. For instance, the quotient of $\mathbf{RP}^3$ by the tetrahedral group acting on the left and a cyclic group of order five acting on the right is the quotient of the space of pentagons congruent to P by the action of the tetrahedral group. No element of the tetrahedral group takes any pentagon exactly to itself, since the group contains no order-five rotations. Therefore, the quotient is an elliptic three-manifold. It can be thought of as the set of all ways to arrange 24 pentagons congruent to P on the surface of the sphere so that they are symmetrical by the tetrahedral group.

Theorem 4.4.14 (classification of elliptic manifolds). *Let M be an elliptic three-manifold.*

(a) *If* $\pi_1(M)$ *is abelian, it is cyclic, and* M *is a lens space.*

Otherwise, M *is the quotient of* $\mathbf{RP}^3$ *by a group* H *of one of the following types:*

(b) $H = H_1 \times H_2$, *where* H_1 *is a dihedral group, the tetrahedral group, the octahedral group or the icosahedral group, and* H_2 *is a cyclic group with order relatively prime to the order of* H_1.

(c) H *is a subgroup of index* 3 *in* $T \times C_{3m}$, *where* m *is odd and* T *is the tetrahedral group.*

(d) H *is a subgroup of index* 2 *in* $C_{2n} \times D_{2m}$, *where* n *is even and* m *and* n *are relatively prime.*

Proof of 4.4.14. Suppose that $\pi_1(M)$ is abelian. According to the plan outlined after Exercise 4.4.3, if M is not a quotient of $\mathbf{RP}^3$, we reduce to the case where it is by considering instead $N = M/\{\pm\mathrm{Id}\}$. Since $-\mathrm{Id}$ is central, $\pi_1(M)$ abelian implies $\pi_1(N)$ abelian, and clearly $\pi_1(N)$ cyclic implies $\pi_1(M)$ cyclic.

Therefore we can assume that M is the quotient of $\mathbf{RP}^3$ by an abelian group $H \subset \mathrm{SO}(3) \times \mathrm{SO}(3)$. Applying Corollary 4.4.11 to H, we find that at least one of its projections to the factors—say Γ_1—is cyclic. By the classification of discrete subgroups of $\mathrm{SO}(3)$ (Problem 4.4.6), Γ_2, if not cyclic, must be $C_2 \oplus C_2$. But this cannot happen, as we see by taking two elements of H whose projections in Γ_2 are distinct and non-trivial, and lifting to $\pi_1(M) \subset \mathrm{SO}(4) = (S^3 \times S^3)/C_2$. The two lifts are of the form (p_1, p_2) and (q_1, q_2), where p_1 and q_1 commute, while p_2 and q_2 are non-central elements of the quaternion group $\{\pm 1, \pm i, \pm j, \pm k\}$ and so anticommute. Therefore the lifts also anticommute, and $\pi_1(M)$ is not abelian.

We now know that Γ_1 and Γ_2 are cyclic, and we show that H is cyclic. If not, H contains a subgroup isomorphic to $C_p \times C_p$, for some prime p. Since all its non-trivial elements have order p, this subgroup is contained in the product of $C_p \subset \Gamma_1$ with $C_p \subset \Gamma_2$, and therefore it equals this product. This means that $C_p \subset H_1$ and $C_p \subset H_2$, contradicting the fact that H acts freely, by Exercise 4.4.5.

If H is cyclic, so is its lift $\pi_1(M)$. For if $(p_1, p_2) \in \mathrm{SO}(3) \times \mathrm{SO}(3)$ generates H, we have

$$\mathrm{lcm}(\mathrm{order}\, p_1, \mathrm{order}\, p_2) = \mathrm{order}\, H.$$

We can take lifts $\tilde{p}_1$ and $\tilde{p}_2$ with twice the order. Then

$$\mathrm{lcm}(\mathrm{order}\, \tilde{p}_1, \mathrm{order}\, \tilde{p}_2) = 2\, \mathrm{order}\, H,$$

so $(\tilde{p}_1, \tilde{p}_2)$ generates $\pi_1(M)$. Since $\pi_1(M)$ is cyclic, M is a lens space, as we saw at the beginning of this section. This proves case (a).

Now suppose that $\pi_1(M)$ is not abelian. If $M = \mathbf{RP}^3/H$, then H is not cyclic. We can assume, from Proposition 4.4.9, that H_1 contains all elements of order two in Γ_1 and that $H_2 = C_m$, where m is odd. So either $H_1 = \Gamma_1$, or Γ_1 is the tetrahedral group with H_1 of index 3, or Γ_1 is cyclic and H_1 is a proper subgroup.

If $H_1 = \Gamma_1$, then also $H_2 = \Gamma_2$, by the proof of Proposition 4.4.8. Therefore $H = H_1 \times H_2$, so $\pi_1(M) = \tilde{H}_1 \times H_2 = H_1 \times \tilde{H}_2$. By Exercise 4.4.5, the orders of H_1 and H_2 must be relatively prime and we have case (b).

If Γ_1 is the tetrahedral group and H_1 has index 3, we get case (c), as in the last case in the proof of 4.4.9.

If Γ_1 is cyclic and $\Gamma_1/H_1 = \Gamma_2/H_2$ is nontrivial, Γ_2 must be dihedral or cyclic, since the groups of the regular polyhedra are not cyclic extensions of cyclic groups. The case Γ_2 cyclic leads back to H cyclic. In the case $\Gamma_2 = D_{2m}$ the only candidate for H_2 (which must be normal, cyclic, and of odd order) is C_m. Therefore m is odd, $\Gamma_1/H_1 = \Gamma_2/H_2 = C_2$, and Γ_1 has order $2n$, where n is even (otherwise there would be an element of order 2 in Γ_1 that is not in H_1). By 4.4.5, m and n are relatively prime, so we have case (d).

Finally, suppose that $\pi_1(M)$ is not abelian and M is not a quotient of $\mathbf{RP}^3$. Then M is a double cover of a manifold $\mathbf{RP}^3/H$ with H of odd order. The above analysis shows that such a quotient of $\mathbf{RP}^3$ does not occur. 4.4.14

Exercise 4.4.15 (characterizations of lens spaces). Give the correspondences between these various descriptions of lens spaces:

(a) $\mathbf{RP}^3/H$, where $H \subset C_m \times C_n \subset \mathrm{SO}(3) \times \mathrm{SO}(3)$, or the double cover of a manifold of this form.

(b) S^3/Γ, where Γ is generated by a map of the form 4.4.1. What condition on p and q classifies lens spaces up to isometry? Compare Problem 1.4.7(c).

(c) The result of gluing together two solid tori $D^2 \times S^1$ by an affine homeomorphism ϕ between their boundaries such that $\phi(\partial D^2)$ is not parallel to ∂D^2.

Problem 4.4.16 (another description of elliptic three-manifolds). Describe the various elliptic three-manifolds classified in Theorem 4.4.14 in terms of the tangent circle bundle of S^2, as explained before the classification. For instance, the manifold of Theorem 4.4.14(c), with $m = 5$, can be described as the space of regular pentagons congruent to P, modulo an action of the tetrahedral group that transforms the pentagons as usual, but then rotates by $0°$, $24°$ or $48°$ according to a homomorphism from the tetrahedral group to C_3. Can you find any better description of this three-manifold?

Problem 4.4.17 (Poincaré dodecahedral space revisited). Show that the quotient of $\mathbf{RP}^3 = UTS^2$ by the action of the icosahedral group is the Poincaré dodecahedral space of Example 1.4.4. (Hint: Use the faces of the dodecahedron to construct an invariant decomposition of UTS^2 into 12 solid tori. The stabilizer of a face is C_5. A fundamental domain for the icosahedral group can be constructed by choosing a vector field on a face as shown in Figure 4.11, then chopping the corresponding solid torus along the five images of this vector field under the face's stabilizer. Show that the resulting pieces of UTS^2 are dodecahedral fundamental domains for the action of the group, and that they meet in the same combinatorial pattern of Example 1.4.4.)

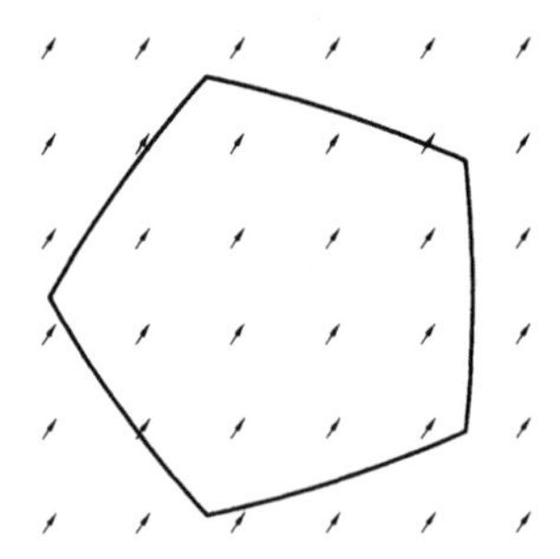

Figure 4.11. A near-constant unit vector field on a spherical pentagon. The pentagon shown is one of the faces of a dodecahedral tiling of the sphere, stereographically projected from the center of the opposite face. The field is constant on the plane. Projecting back to the sphere and normalizing gives the desired vector field.

Problem 4.4.18. The *binary* group $\tilde{\Gamma}$ associated with a subgroup Γ of $\mathrm{SO}(3)$ is the inverse image of Γ under the double cover in $S^3 \to \mathrm{SO}(3)$.

(a) $\tilde{\Gamma}$ equals $\Gamma \times C_2$ if and only if Γ has odd order. (Hint: see the proof of Proposition 4.4.4.)

(b) We saw in Example 1.4.4 that the quotient of S^3 by the binary icosahedral group is the Poincaré dodecahedral space. Give a similar geometric description for the quotients of S^3 by the binary tetrahedral group and the binary octahedral group. Describe fundamental polyhedra for the action, and how they are glued together.

Problem 4.4.19 (Frobenius map). This is a generalization of the fact that a subgroup of isometries of SO(3) lifts to S^3 if and only if it has odd order.

Consider a central group extension $A \to \tilde{G} \to G$, where G has finite order k. Show that the *Frobenius map* $\tilde{G} \to A$ that sends $g \in \tilde{G}$ to g^k is a homomorphism.

(Hint: choose a product structure for $\tilde{G}$ as a set, but respecting the action of A. Reinterpret the map $g \to g^k$ as the total amount by which g "rotates" the cosets of A.)

Use the Frobenius map to show that, if A is finite and k is relatively prime to the order of any element in A, there is a canonical isomorphism between $\tilde{G}$ and $G \times A$. Give a counterexample in the case that A is not finite.

4.5. The Thick-Thin Decomposition

In Sections 4.3 and 4.4 we produced lists of all Euclidean and elliptic three-manifolds, but hyperbolic three-manifolds are another matter: they are much more complicated. In principle, an algorithm can be given that will list all hyperbolic three-manifolds, but this isn't at all the same as an organized description. In fact, a complete analysis of hyperbolic surfaces is already rich and interesting: that is the topic of the next section. In this section we prove a general structural result about hyperbolic manifolds of any dimension.

Let M be a complete hyperbolic manifold of dimension n, possibly with infinite volume. For any point $x \in M$, the lifts of x to the universal cover $\tilde{M} = \mathbf{H}^n$ form a regular array, like atoms in a crystal. Because the group of covering transformations is discrete, there is a minimum distance d between any two lifts of x; this is also the shortest possible length of a homotopically nontrivial loop based at x. A ball of radius $r = \frac{1}{2}d$ about x is embedded, since all its lifts are disjoint, but no larger ball can be embedded. In other words, r is the radius of the largest open ball on which the exponential map $\mathbf{E}^n \to T_x M$ is injective. We call $r = r(x)$ the *injectivity radius* of M at x.

We can decompose M into a *thick part*

$$M_{\geq \varepsilon} = \{x \in M : r(x) \geq \tfrac{1}{2}\varepsilon\}$$

and its complement

$$M_{<\varepsilon} = \{x \in M : r(x) < \tfrac{1}{2}\varepsilon\}.$$

When the particular value of ε is not important, we simply write $M_{\geq}$ and $M_{<}$. We will also be interested in the *thin part* $M_{\leq} = \overline{M_{<}}$, where the bar, as usual, denotes closure. $M_{\leq\varepsilon}$ is usually the same as $\{x \in M : r(x) \leq \frac{1}{2}\varepsilon\}$. The more complicated definition eliminates the undesirable borderline case of a simple closed geodesic of length exactly ε. Along such a geodesic the injectivity radius is $\frac{1}{2}\varepsilon$, but just to the side the injectivity radius is greater.

We shall see that, as a consequence of Corollary 4.1.17 (small discrete hyperbolic almost abelian), $M_{\leq\varepsilon}$ has a standard form, provided ε is less than some constant that depends only on the dimension n.

For a nontrivial element $\gamma \in \pi_1(M)$, let $T_\varepsilon(\gamma)$ (or $T(\gamma)$ if ε is understood) denote the set of points in $\mathbf{H}^n$ moved a distance less than ε by γ. Since, by Theorem 2.5.8, the translation distance $d_\gamma(x) = d(x, \gamma(x))$ is a convex function, each $T(\gamma)$ is convex.

For example, in dimension two, if γ is hyperbolic, $T(\gamma)$ is the region between two curves equidistant from the axis of γ, as in Figure 1.10, since γ commutes with reflection in the axis and with translations along the axis. In dimension three, a hyperbolic element γ that preserves orientation is composed of a translation along an axis l and a rotation about l. This commutes with rotation about l and with translations along l, so $T(\gamma)$ in this case is a cylindrical (that is, translation-invariant) neighborhood of l, with a round cross-section. If γ is hyperbolic but reverses orientation, it is a glide-reflection, composed of a translation along an axis l and reflection in some plane P containing l. Then $T(\gamma)$ is again a cylindrical neighborhood, but it has an oval cross-section, with major axis in P.

Exercise 4.5.1 (shapes of $T(\gamma)$). (a) Describe the shape of $T(\gamma)$ in dimensions two and three when γ is a parabolic or elliptic transformation.

(b) Give an example of an orientation-preserving hyperbolic transformation γ of $\mathbf{H}^3$ such that $T(\gamma)$ is a proper subset of $T(\gamma^{100})$.

Now fix ε and m as in Corollary 4.1.17 (small discrete hyperbolic almost abelian). With these choices, if there is a point x is in the intersection of several $T(\gamma_i)$, the group generated by the γ_i contains a normal abelian subgroup of index at most m. We use this to study the connected components of $M_{<} = M_{<\varepsilon}$ and those of $\tilde{M}_{<}$, the subset of $\mathbf{H}^n = \tilde{M}$ lying above $M_{<}$. Note that $\tilde{M}_{<}$ is the union of $T(\gamma)$, for all nontrivial $\gamma \in \pi_1(M)$.

A component T_0 of $\tilde{M}_{<}$ is a union of convex sets $T(\gamma)$, for γ in some subset S_0 of $\pi_1(M)$. Suppose $T(\alpha)$ intersects $T(\beta)$, for

$\alpha, \beta \in S_0$. Let A be a normal abelian subgroup of finite index in the group generated by α and β, and let ϕ be a nontrivial element of A. Since $\pi_1(M)$ is torsion-free, ϕ is either parabolic or hyperbolic.

If ϕ is parabolic, it fixes a unique point $p \in S_\infty^{n-1}$, so all of A must fix p, by Exercise 2.5.18 (conjugating fixed points). By the same exercise, an element normalizing A must also fix p. Such an element cannot be hyperbolic, for if it were, powers of it or of its inverse would conjugate elements of A to parabolic elements moving any given point an arbitrarily short distance. Therefore the group generated by α and β consists of parabolic elements with fixed point p, and is isomorphic to a subgroup of $\operatorname{Isom} \mathbf{E}^{n-1}$. Any geodesic ray tending toward p meets $T(\alpha)$ (as well as $T(\beta)$) in an infinite half-open interval, since the translation distance $d_\alpha(x)$ approaches zero as x moves out toward p.

Using a similar argument about fixed points, we see that if ϕ is hyperbolic with axis l, all nontrivial elements of the group generated by α and β are likewise hyperbolic with axis l. This group acts on l by translations, and so is isomorphic to a subgroup of $\mathbf{E}^1 \times O(n-1)$. A simple analysis shows that a discrete and torsion-free subgroup of $\mathbf{E}^1 \times O(n-1)$ is cyclic. In particular, α and β commute. Moreover, we have $l \subset T(\alpha)$, since l is the locus of points moved a minimal distance by α (Proposition 2.5.17). This also shows that a generator γ of the cyclic group generated by α and β is in S_0: since γ translates l by no more than α does, we have $l \subset T(\gamma)$.

It follows from this analysis that the type of γ (whether parabolic or hyperbolic) and its fixed points at infinity are the same for all $\gamma \in S_0$—since T_0 is connected, the type and fixed points "propagate" inductively from one element of S_0 to all others. We call the component T_0 parabolic or hyperbolic accordingly. We use this classification to figure out the shape of the component M_0 of $M_<$ that is the image of T_0. Let G_0 by the subgroup of $\pi_1(M)$ generated by S_0, so that $M_0 = T_0/G_0$.

If T_0 is hyperbolic, G_0 is cyclic. T_0 itself is a finite union of convex neighborhoods of the axis l of G_0, and so diffeomorphic to an open ball D^n—or better, to $D^{n-1} \times \mathbf{R}$, in such a way that the diffeomorphism is equivariant under the action of $G_0 = \mathbf{Z}$ (Problem 4.5.3(c)). The component M_0 is diffeomorphic to $D^{n-1} \times S^1$ if the action of G_0 preserves orientation, and to a nonorientable disk bundle over S^1 otherwise. The closure $\bar{M}_0$ is compact, and topologically a manifold-with-boundary.

Exercise 4.5.2. (a) Describe T_0 when the generator of $G_0 \subset \operatorname{Isom} \mathbf{H}^3$ is a glide-reflection.

(b) Give an example of an orientation-preserving cyclic group $G_0 \subset \operatorname{Isom} \mathbf{H}^4$ such that T_0 is nonconvex.

Problem 4.5.3 (open star-shaped is ball). (a) An open subset $U \subset \mathbf{E}^n$ star-shaped with respect to a point $x \in U$ is diffeomorphic to $\mathbf{E}^n$. (Hint: modify the distance-to-the-boundary function on U so it becomes smooth and increases monotonically from 0 to ∞ along each ray from x to the boundary. See [Ber87, 11.3.6.1] if you get stuck; the construction there can be made smooth by "smearing".)

(b) Generalize to $U \subset \mathbf{H}^n$.

(c) If $U \subset \mathbf{H}^n$ is an open set invariant under the action of a cyclic group of hyperbolic transformations and is star-shaped with respect to each point on the axis of the group, U is equivariantly diffeomorphic to $\mathbf{E}^{n-1} \times \mathbf{E}$ (that is, the action on U maps to an action of $\mathbf{Z}$ on $\mathbf{E}^{n-1} \times \mathbf{E}$ by isometries, which preserve the product decomposition.)

(d) Prove an analogous statement if U is an open subset of $\mathbf{H}^n$ that is "star-shaped" with respect to a point x at infinity, invariant under a group of parabolic transformations fixing x, and such that every ray ending at x intersects U.

If T_0 is parabolic, with fixed point $p \in S^{n-1}_\infty$, it is again a union of convex sets that touch infinity at p. Therefore T_0 is "star-shaped" in the sense that the ray from any point $x \in T_0$ in the direction of p lies in T_0. By Problem 4.5.3(d), T_0 is diffeomorphic to upper half-space $\mathbf{E}^{n-1} \times (0, \infty)$. Modding out by the action of the parabolic group G_0, we see that M_0 is diffeomorphic to a Euclidean $(n-1)$-manifold cross $(0, \infty)$; any Euclidean $(n-1)$-manifold can occur in this role. The closure $\bar{M}_0$ is again a manifold-with-boundary, this time noncompact.

We can think of M_0 in this case as a neighborhood of a "point at infinity", or *cusp*, of M. More precisely, consider the set of fixed points of parabolic elements of $\pi_1(M)$, and divide by the action of $\pi_1(M)$. The elements of the resulting set are called cusps of M (of course, cusps are not part of the manifold itself). Each component of $M_<$ whose closure is noncompact is associated with exactly one cusp.

In either the parabolic or hyperbolic case, T_0 is simply connected, so the fundamental group of M_0 is G_0 (up to conjugacy).

Exercise 4.5.4. Based on Exercise 4.5.1 and the paragraph before it, what are the actual possibilities for M_0 in dimensions two and three? (See also Figure 4.12.)

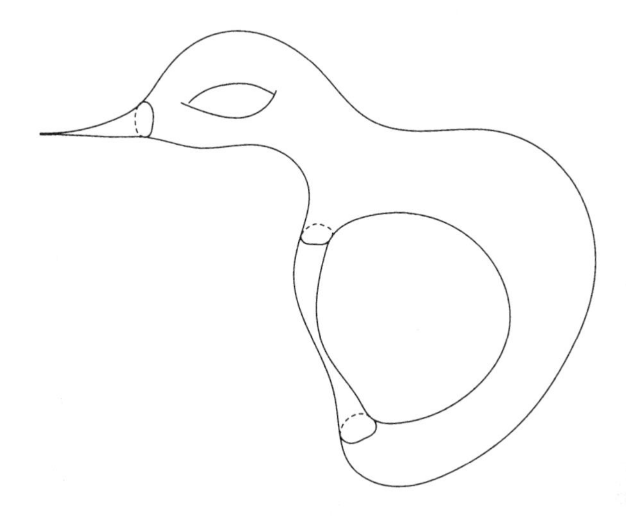

Figure 4.12. The thick-thin decomposition of a surface. The thin part of a hyperbolic surface of finite area consists of neighborhoods of short geodesics, and cusps isometric to pseudospheres.

Exercise 4.5.5 (compact manifolds have no cusps). It follows from this discussion that a compact hyperbolic manifold has no cusps (thus, its fundamental group acts by hyperbolic transformations only). Give a more direct proof of this fact. (Hint: In a compact manifold, the length of the representatives of each nontrivial free homotopy class of curves has a positive lower bound.)

Here is a summary of the discussion so far:

Theorem 4.5.6 (structure of thin part). *For ε less than a certain value, depending only on the dimension, the thin part $M_{\leq\varepsilon}$ of a complete hyperbolic manifold is a union of components of one of the following types: neighborhoods of short geodesics, homeomorphic to disk bundles over the circle; and neighborhoods of cusps, homeomorphic to products of Euclidean manifolds with a half-infinite interval.*

In general we can't say much about $M_{\geq\varepsilon}$ beyond the fact that it is a manifold-with-boundary for ε small. But we have the following result:

Theorem 4.5.7 (thick compact if and only if finite volume). *A complete hyperbolic manifold M has finite volume if and only if $M_{\geq\varepsilon}$ is compact for all $\varepsilon > 0$.*

Proof of 4.5.7. Suppose M has finite volume, and consider a set S of points in $M_{\geq\varepsilon}$ with the property that any two of them are separated by at least ε. The $\frac{1}{2}\varepsilon$-balls in M about the points of S are disjoint, and all have the same volume, so they must be finite in number. On the other hand, if we choose S to be maximal, the closed ε-balls about its points must cover all of $M_{\geq\varepsilon}$. The union of these balls is a compact set; being a closed subset of it, $M_{\geq\varepsilon}$ must be compact.

For the converse, we take ε such that the structure theorem 4.5.6 applies. If $M_{\geq\varepsilon}$ is compact, we must show that each of the finitely many components of $M_{<}$ has finite volume. This is trivial for components with compact closure. If a component M_0 is noncompact, consider a component T_0 of its inverse image in $\mathbf{H}^n$, arranged so the parabolic fixed is the point at infinity in upper half-space. The foliation defined on T_0 by vertical rays is preserved by G_0, and so induces a foliation on M_0, since images of T_0 under elements of $\pi_1(M)$ not in G_0 are disjoint from T_0. Each leaf of the foliation on M_0 ends at a point of $M_{\geq}$. Since $M_{\geq}$ is compact, the action of G_0 on horizontal (Euclidean) planes is cocompact, with fundamental domain P, say. This implies that T_0 is bounded away from the bounding plane $\mathbf{H}^n$: it never dips below a certain height $h > 0$. Thus the volume of M_0 is at most the volume of the half-cylinder $P \times [h, \infty)$, which is easily seen to be finite. $\boxed{4.5.7}$

Thus, in order to verify that M has finite volume, it is enough to check that $M_{\geq\varepsilon}$ is compact for a value of ε that makes Theorem 4.5.6 work. On the other hand, if ε is large, $M_{\geq\varepsilon}$ compact does not necessarily imply $\operatorname{vol} M < \infty$. For instance, if we take a hyperbolic three-manifold that fibers over the circle with fiber a surface, the covering space gotten by unwrapping the circle direction still has bounded injectivity radius, yet it doesn't have finite volume.

4.6. Teichmüller Space

In Section 1.3, we saw that every closed oriented surface of genus $g > 1$ has a hyperbolic structure. Actually, there is a lot of freedom in the construction, with the consequence that there is not just one hyperbolic structure, but many. Let's make a crude dimensional analysis to get some idea of how many.

Consider, for example, a surface of genus two, obtained from gluing opposite sides of an octagon. There is a 16-dimensional space of octagons in $\mathbf{H}^2$, since an octagon is determined by its eight vertices, each of which has two degrees of freedom. Isometric octagons obviously give the same structure, so we subtract three degrees of freedom, the dimension of $\operatorname{Isom} \mathbf{H}^2$.

Not all shapes of octagon will do, however. Opposite sides should have equal length, so they can be glued up; we take away four degrees of freedom since there are four such conditions. In addition, the sum of the angles has to be 2π, or, equivalently, the area must be 2π; we take away an additional degree of freedom for that. We're down to eight.

It sometimes happens that two non-isometric octagons glue together to give isometric surfaces: under the gluing, all vertices are identified and correspond to a single point on the surface. There are two degrees of freedom for the choice of this basepoint, and as it is moved around we get a varying description of the surface as a glued-up octagon. We subtract two more degrees of freedom to account for this multiplicity. Consequently there appears to be a six-dimensional family of essentially distinct hyperbolic structures on a surface of genus two.

This crude analysis ignores many questions. What kinds of octagons are we allowing: Do they have to be convex, and do they have to be embedded? Does the isometry group act freely, so the quotient is a manifold? Are the various conditions independent, so that the solution set is a manifold of the indicated dimension? As you move the basepoint around on the surface, do you get different shapes of octagons both locally and globally? What is the global structure of the space of hyperbolic shapes? These questions can be answered, but instead of formalizing this particular analysis, we will take a different path.

First, though, we must be more precise about what we mean by two hyperbolic structures being the same. There are in fact two important notions of equivalence, giving rise to two spaces: *moduli space* and *Teichmüller space*. Informally, in Teichmüller space, we pay attention not just to what metric a surface is wearing, but also to how it is worn. In moduli space, all surfaces wearing the same metric are equivalent. The importance of the distinction will be clear to anybody who, after putting a pajama suit on an infant, has found one leg to be twisted.

Remark 4.6.1. A *modulus*, in complex analysis, is an invariant associated with a domain in **C**, or, more generally, with a Riemann surface. In 1859, Riemann stated that isomorphism classes of closed Riemann surfaces of genus $g \geq 2$ are parametrized by $3g - 3$ complex parameters, or moduli—in other words, the moduli space of a closed surface of genus g has real dimension $6g - 6$. Because of the uniformization theorem (Example 3.1.17), the study of complex structures on a surface is also, in some sense, the study of its hyperbolic structures. The two points of view have given rise to two main currents in Teichmüller theory.

Riemann's justification for his assertion did not amount to a formal proof. The first rigorous analysis of these questions seems to be due to Fricke and Klein [FK12]. Teichmüller space was so named after the publication of [Tei39, Tei43], which contained important advances (including the introduction of the a metric, discussed on page 267) and rekindled interest in the subject.

Convention 4.6.2. Throughout this section, "surface" will refer to any connected differentiable two-dimensional manifold-with-boundary. "Hyperbolic structure" will refer to a complete hyperbolic structure with geodesic boundary: formally, if $\mathcal{G}$ is the pseudogroup defined in Definition 3.3.11, a hyperbolic structure is a $\mathcal{G}$-stiffening of S that makes S into a complete metric space. Euclidean and elliptic structures should be understood similarly.

Exercise 4.6.3. In Example 3.1.17 we saw that all surfaces without boundary have either a hyperbolic, a Euclidean or a spherical structure. Extend the reasoning given there to cover surfaces with boundary. What surfaces have Euclidean structures? How about spherical structures?

Let S be a surface that admits a hyperbolic structure, and let $\mathcal{S}S$ be the space of such structures.

Definition 4.6.4 (moduli space and Teichmüller space). The moduli space $\mathcal{M}S$ of S is the quotient of $\mathcal{S}S$ by the action of the group $\operatorname{Diff} S$ of diffeomorphisms of S. The Teichmüller space $\mathcal{T}S$ of S is the quotient of $\mathcal{S}S$ by $\operatorname{Diff}_0 S$, the group of diffeomorphisms of S homotopic to the identity by a homotopy that takes the boundary into itself at all times.

A theorem of Baer [Bae28, Eps66] says that, except for S the open disk or the annulus, two diffeomorphisms of a surface S are isotopic if and only if they are homotopic by a homotopy that takes the boundary into itself. Therefore, except in those simple exceptional cases, we can phrase the definition of Teichmüller space more succinctly using isotopy, and $\text{Diff}_0 S$ can be naturally interpreted as the connected component of the identity in $\text{Diff}\, S$ (in particular, $\text{Diff}_0 S$ is normal in $\text{Diff}\, S$). But generally the logic works more smoothly with homotopy as the basis of the equivalence relation, rather than isotopy—see Proposition 4.6.10, for example.

Exercise 4.6.5. In Baer's theorem, the condition that the homotopy take the boundary to itself is unnecessary except when S is the closed disk or the closed annulus, or when S has boundary components that are intervals. Find counterexamples for the statement in each exceptional case.

Another common reformulation of Definition 4.6.4 is illustrated in Figure 4.13. A pair (f, M), where M is a hyperbolic surface and $f : S \to M$ is a diffeomorphism, is called a *marked surface* (see also Problem 4.6.26). Now consider all surfaces marked by S. To get $\mathcal{T}S$, identify (f, M) and (f', M') if and only if $f' \circ f^{-1}$ is isotopic to an isometry. To get $\mathcal{M}S$, erase the information about the marking, that is, identify isometric surfaces M and M' regardless of f and f'. We will generally dispense with f and f', but we will use $M, M', \ldots$ for stiffenings of S.

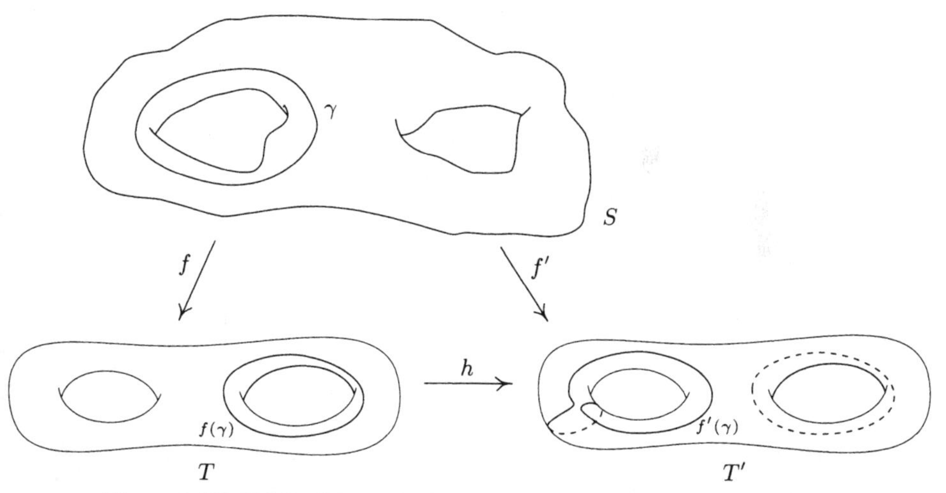

Figure 4.13. Teichmüller space is finer than moduli space. Here (f, M) and (f', M') coincide in $\mathcal{M}S$, because there is an isometry h—schematically depicted as translation in $\mathbf{R}^3$—between M and M'. But the two marked surfaces give different points in $\mathcal{T}S$, because $f' \circ f^{-1}$ is not isotopic to h, nor to any isometry. (If $f' \circ f^{-1}$ were isotopic to h, the dotted curve $h(f(\gamma))$ would be isotopic to $f'(\gamma)$.)

Let's introduce one more important concept before looking at some examples. Almost by definition, we have

$$\mathcal{M}S = \mathcal{T}S / \operatorname{mcg} S,$$

where $\operatorname{mcg} S = \operatorname{Diff} S / \operatorname{Diff}_0 S$ is the *mapping class group* or *modular group* of S.

Remark 4.6.6. In the complex analytic setting, surfaces come with a built-in orientation, so the traditional definitions for the concepts of this chapter differ from the ones given here in that the starting point is an *oriented* surface S and the group $\operatorname{Diff}_+ S$ of orientation-preserving diffeomorphisms is used instead of $\operatorname{Diff} S$. The space of stiffenings and Teichmüller space don't change (why?), but the mapping class group and moduli space generally do. We will write $\operatorname{mcg}_+ S$ and $\mathcal{M}_+ S$ for the orientation-preserving mapping class group and moduli space, so we still have

$$\mathcal{M}_+ S = \mathcal{T}S / \operatorname{mcg}_+ S.$$

Example 4.6.7 (the pair of pants). The surface P of Figure 4.14 is called a *pair of pants*. Topologically, it is the sphere minus three open disks. Let's compute its Teichmüller space and moduli space.

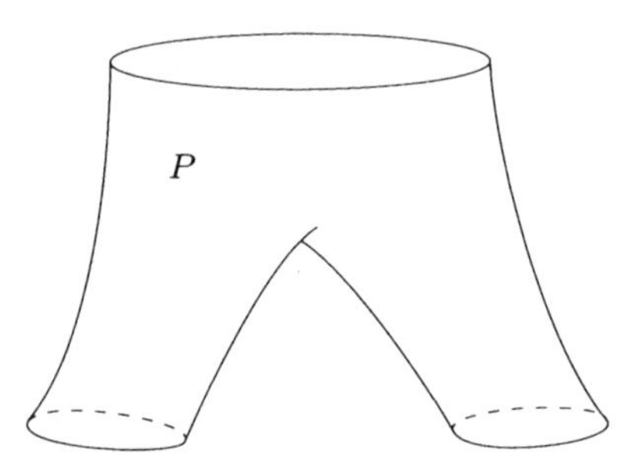

Figure 4.14. A pair of pants.

To each point in $\mathcal{T}P$ we can associate an ordered triple of positive numbers, the lengths of the three boundary components. (Note that we only get an ordered triple for Teichmüller space, not for moduli space, since there are elements of the mapping class group that permute the boundary components. Likewise, for a point in Teichmüller space, but not for one in moduli space, we can talk about the free homotopy class of a loop, and ask, for example, what is the length of its shortest representative.) We will show that the resulting map $\mathcal{T}P \to (0,\infty)^3$ is a bijection.

Indeed, consider some hyperbolic structure on P. Each pair of boundary components is joined by a seam, or geodesic arc orthogonal to the boundary, and these arcs are disjoint and uniquely determined. To see this, draw any three disjoint arcs joining the three pairs of boundary components; up to isotopy, there is no choice in this construction. Take the double $P \cup P'$ of P (Definition 3.3.11): the arcs become closed curves. We will see in Proposition 4.6.20 that these curves can be isotoped to give disjoint simple geodesics, which are uniquely determined. These geodesics must be orthogonal to the boundary, otherwise a symmetry interchanging P and P' would contradict uniqueness. Restrict back to P to get the seams.

The two hexagons arising from cutting P along the seams are congruent, because, by Exercise 2.4.11, a right-angled hyperbolic hexagon is determined up to isometry if one knows the lengths of every other side (see also Figure 4.15 for a more visual proof).

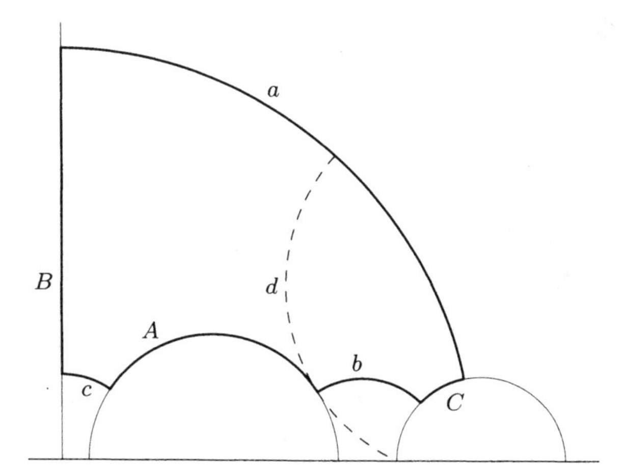

Figure 4.15. There is exactly one right-angled hexagon $aCbAcB$ **with given** $a, b, c > 0$. In this construction (taking place in the upper half-plane), side a and the length of side b are fixed. As C varies, one endpoint of b describes the dashed curve d, a Euclidean arc of circle (compare Figure 1.10). The line supporting A is tangent to d at that point. For C greater than some C_0, the supports of A and B are ultraparallel, and c does exist. As C gets longer, the support of A gets squeezed between d and S^1_∞, so the Euclidean radius of the support of c increases and B gets shorter. By the hexagonal law of sines (Equation 2.4.10), this means that c is a strictly monotonic function of C, going from 0 to ∞ as C goes from C_0 to ∞. Thus every value of c occurs exactly once.

It follows that the seams bisect the boundary components of P, and that the boundary lengths determine the hyperbolic structure on P. Moreover, any triple of positive numbers occurs as the boundary lengths, again by Exercise 2.4.11 or Figure 4.15.

The moduli space $\mathcal{M}P$ is the set of unordered triples of positive reals. This is because $\mathrm{mcg}_+ P$ is the symmetric group S_3, acting by permutations on the boundary components, and $\mathcal{M}P = \mathcal{M}_+P$ since every hyperbolic structure on P admits an orientation-reversing isometry. Note that P is quite exceptional in this way: for most surfaces it is a rare event to have orientation-reversing symmetries, and $\mathcal{M}_+P$ is a double (branched) cover of $\mathcal{M}P$.

Exercise 4.6.8. Describe Teichmüller space and moduli space for the thrice-punctured sphere, that is, the boundaryless counterpart of P. (Hint: Each boundary component of P can be replaced by a cusp or by a hyperbolic "flare". We will see later that the two types of structures are fundamentally different: see Problem 4.6.16.)

Exercise 4.6.9. Describe Teichmüller space and moduli space for the open annulus, the open Möbius strip, and the surface obtained by removing an open disk from a closed half-plane.

Suppose for a moment that S has empty boudary. From the discussion in Section 3.5 we see that moduli space, as the space of complete $(\mathrm{Isom}\,\mathbf{H}^2, \mathbf{H}^2)$-structures on S, can be identified with the space of conjugacy classes of discrete subgroups of $\mathrm{Isom}\,\mathbf{H}^2$ acting on $\mathbf{H}^2$ with quotient S. Teichmüller space has a similar algebraic interpretation, in terms of *faithful representations* (injective homomorphisms) $\pi_1(S) \to \mathrm{Isom}\,\mathbf{H}^2$ whose image is discrete:

Proposition 4.6.10 (holonomy defines structure). *If S has empty boundary, two elements of $\mathcal{S}S$ are equivalent in moduli space if and only if they have the same holonomy group, and they are equivalent in Teichmüller space if and only if they have the same holonomy map.*

Of course, since the holonomy is only defined up to conjugacy, "the same" should be interpreted accordingly.

Proof of 4.6.10. The statement about moduli space is just Corollary 3.5.12. To prove the statement about Teichmüller space, let M and M' be hyperbolic structures on S, and consider locally isometric covering projections $p : \mathbf{H}^2 \to M$ and $p' : \mathbf{H}^2 \to M'$. Lift the identity map $M \to M'$ to a map $\phi : \mathbf{H}^2 \to \mathbf{H}^2$. This map need not be the identity, or even an isometry.

For any $\sigma \in \pi_1(S)$, conjugation by ϕ takes the deck transformation $H(\sigma)$ over M to the deck transformation $H'(\sigma)$ over M', where H and H' are the holonomy maps of M and M'—in other words, ϕ conjugates H to H'. Saying that $H = H'$ is saying that ϕ commutes with the holonomy. In that case the straight-line homotopy joining ϕ to the identity map $\mathbf{H}^2 \to \mathbf{H}^2$ also commutes with the holonomy, and therefore yields a homotopy downstairs. The end stage of the downstairs homotopy is an isometry, since its lift is an isometry of $\mathbf{H}^2$. This shows that M and M' represent the same point in Teichmüller space.

Conversely, given a homotopy downstairs joining the identity map $M \to M'$ to an isometry, we can lift to a homotopy between ϕ and the identity map $\mathbf{H}^2 \to \mathbf{H}^2$, so M and M' have the same holonomy. 4.6.10

A similar statement can be made about $\mathcal{T}S$ when $\partial S \neq \varnothing$: then two elements of $\mathcal{S}S$ are equivalent if and only if the holonomy maps of their doubles are the same. This follows by applying the argument above to the double of S, and observing that the lifts of the boundary components are geodesics that coincide when $H = H'$.

In general, it is difficult to isolate, among the representations of $\pi_1(S)$ in a Lie group, those that are discrete and have quotient S. But here is an elementary case where the algebraic approach works well:

Example 4.6.11 (Teichmüller space of the torus). If S admits Euclidean structures, we define the Euclidean moduli space and Teichmüller space of S just as in Definition 4.6.4, except that $\mathcal{S}S$ is the space of Euclidean stiffenings of S modded out by scale changes. (If S is compact we can normalize so the total area is one.)

The most interesting case is the torus T^2. We must look at discrete representations $\pi_1(T^2) = \mathbf{Z}^2 \to \operatorname{Isom} \mathbf{E}^2$. Actually we already know that the holonomy of a torus consists of translations. By conjugation, we can assume that the image of the generator $(1,0)$ of $\mathbf{Z}^2$ is the translation $(1,0)$. The image of the other generator $(0,1)$ can be any translation (x,y) with $y \neq 0$, but, again by conjugation, we can assume $y > 0$. We conclude that Teichmüller space for the torus is the upper half-plane $y > 0$.

Exercise 4.6.12 (the classical modular group). Show that $\operatorname{mcg} T^2$ is isomorphic to $\mathrm{PGL}(2, \mathbf{Z})$. How does $\operatorname{mcg} T^2$ act on $\mathcal{T}T^2$? What does the quotient look like? (Hint: see Figure 4.16.) The name *modular group* by itself refers specifically to $\mathrm{mcg}_+ T^2 = \mathrm{PSL}(2, \mathbf{Z})$.

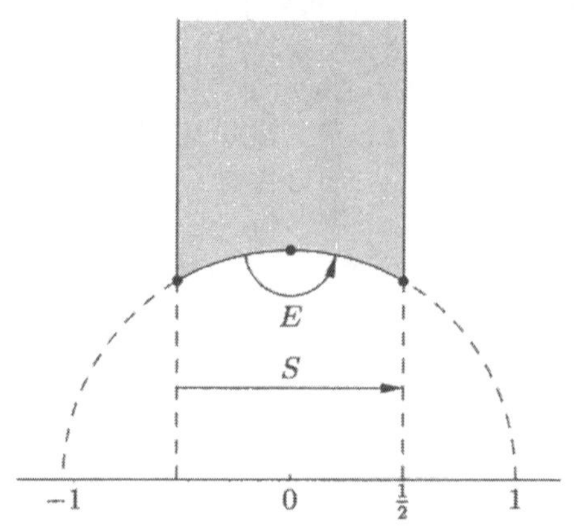

Figure 4.16. The Teichmüller space of the torus and the action of the modular group. The group $\mathrm{mcg}\, T^2 = \mathrm{PGL}(2, \mathbf{Z})$ acts on the Teichmüller space of the torus by hyperbolic isometries. The shaded region is a fundamental domain of the action of $\mathrm{mcg}_+ T^2 = \mathrm{PSL}(2, \mathbf{Z})$, with generators $S = \left(\begin{smallmatrix} 1 & 1 \\ 0 & 1 \end{smallmatrix}\right)$ and $E = \left(\begin{smallmatrix} 0 & 1 \\ -1 & 0 \end{smallmatrix}\right)$, in the notation of Equation 2.6.1.

Exercise 4.6.13. Describe the Euclidean Teichmüller space and moduli space for the closed annulus $S^1 \times [0, 1]$, the open annulus $S^1 \times (0, 1)$, the open and closed Möbius strips, and the Klein bottle.

Exercise 4.6.14. The (spherical) Teichmüller space of a surface admitting spherical structures is a point.

We now give $\mathcal{T}S$ a natural topology, where nearby points represent "almost the same" hyperbolic structure. There are many ways to define the topology, one of the most obvious being through a metric. For $K \geq 1$, call a homeomorphism between two metric spaces a K-*quasi-isometry* if it distorts distances by a factor at most K either way. For a diffeomorphism between Riemannian manifolds, this is the same as saying that tangent vectors stretch or shrink at most K times. Then, if M and M' are stiffenings of S, set

$$d_{\mathrm{qi}}(M, M') = \tfrac{1}{2} \log \inf K,$$

where the infimum is taken over all K such that there is a K-quasi-isometry $M \to M'$ in the isotopy class of the identity. By construction, d_{qi} is defined on pairs of points of $\mathcal{T}S$. Because it is invariant under the action of the mapping class group, d_{qi} is defined on pairs of points of $\mathcal{M}S$ as well.

It is easy to see that d_{qi} satisfies the axioms for a metric, except that $d_{\mathrm{qi}}(M, M')$ can be infinite. When S is compact any diffeomorphism is a quasi-isometry, so this doesn't happen. Otherwise it does happen:

Exercise 4.6.15 (inequivalent cylinders). Let M_a be the hyperbolic cylinder with girth a, obtained, for example, as the quotient of $\mathbf{H}^2$ by the translation expressed as $z \mapsto ze^a$ in the upper half-plane model. If M_b is the cylinder with girth b, compute $d_{\mathrm{qi}}(M_a, M_b)$ by finding explictly the best quasi-isometry between M_a and M_b.

Show that there is no quasi-isometry between M_a and M_0, the quotient of $\mathbf{H}^2$ by a parabolic transformation; therefore $d_{\mathrm{qi}}(M_a, M_0) = \infty$, and the Teichmüller space of the open cylinder is disconnected.

Problem 4.6.16 (inequivalent surfaces). Let M be a connected manifold and cover it with compact sets $K_1 \subset K_2 \subset \cdots$ such that $K_i \subset \mathring{K}_{i+1}$. An *end* of M is a choice, for each i, of a connected component A_i of $M \setminus K_i$, subject to the condition $A_i \subset A_j$ for $i > j$.

(a) Show that this definition is robust with respect to changes in the sequence K_i.

(b) Let S be a surface with no boundary, and let M be a hyperbolic structure on S, with holonomy group Γ. Show how to associate to each end of S an isometry of $\mathbf{H}^2$, defined up to conjugacy by elements of Γ, and a Γ-invariant subset of the circle at infinity. The quotient of this subset by Γ is either a point or a circle. In the first case the end is *parabolic* (a cusp), in the second case *hyperbolic* (a flare). Punctures and flares constitute the *ideal boundary* of M.

(c) Two hyperbolic structures M and M' on S cannot be quasi-isometric if corresponding ends in M and M' are of different types.

(d) (Harder.) If S is *of finite type* ($\pi_1(S)$ is finitely generated), ends of different types are the only obstruction to the existence of a quasi-isometry between M and M'.

Closely related to d_{qi} is the metric originally introduced by Teichmüller [Tei39, Tei43]. The Teichmüller metric is defined just like d_{qi}, but using quasiconformal maps rather than quasi-isometries. A diffeomorphism is K-*quasiconformal* if it changes the lengths of tangent vectors *at the same point* by a *relative* factor at most K; in other words, a small circle on the tangent space to a point in the domain is mapped to an ellipse whose axes have ratio at most K.

Thus the Teichmüller distance records the amount of conformal distortion, while d_{qi} records the metric distortion. Obviously we have $d_{\mathrm{qc}} \leq d_{\mathrm{qi}}$. It is also true—but far from obvious—that $d_{\mathrm{qi}} \leq C d_{\mathrm{qc}}$ for some constant C [DE86]; this shows that the two metrics determine the same topology.

One of Teichmüller's contributions was finding the most economic way to deform a surface: he described the geodesics of Teichmüller space in the Teichmüller metric, and proved their uniqueness (see [Abi80] for a good exposition). Later Royden [Roy71] showed that all isometries of the Teichmüller metric are given by the action of the mapping class group. Ahlfors [Ahl60] proved that Teichmüller space has a natural complex manifold structure, and that through any two points there is a unique complex geodesic (an embedded copy of the hyperbolic plane). Royden showed that the canonical (Kobayashi) metric arising from the complex structure is the same as the Teichmüller metric. All these results were proved first for compact surfaces, but they generalize naturally to arbitrary surfaces (see the next remark).

Remark 4.6.17. Because two hyperbolic (or complex) structures on an open surface S can be infinitely far apart, it is usual in the literature to associate a Teichmüller space not with S, but with a particular hyperbolic (or complex) structure M on S. The Teichmüller space of M then classifies only those structures that are quasi-isometric (or quasiconformally equivalent) to M, namely those that sit in the same connected component of $\mathcal{T}S$ as M.

Moreover, when M has flares, there is another interesting group of diffeomorphisms that can be used instead of $\operatorname{Diff}_0 T$ in the definition of Teichmüller space: namely, diffeomorphisms that are isotopic to the identity and equal the identity on the ideal boundary. The quotient of the space of structures quasi-isometric to M by this smaller group is what complex analysts call the Teichmüller space of M, while the quotient by $\operatorname{Diff}_0 T$ is called *reduced Teichmüller space*. The "unreduced" space is nicer than the reduced one in many ways—in particular, it is to it that the results mentioned above can be extended—but it is infinite-dimensional, unless M is compact or has only cusps, in which case the distinction collapses.

Problem 4.6.18. By Proposition 4.6.10, $\mathcal{T}S$ can be seen as a subspace (modulo conjugacy) of the space of representations of the fundamental group of S (or of its double). The space of representations, in turn, is an algebraic subset of $G^{|A|}$, where $G = \operatorname{Isom} \mathbf{H}^2$ and A is a set of generators of the fundamental group. How does the topology inherited by $\mathcal{T}S$ from $G^{|A|}$ relate to the topology defined by the Teichmüller metric?

In the rest of this section, we investigate the Teichmüller space of *compact* surfaces S, by first chopping S into pairs of pants, then reconstructing $\mathcal{T}S$ from the Teichmüller space of the pairs of pants.

Exercise 4.6.19 (cutting up surfaces). Let S be a compact surface.

(a) S is homeomorphic to a closed surface minus a finite number of disjoint open disks. S is characterized by the Euler number, the number of boundary components, and whether or not it is orientable.

(b) A simple closed curve on S is *essential* if it is neither null-homotopic nor freely homotopic to a boundary component. What surfaces S admit essential curves? Show that, if S is cut along an essential curve, the resulting surface or surfaces are compact and less complex than S, in a suitable sense.

(c) It follows from (b) that, if S has an essential curve, we can repeatedly cut along essential curves until only pairs of pants and annuli are left (Figure 4.17). When can annuli occur? When are they unavoidable? Show that the number of pairs of pants in the decomposition is an invariant.

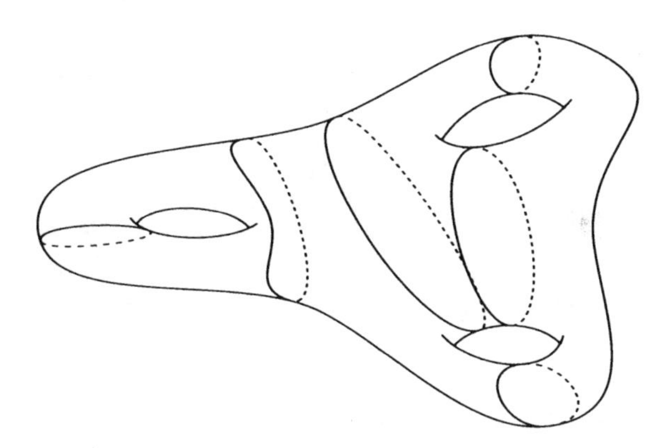

Figure 4.17. A pants decomposition of a surface

This gives a topological decomposition. For a geometric decomposition we need to be able to straighten simple closed curves in S into geodesics. Part of the straightening-up works in any dimension:

Proposition 4.6.20 (straightening closed curves). *Let M be a complete, compact hyperbolic manifold, possibly with geodesic boundary, and let $\alpha : S^1 \to M$ be a closed curve that is not null-homotopic.*

Then the free homotopy class of α is represented by a closed geodesic, unique up to reparametrization, and either disjoint from the boundary or contained in it.

Proof of 4.6.20. First suppose that M has no boundary, and fix a covering projection $\mathbf{H}^n \to M$. Fix a lift $\tilde{\alpha} : \mathbf{R} \to \mathbf{H}^n$ of α, and call T_α the holonomy of α corresponding to this choice of $\tilde{\alpha}$ (see Exercise 3.4.4). By Exercise 4.5.5, T_α is a hyperbolic transformation. Let $\tilde{\gamma} : \mathbf{R} \to \mathbf{H}^n$ be its axis, parametrized at such a speed that the action of T_α corresponds to a unit translation of the domain under $\tilde{\gamma}$, as it does under $\tilde{\alpha}$, by construction (Figure 4.18, left). Then the straight-line homotopy between the two curves is equivariant under the action of the group generated by T_α, and so projects to a free homotopy between α and γ.

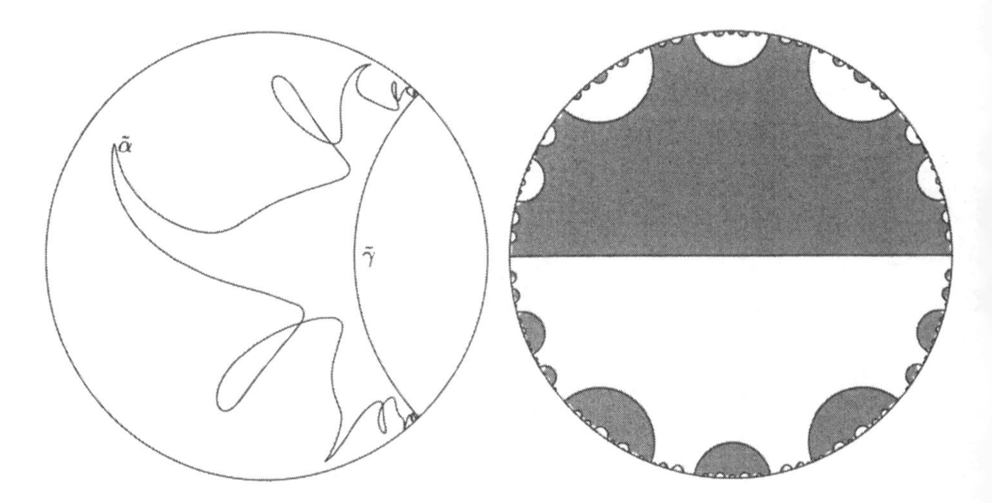

Figure 4.18. Straightening a closed curve. Left: To find a geodesic freely homotopic to a closed curve α in a closed manifold, lift α to the universal cover, take the axis of the holonomy and project back to the manifold. Right: The genus-two surface obtained by gluing a regular hyperbolic octagon (Figure 1.13) is the double of a surface M homeomorphic to $T^2 \setminus D^2$ and having geodesic boundary. Each shaded region represents an embedding of the universal cover of M in $\mathbf{H}^2$.

Conversely, if α is freely homotopic to a geodesic γ, the homotopy between the lifts $\tilde{\alpha}$ and $\tilde{\gamma}$ moves each point a bounded hyperbolic distance, so $\tilde{\alpha}$ converges at each end to the corresponding endpoint of $\tilde{\gamma}$. Therefore $\tilde{\gamma}$ must coincide with the axis of T_α.

This concludes the proof when M is closed. When M has boundary, we first form its double. We look at the universal cover $\tilde{M}$ of M, which can be regarded as a connected component of the inverse image of M in the universal cover $\mathbf{H}^n$ of the double (Figure 4.18, right). Being an intersection of half-spaces, $\tilde{M}$ is convex, so the previous argument still applies: a lift of α contained in $\tilde{M}$ can be straightened to a geodesic also contained in $\tilde{M}$, and this geodesic is either contained in the boundary or disjoint from it. 4.6.20

In the two-dimensional case, we can say more.

Proposition 4.6.21 (straightening preserves embedding). *Let M be a compact surface with a hyperbolic structure, and let $\alpha : S^1 \to M$ be a simple closed curve that is not null-homotopic. Then the geodesic representative of α is also a simple closed curve, possibly described more than once.*

Proof of 4.6.21. A closed curve in M is a multiple of a simple curve if and only if any two of its lifts are disjoint or coincide. Applying this criterion to two lifts $\tilde{\alpha}_1$ and $\tilde{\alpha}_2$ of α, and looking at the relative positions of their endpoints on S^1_∞, we see that the axes of the corresponding transformations are disjoint or coincide. All lifts of the straightening of α are axes coming from lifts of α, so another application of the criterion shows that the straightening is a multiple of a simple geodesic γ. 4.6.21

Exercise 4.6.22 (multiplicity). In the situation of the preceding proposition, show that the multiplicity of α with respect to its straightening γ is at most two, and that it is two if and only if α bounds a Möbius strip. (Hint: Consider the possibilities for the intermediate cover $\mathbf{H}^2/G$, where G is the group generated by the holonomy of γ.)

How else can you conclude that a curve that bounds a Möbius strip must collapse to a double geodesic when straightened up? See also the question about annuli in Exercise 4.6.19(c).

Note that, in all these results (4.6.20–4.6.22), compactness can be replaced by the weaker assumption that the manifold (or its double) has no cusps.

Theorem 4.6.23 (Teichmüller space of compact surfaces). *The Teichmüller space of a compact surface S that admits a hyperbolic structure is homeomorphic to $\mathbf{R}^{3|\chi(S)|}$. In the particular case of S closed and orientable of genus g, we get $\mathcal{T}_g = \mathcal{T}S = \mathbf{R}^{6g-6}$.*

Proof of 4.6.23. We first give an informal description of the correspondence, and later make it more rigorous. If you did Exercise 4.6.19(c), you know that S can be cut into $|\chi(S)|$ pairs of pants by a collection of disjoint simple closed curves. When S is given a hyperbolic structure, these curves correspond to closed geodesics that cut the surface into $|\chi(S)|$ hyperbolic pairs of pants. Characterizing a point of $\mathcal{T}S$ amounts to characterizing these pairs of paints, and how they are to be glued together in the prescribed combinatorial pattern.

The pairs of pants have a total of $3\,|\chi(S)|$ boundary components. Each boundary component can match up with another, if it comes from a two-sided curve, or with itself, if it comes from a one-sided curve. If it matches up with itself, there is one free parameter, the length of the curve. If it matches up with another, there are two free parameters associated with the pair: the common length, and a *twist parameter* giving the relative displacement of the two sides along the common geodesic. The twist parameter is an arbitrary real number, representing a signed distance measured in multiples of the geodesic's length. (That the twist parameter takes values in $\mathbf{R}$, rather than in S^1, tends to be a confusing issue, because twist parameters that are the same modulo 1 result in surfaces that are isometric. But, remember, to determine a point in Teichmüller space we need to consider how many times the leg of the pajama suit is twisted before it fits onto the baby's foot. Hopefully as we fill in the details this point will become clear.) In any case, we are left with $3\,|\chi(S)|$ free parameters taking arbitrary values in $(0,\infty)$ or $(-\infty,\infty)$. These parameters are known as the *Fenchel–Nielsen coordinates* of Teichmüller space.

Now for the formalization. Fix disjoint, oriented, simple closed curves γ_i cutting S into $|\chi(S)|$ pairs of pants, and in each pair of pants fix three disjoint arcs connecting the boundary components pairwise, with the condition that, for each closed curve γ_i, the two endpoints of the arcs in the pair of pants on one side of γ_i match with the endpoints of the arcs in the pair of pants on the other side. Their directions should also match, and be transverse to γ_i. Finally, choose a preferred side for each two-sided γ_i—if S is oriented, it's natural to make the choice based on the orientations of S and γ_i. Different choices for any of the objects above will generally result in different sets of Fenchel–Nielsen coordinates for $\mathcal{T}S$, unless of course the difference can be accounted for by an isotopy of S.

Now let M be S with some hyperbolic structure. On a surface, two simple closed curves that are freely homotopic can be taken to one another by an isotopy. Therefore, by repeated application of Proposition 4.6.21 and Exercise 4.6.22, M can be isotoped so that the γ_i become simple closed geodesics. Each arc is dragged along by the isotopy, but remains a simple arc joining two boundary components of the pair of pants it's in, as in Figure 4.19 (left). Further isotope the interior of each pair of pants so that each arc acquires the standard form shown in Figure 4.19 (right), that is, so that it coincides with the corresponding geodesic seam (Example 4.6.7) except in annular neighborhoods of the two boundary components it joins. In these neighborhoods, make the arc spiral according to some formula, so the picture depends only on the amount of spiraling and the length of the boundary circle.

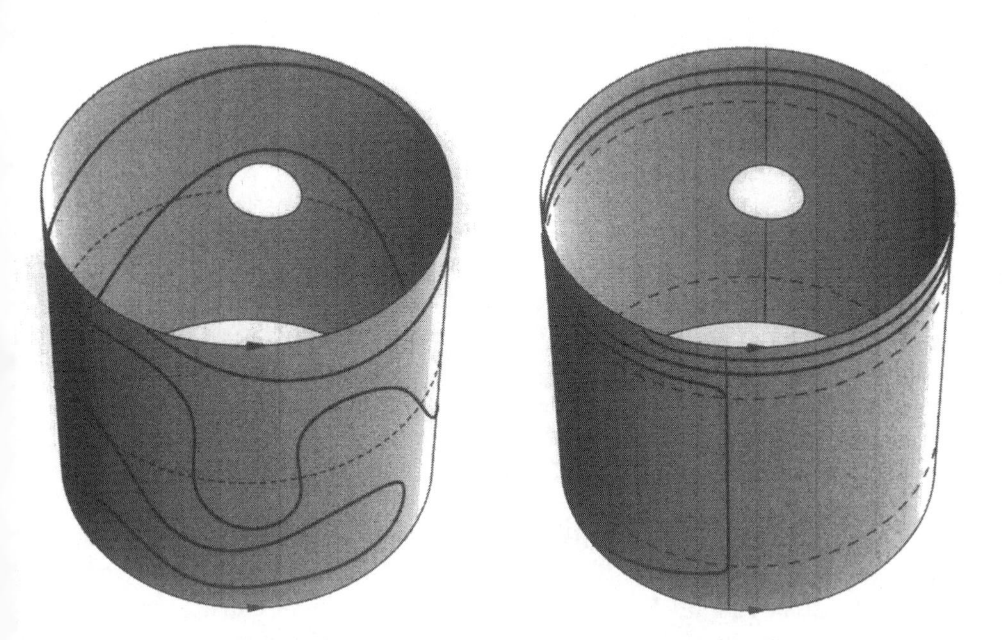

Figure 4.19. Putting an arc in standard form. A simple arc joining fixed points on two boundary components of a hyperbolic pair of pants (left) can be deformed so as to coincide with the corresponding geodesic seam, except for some spiraling at each end (right). The amount of spiraling is well defined: in this picture it is approximately -2.1 turns at the top and $-.15$ at the bottom. (To show it's well defined, you can start by considering the intersection number of the arc with the middle curve shown dotted on the left.)

Taking into account the orientation of the γ_i, this gives for each arc two real numbers, representing the amount of spiraling as one goes from the middle to either end. Now subtract the numbers associated with the endpoints of two arcs that match up at γ_i, and call the difference the *twist parameter* at γ_i. The twist parameter does not depend on which pair of arcs is used in the computation (Figure 4.20), and it equals zero if the curve is one-sided (why?).

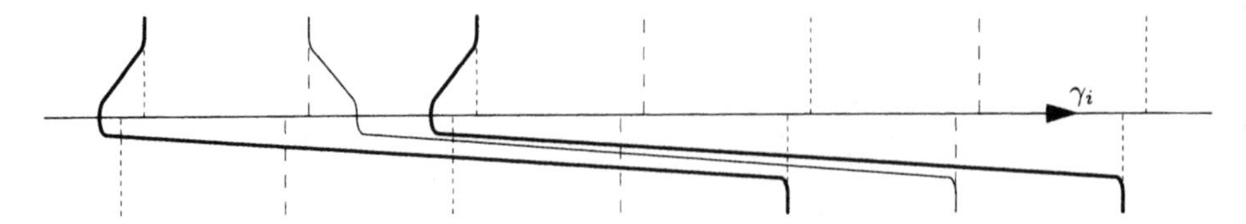

Figure 4.20. Computing the twist parameter. The annular neighborhood of a closed geodesic γ_i is shown here unfolded. The thick curve comes from two matching arcs in standard position within their respective pairs of pants, and the thin curve from the other two. (Compare with Figure 4.19: the drawing here comes from gluing the pair of pants shown there to itself, preserving orientation.) Because the two seams (dashed lines) on each side of γ_i hit γ_i at diametrically opposite points, the thick and thin curves determine the same twist parameter.

From the construction, it's clear that the twist parameters are the same for two hyperbolic structures M and M' related by an isometry isotopic to the identity. Of course, the same is true about the lengths of the γ_i and of the boundary components. The upshot is that we have a map $\mathcal{T}S \to (0,\infty)^{s+b} \times \mathbf{R}^t$, where s is the number of closed curves in the decomposition of S, t is the number of two-sided such curves, and b is the number of boundary components. As already seen, $s+b+t = 3\,|\chi(S)|$. We now must show that this map is bijective.

Surjectivity is easy: given a set of Fenchel–Nielsen coordinates, take pairs of pants with the right measurements, draw on them standard arcs with the right amount of spiraling, then define diffeomorphisms from the topological pairs of pants in the decomposition of S to these geometrical pairs of pants (start with the γ_i's and boundary components, then extend to the arcs, and finally to the hexagons). With a bit of care to make everything match up, this gives a hyperbolic structure on S.

For injectivity, suppose M and M' are hyperbolic structures on S that have the same Fenchel–Nielsen coordinates. We can assume after applying isotopies that the γ_i are geodesic in M and M'. The isometries between corresponding pairs of pants piece together to give an isomety $f : M \to M'$; this uses the fact that the twist parameters are the same modulo 1. We still must show that f is isotopic to the identity. First we adjust f so it is the identity along the γ_i—this is done as in Exercise 1.3.1. Then we put the arcs in standard position, and adjust f so it is the identity on the arcs as well. There remains to adjust f in the interior of the hexagons, which we can do because a diffeomorphism of the closed disk that is the identity on the boundary is isotopic to the identity.

Once you understand how Fenchel–Nielsen coordinates work, the proof that they parametrize $\mathcal{T}S$ homeomorphically is straightforward, and we won't spell it out. 4.6.23

When two hyperbolic structures differ only in the twist parameter along some γ_i, and the amount of the difference is ± 1, we say they're obtained from one another by a Dehn twist. More generally, suppose S is an oriented surface and γ is an essential simple loop on S. A *Dehn twist* about γ is the mapping class corresponding to the operation of cutting S along γ and gluing back together after a full twist of the two sides, relative to one another. The twist is *positive* if it can be made to look like Figure 4.21 in a collar neighborhood of γ, while being the identity outside that neighborhood. We denote by τ_γ a positive Dehn twist about γ. Sometimes any power τ_γ^n is also called a Dehn twist.

Exercise 4.6.24. Let M be a closed oriented hyperbolic surface, and γ a simple closed geodesic.

(a) Sketch the effect of the Dehn twist τ_γ on the universal cover of M. (Assume τ_γ is the identity outside a collar neighborhood of γ, and lift τ_γ so that it is still the identity on some open subset of $\mathbf{H}^2$. You can use the right side of Figure 4.18 as a model.)

(b) Show that a countable set of points on S^1_∞ is kept pointwise fixed, but that other points on S^1_∞ are moved. Show that this implies that τ_γ is not homotopic to an isometry, and therefore that a Dehn twist acts on Teichmüller space without fixed points.

(c) Generalize to a composition of multiple Dehn twists about several simple closed geodesics. This confirms that sets of Fenchel–Nielsen coordinates whose twist parameters differ by integers do map to different elements in Teichmüller space (unless all the integers are zero).

For a closed orientable surface, the orientation-preserving mapping class group is generated by Dehn twists about a finite number of

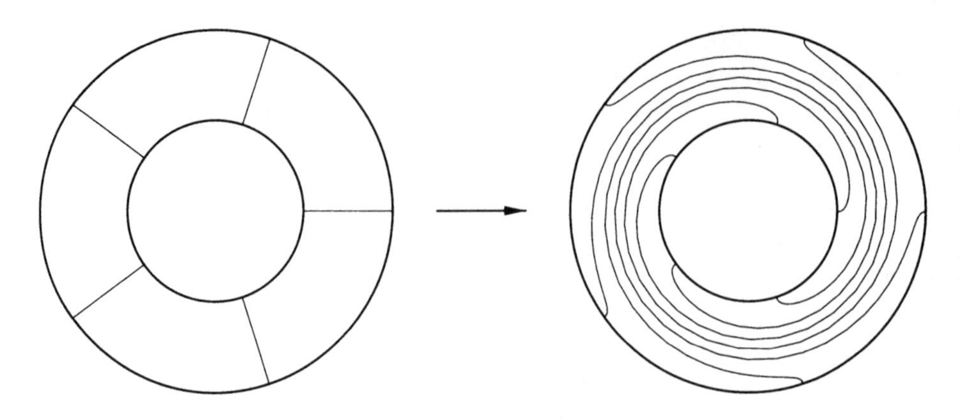

Figure 4.21. A model Dehn twist. A positive Dehn twist is one that is conjugate to the map $A \to A$ shown here, where the annulus A has the standard orientation induced from $\mathbf{R}^2$. It makes no difference how the conjugating map embeds A in the surface, so long at it preserves orientation; this is the same phenomenon as a right-handed bolt remaining so when turned end-for-end.

curves. This fact, which we will revisit in a later chapter, was first proved by Dehn, in a long paper [Deh38] that remained very little known. It was rediscovered independently in [Lic62]. Dehn also appears to be the author of the following result, first published by Nielsen (who nonetheless attributes it to Dehn):

Theorem 4.6.25 (Dehn–Nielsen). *The mapping class group of a closed surface S is isomorphic to the outer automorphism group of $\pi_1(S)$.*

(An "outer automorphism" of a group G is a conjugacy class of automorphisms of G; note that the action of a diffeomorphism $S \to S$ on $\pi_1(S)$ is defined only up to conjugacy, because there is no basepoint.) For a clear proof and historical details, see [Sti87].

Problem 4.6.26 (markings). The name "marked surface" comes from the following setup. Let M be an orientable, closed, hyperbolic surface of genus g, and mark M with $2g$ disjoint loops $\{a_i, b_i\}$ based at the same point, and cuting M into a $4g$-gon in some standard combinatorial way (for instance, such that

$$a_1^{-1} b_1^{-1} a_1 b_1 \dots a_g^{-1} b_g^{-1} a_g b_g$$

is null-homotopic). Call two marked surfaces $(M, \{a_i, b_i\})$ and $(M', \{a'_i, b'_i\})$ equivalent if there is an isometry $f : M \to M'$ such that the system of curves $\{f(a_i), f(b_i)\}$ can be homotoped to $\{a'_i, b'_i\}$. Define a map Φ

from Teichmüller space into the set of marked surfaces $(M, \{a_i, b_i\})$ modulo equivalence. Use Theorem 4.6.25 (among other things) to show that Φ is a bijection.

Remark 4.6.27. The literature on Teichmüller space is huge, but much of it assumes familiarity with some high-power complex analytic tools. As a book that requires little background and does not neglect the geometric point of view, you might try [IT92]. Other standard textbooks are [Gar87, Leh87, Nag88]; see also the survey [Ber81].

4.7. Three-Manifolds Modeled on Fibered Geometries

In Sections 4.3 and 4.4, we have seen that Euclidean and elliptic three-manifolds have a close relation to discrete groups acting in one and two dimensions. It is not surprising that there are similar descriptions for manifolds modeled on any of the five anisotropic geometries of Theorem 3.8.4(b–c), since each of these geometries fiber over a lower-dimensional geometry in a natural way.

In this section we will analyze the properties of manifolds modeled on these five geometries. Our proofs will be in many cases sketchier than those of preceding sections. As a result of this analysis, it will turn out that discrete cocompact (or cofinite) groups of any one of the eight model geometries can be distinguished from corresponding groups of any other geometry by means of purely algebraic properties (Theorem 4.7.8). In particular, a compact three-manifold can be given at most one among the eight types of three-dimensional geometries.

Exercise 4.7.1 (classification of $S^2 \times \mathbf{E}^1$-manifolds). If ϕ is an isometry of S^2, the mapping torus M_ϕ of ϕ (Example 3.6.5) is an $S^2 \times \mathbf{E}^1$-manifold. In fact, it is the quotient of $S^2 \times \mathbf{E}^1$ by the discrete group generated by the transformation $(v, t) \mapsto (\phi v, t + 1)$, where $v \in S^2$. The manifold is diffeomorphic to $S^2 \times S^1$ when ϕ preserves orientation, and to a nonorientable manifold otherwise. What other manifolds admit $S^2 \times \mathbf{E}^1$-structures?

(a) Any discrete subgroup of isometries of $S^2 \times \mathbf{E}^1$ acts discretely (but not necessarily freely or effectively) on $\mathbf{E}^1$.

(b) A discrete group of isometries of $\mathbf{E}^1$ is isomorphic to $\mathbf{Z}$ or to the free product $C_2 * C_2$, where C_2 is the cyclic group of order 2.

(c) There are only three closed three-manifolds, up to diffeomorphism, that admit $S^2 \times \mathbf{E}^1$-structures. Two are orientable and one is not.

We will now analyze the remaining four cases: $\mathbf{H}^2 \times \mathbf{E}^1$, $\widetilde{\mathrm{PSL}}(2,\mathbf{R})$, nilgeometry and solvegeometry. As in Theorem 3.8.4, X will denote the model space, G the group of all isometries of this space, and G' the connected component of the identity in G.

It will be convenient to say that a group is *almost* P, where P is some adjective, if it contains a subgroup of finite index that is P. For instance, discrete groups of Euclidean motions are almost abelian (Corollary 4.1.13).

First we study $\mathbf{H}^2 \times \mathbf{E}^1$ and $\widetilde{\mathrm{PSL}}(2,\mathbf{R})$, the two geometries that fiber over $\mathbf{H}^2$. Let $p : G \to \operatorname{Isom} H^2$ be projection onto the group of isometries of the base. Any element of G' that projects to the identity is central in G, so the commutator of two elements of G' depends only on their images under p.

Proposition 4.7.2 (image in $\operatorname{Isom}\mathbf{H}^2$ discrete or almost abelian). *If Γ is a discrete group of isometries of $\mathbf{H}^2 \times \mathbf{E}^1$ or $\widetilde{\mathrm{PSL}}(2,\mathbf{R})$, its image $p(\Gamma)$ in $\operatorname{Isom}\mathbf{H}^2$ either is discrete or contains an abelian subgroup with finite index.*

Proof of 4.7.2. Since either criterion is preserved when we pass to a subgroup of finite index, we may as well assume $\Gamma \subset G'$.

Choose a neighborhood U of the identity in G_0 such that $[U,U] \subset U$ and $U \cap \Gamma = \{1\}$. Two elements of Γ that project to $p(U)$ commute, because their commutator lies in Γ and in U (move the elements along their fibers until they are inside U, and apply the observation made just before the proposition). Therefore the group H generated by $p(\Gamma) \cap p(U)$ is abelian.

Similarly, given $\gamma \in \Gamma$, choose a neighborhood U_γ of 1 in G_0 satisfying $[\gamma, U_\gamma] \subset U$; then γ commutes with any element of Γ that projects to $p(U_\gamma)$.

If $p(\Gamma)$ is not discrete, the closure $\bar{H}$ of H is a closed abelian subgroup of $\operatorname{Isom}\mathbf{H}^2$ with no isolated points, so it's a one-parameter group. For $\gamma \in \Gamma$, the projection $p(\gamma)$ commutes with elements in $p(\Gamma) \cap p(U_\gamma)$, as we have seen. The group generated by such elements is dense in $\bar{H}$, so $p(\gamma)$ centralizes $\bar{H}$. But the centralizer of a one-parameter subgroup of $\operatorname{Isom}\mathbf{H}^2$ is abelian, so $p(\Gamma)$ is abelian. $\boxed{4.7.2}$

Corollary 4.7.3 (discrete groups of isometries of $\mathbf{H}^2 \times \mathbf{E}^1$ and $\widetilde{\mathrm{PSL}}(2,\mathbf{R})$). *A discrete group Γ of isometries of $X = \mathbf{H}^2 \times \mathbf{E}^1$ or $X = \widetilde{\mathrm{PSL}}(2,\mathbf{R})$ is cofinite if and only if its action on $\mathbf{H}^2$ is discrete, cofinite, and has infinite kernel.*

Proof of 4.7.3. If the action on $\mathbf{H}^2$ is discrete—that is, if $p(\Gamma)$ is a discrete subgroup of $\operatorname{Isom}\mathbf{H}^2$—we have

$$\operatorname{vol}(X/\Gamma) = \operatorname{area}(\mathbf{H}^2/p(\Gamma))\,\operatorname{length}(\mathbf{E}^1/(\Gamma \cap \operatorname{Ker} p)),$$

so in this case Γ cofinite is equivalent to $p(\Gamma)$ cofinite and $\Gamma \cap \operatorname{Ker} p$ infinite.

On the other hand, if $p(\Gamma)$ is not discrete, the proof of Proposition 4.7.2 shows that it preserves a line, point or horocycle in $\mathbf{H}^2$, so Γ preserves and acts discretely on the preimage of the line, point or horocycle in X. It is easy to see in this case that X/Γ has infinite volume. 4.7.3

Exercise 4.7.4. Justify the equation in the preceding proof. How wide is its validity?

Proposition 4.7.5 (discrete groups of isometries of nilgeometry). *A discrete group* Γ *of isometries of nilgeometry projects to a group of isometries of* $\mathbf{E}^2$ *which is either discrete, or preserves some line or some point in* $\mathbf{E}^2$.

Proof of 4.7.5. Let $p : G \to \operatorname{Isom}(\mathbf{E}^2)$ be the projection of G to its action on $\mathbf{E}^2$. Let us first suppose that Γ is contained in the identity component G_0 of G, the group of isometries of nilgeometry. Then the commutator subgroup of Γ acts by translations on $\mathbf{E}^2$, since the commutator of any two orientable isometries of $\mathbf{E}^2$ is a translation. If γ_1 and γ_2 are elements of $[\Gamma, \Gamma]$, then $[\gamma_1, \gamma_2]$ is a vertical translation t of nilgeometry by a distance equal to the area of the parallelogram spanned by $p(\gamma_1)(0) - 0$ and $p(\gamma_2)(0) - 0$ in $\mathbf{E}^2$.

Suppose that there is some such parallelogram with nonzero area. Let T be the group of all vertical translations in Γ. T is central in G_0, and the group G_0/T acts on $\mathbf{E}^2$ with compact stabilizers. Therefore, by Proposition 3.5.8 (proper map preserves proper discontinuity), Γ/T acts properly discontinuously on $\mathbf{E}^2$.

If on the other hand all translations of $\mathbf{E}^2$ in $p([\Gamma, \Gamma])$ are linearly dependent, then either $p([\Gamma, \Gamma])$ acts trivially on $\mathbf{E}^2$—in which case either some element of $p(\Gamma)$ has nontrivial rotation and thus $p(\Gamma)$ fixes a point or $p(\Gamma)$ consists entirely of pure translations and is thus discrete—or $p([\Gamma, \Gamma])$ acts by translations along some line. Then the rotational parts of the action of $p(\Gamma)$ can only be the identity or $180°$ rotations, and one sees that $p(\Gamma)$ must preserve some line in $\mathbf{E}^2$.

The general case, when $\Gamma \not\subset G_0$, follows readily. 4.7.5

Corollary 4.7.6 (cofinite implies cocompact in nilgeometry). *A cofinite group* Γ *of isometries of nilgeometry is cocompact, and its image in* $\mathrm{Isom}(\mathbf{E}^2)$ *is discrete.*

Finally, consider three-dimensional solve-geometry. Denote the space by X, its group by G and the identity component of G by G_0. Recall the structure of G_0 as an extension $\mathbf{R}^2 \to G_0 \to \mathbf{R}$.

Proposition 4.7.7 (discrete groups of isometries of solvegeometry). *If* $\Gamma_0 \subset G_0$ *is any discrete subgroup, then* $\Gamma_0/(\Gamma_0 \cap \mathbf{R}^2)$ *acts discretely on* $\mathbf{R}$. *Any discrete subgroup* Γ *of* G *acts as a discrete group on the line* $\mathbf{R}$, *and the kernel of the action on the line is a discrete group of isometries of the Euclidean plane.*

Proof of 4.7.7. Any element γ in Γ_0 which is not in $\mathbf{R}^2$ acts as an affine map in $\mathbf{R}^3$ which preserves and translates along some unique vertical line (since γ acts on $\mathbf{R}^2$ without any eigenvalue of 1).

If Γ_0 is abelian, it is either contained in $\mathbf{R}^2$, or all of Γ_0 must act as translations on a single vertical line. The proposition follows in either case.

If Γ_0 is not abelian, its commutator subgroup is contained in $\mathbf{R}^2$. If $t \in \Gamma_0 \cap \mathbf{R}^2$ and $\gamma \in \Gamma_0 - \mathbf{R}^2$, then t and $\gamma t \gamma^{-1}$ are linearly independent translations, so $\Gamma_0 \cap \mathbf{R}^2$ is cocompact. The action of Γ_0 on $\mathbf{R}$ factors through the action on $X/(\Gamma_0 \cap \mathbf{R}^2)$, so by 3.5.8 the image of Γ_0 in $\mathbf{R}$ is discrete.

The statements concerning Γ follow readily. 4.7.7

We now have enough information to be able to find algebraic properties of cocompact groups in any of the eight geometries which identify the geometry.

Theorem 4.7.8 (distinguishing geometries of cocompact groups). *A discrete cocompact group* Γ *of automorphisms of any one of the eight basic three-dimensional geometries is not isomorphic to a cocompact group in any of the others.*

The flowchart in Figure 4.22 reconstructs the geometry.

Proof of 4.7.8. We already have most of the information, so we only need to consider a few extra points. From Exercise 4.7.1, Corollary 4.1.13, Corollary 4.7.3, and Corollary 4.7.6, it is clear that a group always takes a yes branch when the flowchart says it should. Proposition 4.7.7 proves the stronger statement that a discrete cocompact group of isometries of solvegeometry is almost polycyclic.

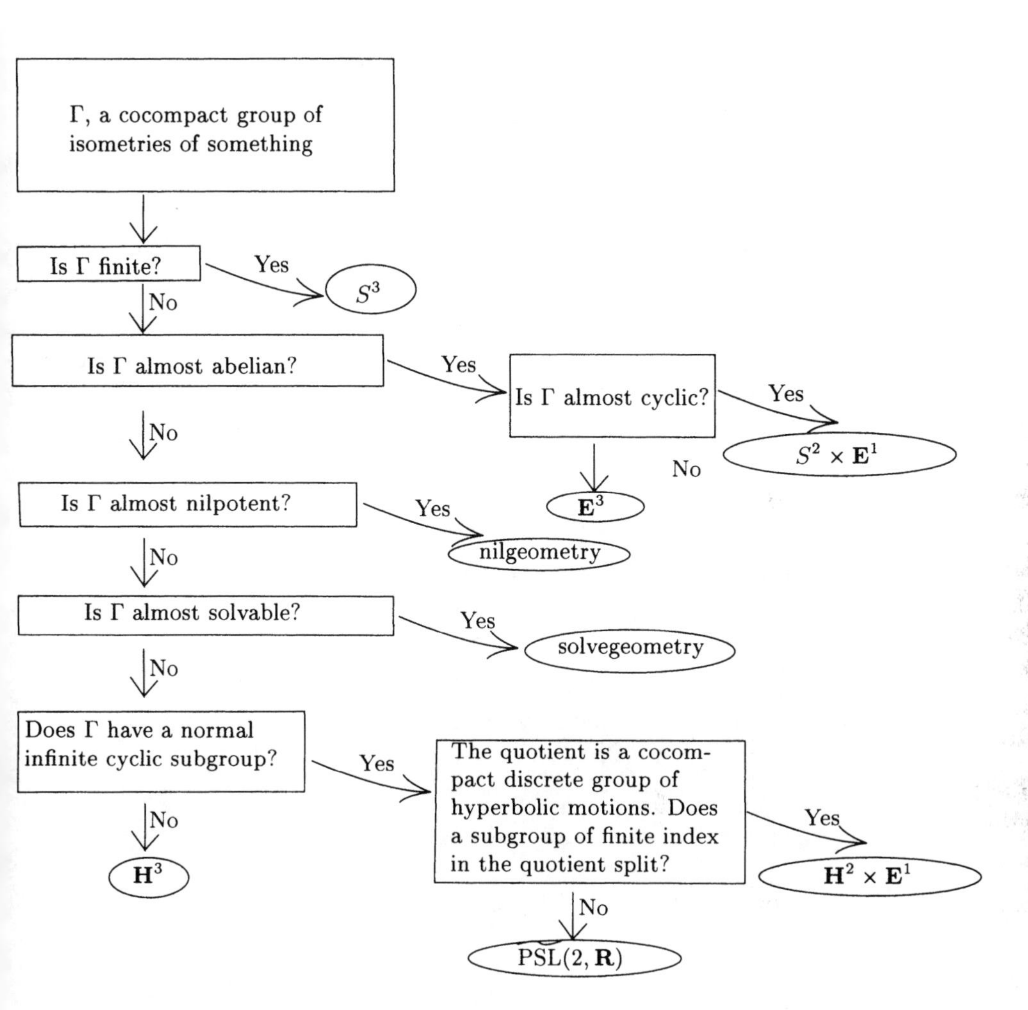

Figure 4.22. Flowchart for cocompact groups. If Γ is a discrete cocompact group of automorphisms of one of the eight geometries, it cannot be a discrete cocompact group of automorphisms of any of the other of the eight geometries. This flowchart shows how to tell which of the eight geometries it acts on.

Since Γ is cocompact, an infinite cyclic normal subgroup in the case of $\widetilde{\mathrm{PSL}}(2,\mathbf{R})$ or $\mathbf{H}^2 \times \mathbf{E}^1$ is $\Gamma \cap Z_0$, where Z_0 is the center of the identity component of the group. The only problem is to make sure that no group takes a yes branch when the flowchart says it shouldn't.

A nontrivial solvable group has a nontrivial normal abelian subgroup. Since cocompact groups of isometries of $\mathbf{H}^3$ have no nontrivial normal abelian subgroups, they cannot take any yes branch. If Γ is a discrete cocompact group of isometries of $\mathbf{H}^2 \times \mathbf{E}^1$ or $\widetilde{\mathrm{PSL}}(2,\mathbf{R})$, then by Corollary 4.7.3 the image in $\mathrm{Isom}(\mathbf{H}^2)$ is discrete and cocompact. The image cannot be almost solvable by the above argument, so Γ cannot be almost solvable. A cocompact subgroup of $\mathrm{PSL}(2,\mathbf{R})$ does not lift to $\widetilde{\mathrm{PSL}}(2,\mathbf{R})$ (Exercise 4.7.9), so if Γ is a cocompact group of isometries of $\widetilde{\mathrm{PSL}}(2,\mathbf{R})$ it cannot take the final yes branch. It is elementary to check the other cases. 4.7.8

Exercise 4.7.9 (cocompact subgroups of PSL$(2,\mathbf{R})$ do not lift). A cocompact subgroup of $\mathrm{PSL}(2,\mathbf{R})$ does not lift to $\widetilde{\mathrm{PSL}}(2,\mathbf{R})$.

The noncocompact but cofinite case is somewhat different.

Theorem 4.7.10 (distinguishing geometries of cofinite groups). *Non-cocompact cofinite groups of automorphisms exist only for $\mathbf{H}^3$, $\mathbf{H}^2 \times \mathbf{E}^1$ and $\widetilde{\mathrm{PSL}}(2,\mathbf{R})$ (out of the eight basic geometries). Any such group of automorphisms of $\mathbf{H}^2 \times \mathbf{E}^1$ acts also as a cofinite group of automorphisms of $\widetilde{\mathrm{PSL}}(2,\mathbf{R})$, and vice versa. Otherwise, groups are distinguished by this flowchart*:

The separation of the case of $\mathbf{H}^3$ from the other two cases is exactly the same as for cocompact groups. We will not prove the rest of 4.7.10.

Discrete cocompact groups in nilgeometry and solvegeometry can be completely characterized in a way closely related to the characterization of discrete cocompact Euclidean crystallograph groups given by the Bieberbach theorems. We'll begin with nilgeometry.

There is a somewhat more symmetric way to map nilgeometry into $\mathbf{R}^3$ than the one we've been using so far. In section Section 3.7, we defined the standard contact structure in $\mathbf{R}^3$ to be the plane field τ spanned by the vectors $(1,0,0)$ and $(0,1,x)$ at the point $(x,y,z) \in \mathbf{R}^3$, and in Section 3.8, we defined nilgeometry to be $\mathbf{R}^3$ with the group of automorphisms consisting of contact automorphisms that preserve the fibration over the (x,y)-plane by lines parallel to the z-axis and project to isometries of the (x,y)-plane.

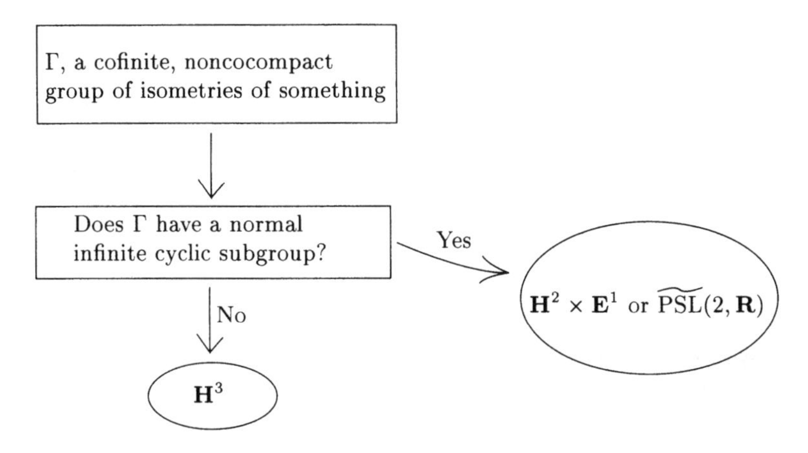

Figure 4.23. Flowchart for cofinite groups. Non-cocompact cofinite groups of autmorphisms exist for only three of the eight geometries. This flowchart shows how to distinguish them.

Instead of τ, let's use the plane field σ spanned by the vectors $(1, 0, -\frac{1}{2}y)$ and $(0, 1, \frac{1}{2}x)$. The plane field σ still defines a connection of curvature 1, and it defines a contact structure isomorphic to that defined by τ: in fact, the diffeomorphism

$$(x, y, z) \mapsto (x, y, z - \tfrac{1}{2}x)$$

sends τ to σ. Any line that meets the z-axis perpendicularly is a Legendrian curve for σ, and this condition, together with the condition that the curvature of σ is 1, determines σ.

One advantage of σ is that it is rotationally symmetric about the z-axis. Furthermore, if we take any area-preserving linear map of the plane and lift it to $\mathbf{R}^3$ so as to fix the z-axis and preserve σ, the result is a linear map. A linear transformation of the plane that expands area by a factor of c lifts to a linear transformation of $\mathbf{R}^3$ that expands the z-axis by a factor of c^2. Also, translations of the plane lift to affine maps of $\mathbf{R}^3$ that preserve σ.

Proposition 4.7.11 (affine nilgeometry). *Nilgeometry has an affine structure such that all nilgeometry automorphisms are automorphisms of the affine structure. If H denotes the Heisenberg group (contained in the group of automorphisms of nilgeometry), then all continuous automorphisms of H are automorphisms of the affine structure.*

Proof of 4.7.11. The only thing still to check is that automorphisms of the Heisenberg group induce automorphisms of the affine structure.

Any automorphism ϕ of H induces an automorphism of the quotient $\mathbf{R}^2$ of H by its center Z. We have seen that such a linear map lifts to an affine automorphism ϕ' preserving σ. Since the Heisenberg group consists of lifts of translations of the plane that preserve σ, ϕ' is necessarily an automorphism of H.

Consider the function $f = \phi'^{-1} \cdot \phi : H \to Z$. This function is actually a homomorphism of H to Z (which is essentially, a linear map from $\mathbf{R}^2 \to \mathbf{R}$). Thus, the automorphism ϕ can be written $\phi' \cdot f$, so in particular, it is an affine transformation. $\boxed{4.7.11}$

Theorem 4.7.12 (cocompact groups in nilgeometry). *A group Γ is a cocompact discrete group of automorphisms of nilgeometry if and only if Γ contains a subgroup of finite index isomorphic to the integral Heisenberg group*

$$H = \langle a, b \mid [a,[a,b]] = [b,[a,b]] = 1 \rangle\,,$$

and such that the centralizer of H is infinite cyclic.

Proof of 4.7.12. If Γ is a discrete cocompact group of isometries of nilgeometry, then Γ acts on the plane as a discrete cocompact group of isometries. Let a and b be two elements of Γ which act on the plane as linearly independent translations. The subgroup they generate is isomorphic to H. Since the commutator $[a,b]$ is a nonzero vertical translation, $\mathbf{E}^3/H$ must be compact, so H is of finite index.

Conversely, suppose Γ is a group which contains H with finite index. Let H_0 be a subgroup with finite index in H which is normal in Γ (take for example the intersection of the conjugates of H).

Then H_0 is a discrete cocompact group of isometries of nilgeometry. In order for the action of H_0 on nilgeometry to extend to Γ, it is necessary that the automorphisms of H_0 induced by conjugacy by elements $\gamma \in \Gamma$ are induced by isometries of nilgeometry.

To arrange this, first we observe that any automorphism of H_0 extends uniquely to a continuous, affine automorophism of H, as in factors through $\mathbf{Z}^2$ (H_0 modulo its center Z_0). Conjugacy by an element $\gamma \in \Gamma$ induces an automorphism of $\mathbf{Z}^2$, which comes from a linear transformation of the plane. Thus, Γ/H_0 acts as a group of linear transformations of the plane. We can choose a metric on $\mathbf{E}^2$ invariant by this action, as in the proof of Theorem 4.2.6.

The action on the plane does not determine the action in space. in fact, for any linear map $f : \mathbf{R}^2 \to \mathbf{R}$, the linear map L_f of $\mathbf{R}^3$ into itself that sends (X, z) to $(X, z + f(X))$ (where $X \in \mathbf{R}^2$), induces an automorphism of the Heisenberg group. Two actions of H_0 as isometries of nilgeometries which give rise to the same action in the plane differ by some map L_f; note, in particular, that the action of $[H_0, H_0]$ is always the same, being determined by the areas of parallelograms in the plane. A finite group of automorphisms of H_0 which respects the action of H_0 on $\mathbf{E}^2$ acts by affine transformations on the set of choices of actions of H_0 in nilgeometry. A finite group of affine transformations has a fixed point—the center of mass of any orbit—so the action of H_0 can be chosen so the finite group comes from isometries of nilgeometry.

Finally, to realize Γ we consider the induced representation of Γ as a group of affine transformations of $N^{(\Gamma/H_0)}$. This space is constructed by considering

$$(\Gamma/H_0) \times N = (\Gamma \times N)/(gh, x) \sim (g, p(h)x)$$

where N is nilgeometry and p is the representation of H_0 as a group of isometries of nilgeometry, and taking the cartesian product of the copies of N. There is a foliation of $N^{(\Gamma/H_0)}$ by copies of N, parallel to a kind of diagonal embedding $x \to ((g_1, x), \ldots, (g_n, x))$ where $i = 1, \ldots, n$ is an index for the cosets of H_0 in Γ and each g_i is a choice of an automorphism of nilgeometry which realizes the automorphism of H_0 defined by the i-th coset. This foliation is invariant by the action of Γ, and each leaf is invariant by H_0. As in the proof of Proposition 4.2.5, there must be some leaf invariant by Γ. The action of Γ on this leaf gives the desired action on N.

The condition for the action of Γ to be effective is easily checked. An element γ cannot possibly act trivially on N unless it centralizes H (or H_0). The group which centralizes H_0 in the full group of isometries of N is $\mathbf{R}$ acting as pure vertical translations), so the centralizer of H acts effectively if and only if it is infinite cyclic.

4.7.12

Theorem 4.7.13 (cocompact groups in solvegeometry). *A group Γ is a cocompact discrete group of automorphisms of solvegeometry if and only if Γ contains a subgroup H of finite index whose centralizer is trivial and that is an extension of the form*

$$\mathbf{Z}^2 \to H \to \mathbf{Z}$$

where the action of $\mathbf{Z}$ *on* $\mathbf{Z}^2$ *is generated by a hyperbolic element of* $\mathrm{SL}(2, \mathbf{Z})$.

Proof of 4.7.13. If Γ is a discrete cocompact group of automorphisms of solvegeometry, then $\Gamma \cap G_0$ has the form indicated for H, where G_0 is the identity component of G (using 4.7.7).

If Γ is a group having a subgroup H of finite index of the form described in 4.7.13, then we can construct a discrete cocompact action of H on X (by Exercise 3.8.10). We may as well assume that H is normal in Γ. The next claim is that every possible automorphism of H is induced by conjugation by some element of G which normalizes the action of H on X. Indeed, any automorphism must preserve the $\mathbf{Z}^2$ subgroup, and it must preserve or interchange the two eigenspaces in $\mathbf{R}^2 = \mathbf{Z}^2 \otimes \mathbf{R}$. The automorphism restricted to $\mathbf{Z}^2$ can therefore be realized by an element of G of finite order, followed, if necessary, by some translation in the (vertical) t-direction to adjust the relative scaling factors of the eigenspaces. The full group H has one remaining generator, which is a translation along some line. After the automorphism, it becomes a translation along a new line, in the same or opposite sense depending on whether the eigenspaces in $\mathbf{R}^2$ are preserved or interchanged. A horizontal translation now adjusts this final generator, without changing the action of the others.

Note that two different elements of G can never induce the same automorphism of H. Therefore the construction above gives a group homomorphism $Aut(H) \to G$, which extends the given embedding of H in G. The composition $\Gamma \to Aut(H) \to G$ completes the proof of 4.7.13. $\boxed{4.7.13}$

Challenge 4.7.14 (nonfibered groups of fibered geometries). Show that there are interesting discrete groups of isometries of $\mathbf{H}^2 \times \mathbf{H}^2$ which project to nondiscrete groups on both factors. Do Challenge 4.3.16 first.

(a) Now consider the group G with presentation

$$\langle \alpha, \beta, \gamma \mid \alpha^2 = \beta^2 = \gamma^2 = (\alpha\beta)^7 = (\beta\gamma)^7 = (\gamma\alpha)^7 = 1 \rangle .$$

Consider three triangles: two in the hyperbolic plane with angles $\pi/7, 2\pi/7$ and $2\pi/7$, and $3\pi/7, \pi/7$ and $\pi/7$, the third on S^2 with angles $2\pi/7, 3\pi/7$ and $3\pi/7$. This gives three homomorphisms of G into isometries of some geometry. Show that the diagonal homomorphism, into $\mathrm{Isom}(\mathbf{H}^2 \times \mathbf{H}^2 \times S^2)$, has discrete image, and consequently the projection to $\mathrm{Isom}(\mathbf{H}^2 \times \mathbf{H}^2)$ is also discrete.

[Hint: write down matrices for the quadratic form descriptions of $\mathbf{H}^2$ and S^2 based on these triangles, as in Section 2.4. Note that all coefficients are algebraic integers. Describe a lattice in $\mathbf{R}^9$ invariant by the diagonal action of G, and conclude that G is discrete.

(b) Prove that the image in $\mathrm{Isom}(\mathbf{H}^2)$ is indiscrete by showing that the two images are isomorphic, and that there is a subgroup which is elliptic in one image but hyperbolic in the other.

(c) Show that these homomorphisms of G are not faithful.

Glossary

action [Ber87, vol. I, p. 5]. If G is a group and X is a topological space, an *action* of G on X (or G-action) is a homomorphism from G into the group of homeomorphisms of X; we denote the image of $x \in X$ under the homeomorphism associated to $g \in G$ by $g(x)$. A G-action defines an equivalence relation on X, whereby $x, y \in X$ are equivalent if and only if $y = g(x)$ for some $g \in G$. The equivalence classes of this equivalence are the *orbits* of G. The space of orbits, with the quotient topology, is the *quotient* of X by the G-action.

ambient. When we talk about two spaces (or manifolds, etc.), one contained in the other, the containing space is sometimes called the *ambient space*. A homeomorphism (or diffeomorphism, etc.), of the containing space is then called an *ambient homeomorphism*.

barycentric subdivision. The barycentric subdivision of a triangle is obtained by joining the triangle's barycenter—the average of its three vertices—to the vertices and to the midpoints of the sides. The subdivision of a triangulation is what you get by taking the barycentric subdivision of each triangle in the triangulation.

basepoint. See †fundamental group.

base space. See †fiber bundle.

bundle. See †fiber bundle.

C^r, C^∞. See †differentiable.

compact-open topology. In the *compact-open topology* on a set $\mathfrak{F}$ of continuous mappings $X \to Y$, a neighborhood basis of an element $f \in \mathfrak{F}$ is given by finite intersections of sets of the form $\{f' \in \mathfrak{F} : f'K \subset U\}$, for all $K \subset X$ compact and $U \subset Y$ open such that $fK \subset U$. If X and Y are Hausdorff, this is the coarsest topology on $\mathfrak{F}$ for which the evaluation map $(f, x) \mapsto f(x)$ is continuous.

cone on a torus. The *cone* on a topological space X is the product $X \times [0, 1]$, modulo the equivalence relation that identifies all points in $X \times \{0\}$.

contractible. A topological space X is *contractible* if the identity map on X is homotopic to a constant map $X \mapsto x_0 \in X$.

covering, covering map, covering space, covering transformation, covering group [Mun75, p. 331–341, 398]. A continuous map $p : \tilde{X} \to X$ between path-connected topological spaces is a *covering map* if every point of X has a neighborhood V such that every connected component of $p^{-1}(V)$ is mapped homeomorphically onto V by p. In this case $\tilde{X}$ is called a *covering space* of X, and $(\tilde{X}, p, X)$ is called a *covering*. A map $\phi : \tilde{X} \to \tilde{X}$ such that $p \circ \phi = p$ is a *covering transformation* (or *deck transformation*) for this covering; covering transformations form a group, called the *covering group*. If $\tilde{X}$ is simply connected, it is called the universal cover of X—"the" because it is unique up to homeomorphism.

deck transformation. See †covering.

deformation retract. If $A \subset X$ are topological spaces, A is a *deformation retract* of X if there a homotopy $F : X \times [0,1] \to X$ such that F_0 is the identity on X, F_1 maps X into A, and F_t fixes every point of A for every t.

diffeomorphism, diffeomorphic. If f is bijective and both f and f^{-1} are differentiable, f is called a *diffeomorphism*, and X and Y are said to be *diffeomorphic*.

differentiable map [Hir76, p. 9, 15]. A map $f : U \to \mathbf{R}^m$, where U is open in $\mathbf{R}^m$, is *differentiable of class* C^r (or a C^r*-map*) if f has continuous partial derivatives of order up to r. It is *smooth*, or *of class* C^∞, if it has continuous partial derivatives of all orders. A map $f : X \to \mathbf{R}^m$, where $X \subset \mathbf{R}^n$ is arbitrary, is of class C^r if it can be extended to a C^r map on a neighborhood of X.

A map $f : X \to Y$ between differentiable manifolds or manifolds-with-boundary is *of class* C^r if, for every point of $x \in X$, the expression of f in local coordinates in a neighborhood of x is of class C^r.

dihedral angle. The *dihedral angle* between two intersecting planes in space is the angle between the lines determined on the two planes by a third plane orthogonal to both.

effective group action. See †faithful.

embed, embedding. See †immersion.

Euclidean [Ber87, p. 153, 202]. We get *Euclidean n-space* $\mathbf{E}^n$ (*plane* if $n = 2$) by giving $\mathbf{R}^n$ the *Euclidean metric* $d(x,y) = \left(\sum_i x_i y_i\right)^{1/2}$, where $x = (x_1, \ldots, x_n)$ and $y = (y_1, \ldots, y_n)$ are points in $\mathbf{R}^n$. For $n = 3$ this is the space of our everyday experience.

Euclid's parallel axiom. "Given a line and a point outside the line, there is exactly one parallel to the line through the point." This axiom, which holds in a Euclidean space, has numerous equivalent formulations, such as "The sum of the angles of a triangle equals 180°."

evenly covered [Mun75, p. 331]. If $p : X \to Y$ is a continuous, surjective map between topological spaces and $U \subset Y$ is open, U is *evenly covered* by p if $p^{-1}(U)$ is a disjoint union of open sets that map homeomorphically onto U. Thus, p is a covering map if every point of Y has a neighborhood that is evenly covered by p.

faithful group action. Let the group G act on the space X. The action is said to be *faithful* if the only element of G which acts as the identity on X is the identity element of G. A synonym for this is an *effective* group action.

fiber, fiber bundle [Spa66, p. 90]. A surjective map $p : E \to B$ between topological spaces is a *bundle projection* if it looks locally like a product: that if, if there is a space F such that, for every point $x \in B$, there is an open neighborhood U of x and a map $\phi_U : U \times F \to p^{-1}(U)$ with $p \circ \phi_U = \pi_1$, where $\pi_1 : U \times F \to U$ is the projection onto the first factor. We call the whole setup (E, B, F, p) a *fiber bundle*, with F the *fiber*, E the *total space* and B the *base space*. Sometimes we also say that E itself is a *fiber bundle*, or that E *fibers* over B.

Let G be a group of homeomorphisms of F. A fiber bundle (E, B, F, p) is called a G-*bundle* if we can choose the sets U and maps ϕ_U above with the following additional property: for any $x \in B$ contained in two such sets U and V, the homeomorphisms $F \to p^{-1}(x)$ induced by ϕ_U and ϕ_V differ by an element of G.

flow. If X is a topological space, a family of continuous maps $f_t : X \to X$, for $t \in \mathbf{R}$, is a *flow* if $f_0 = \mathrm{Id}_X$ is the identity and $f_{s+t} = f_s \circ f_t$ for every $s, t \in \mathbf{R}$.

frame, frame bundle. An r-*frame* for a vector space V is an ordered set of r linearly independent vectors in V. Thus, if V has dimension n, an n-frame (also called just a *frame*) is the same as an ordered basis. If M is a differential manifold, the set of r-frames of tangent vectors at x, for every $x \in M$, forms a fiber bundle of M, called the r-*frame bundle* of M, or simply the *frame bundle* if r is the dimension of M.

fundamental group [Mun75, p. 326]. If X is a connected topological space, the set of homotopy classes of loops beginning and ending at a fixed point of X (the *basepoint*) is a group under concatenation of paths. It is called the *fundamental group* of X. If the fundamental group of X is trivial, we say that X is *simply connected*.

geodesic [dC76, p. ??]. Geodesics are curves that are as straight as possible. More precisely, given a Riemannian manifold X and an interval $I \subset \mathbf{R}$, a curve $\gamma : I \to X$ is a *geodesic* if, for $x, y \in I$ close enough, the shortest path joining $\gamma(x)$ and $\gamma(y)$ in X coincides with $\gamma([x, y])$. Normally we also require that γ be parametrized at constant speed (which can be zero). Often the image $\gamma(I)$, too, is called a geodesic.

geometry. Sometimes we use "geometry" interchangeably with "metric." Other times, "geometry" refers to the set of properties of a space that depend on the metric. Other times yet, it refers to the properties of a space that are invariant under a group of transformations of the space, which depends on the context. Thus projective geometry, Euclidean geometry, hyperbolic geometry.

homothety [Ber87, vol. I, p. 39]. A *homothety* centered at a point $p \in \mathbf{E}^n$ is the map that fixes p and "blows up" or "shrinks down" all of space around p. More precisely, it takes q to $k(q - p) + p$, for $k \in \mathbf{R} \setminus \{0\}$ fixed.

homotopy, homotopic, homotope [Mun75, p. 318–319]. Two continuous maps $f, g : X \to Y$ between topological spaces are *homotopic* if they can be continuously deformed into one another, that is, if there is a continuous map $F : X \times [0, 1] \to Y$ such that $F|_{X \times \{0\}} = f$ and $F|_{X \times \{1\}} = g$. Such a map is called a *homotopy* between f and g, and we say that f can be *homotoped* into g through F. The restrictions $F|_{X \times \{t\}}$, for $t \in [0, 1]$, are the *stages* of the homotopy, and are generally denoted by $F_t : X \to Y$.

immersion [Hir76, p.21]. A differentiable map $f : M \to N$ between differentiable manifolds is an *immersion* if the derivative df has maximal rank at every point $p \in M$. If, in addition, f is a homeomorphism onto its image, it is called an *embedding*, and we say that M is *embedded* in N. In particular, an embedding is a one-to-one immersion.

intermediate value theorem. A continuous function $f : [a, b] \to \mathbf{R}$ that is positive at a and negative at b must be zero somewhere in (a, b). A trivial consequence of the connectedness of an interval.

invariance of domain, theorem on the. If a subset $A \subset \mathbf{R}^n$ is homeomorphic to an open subset of $\mathbf{R}^n$, it is itself open. From this it easily follows that an m-dimensional manifold cannot have a subset homeomorphic to $\mathbf{R}^n$, for $n > m$.

isometry, isometric. A one-to-one map between metric spaces is an *isometry* if it preserves distances. If there is an isometry between two spaces they are said to be *isometric*. A map $f : X \to Y$ is a *local isometry* if any $x \in X$ has a neighborhood restricted to which f is an isometry.

isotopy. An *isotopy* between two maps $f, g : X \to Y$ is a homotopy $F : X \times [0, 1] Y$ between f and g such that every stage $F_t : X \to Y$, for $t \in [0, 1]$, is a homeomorphism onto its image.

isotropy subgroup, isotropic. If a group G acts on a space X, the *stabilizer* or *isotropy subgroup* of a point $x \in X$ is the subgroup G_x of G that leaves x invariant. If X is manifold, we say that X is *isotropic* (under the action of G) if the stabilizer of every point acts transitively on the tangent space at that point.

nilpotent group. Let the center of a group G be denoted by $Z(G)$. The *upper central series*

$$Z_0 \subset Z_1 \subset \cdots \subset Z_i \subset \cdots$$

of G is defined inductively by $Z_0 = \{e\}$ (where e is the identity) and $Z_i/Z_{i-1} = Z(G/Z_{i-1})$. We say that G is *nilpotent* (of class c) if $Z_c = G$. Thus a group is abelian if and only if it is nilpotent of class 1.

$O(\,\cdot\,)$. If f and g are real-valued functions of a real or integer variable x, we say that f is $O(g)$ if f is bounded by a multiple of g for x large enough. For instance, f *grows polynomially* if f is $O(x^n)$ for some n, that is, if f is bounded by a polynomial of order n.

orbit. See †action.

order. The *order* of an element g of a group is the smallest positive integer n (if one exists) such that g^n is the identity.

orientation, oriented, orientable, orientation-preserving, orientation-reversing [Mil65, p. 27]. An *orientation* for a finite-dimensional vector space is an equivalence class of (ordered) bases that can be taken to one another by linear transformations with positive determinant. A linear transformation between oriented vector spaces is *orientation-preserving* or *orientation-reversing* depending on whether its determinant is positive or negative.

A manifold M is *orientable* if the tangent spaces to M at all points can be oriented consistently. This means that M can be covered by coordinate patches such that the derivative map of any coordinate map at any $x \in M$ is an orientation-preserving map between T_xM and $\mathbf{R}^n$ with its standard orientation. We say that M is *oriented* if such a choice of orientations has been made. A local diffeomorphism between oriented manifolds is *orientation-preserving* or *orientation-reversing* according to what its derivative is at each point.

polygonal region. A closed set on the plane bounded by a polygon.

positive definite, positive semidefinite. See †signature.

proper map. A map between topological spaces is *proper* if the inverse image of every compact set is compact.

quotient by a group action. See †action.

quotient topology [Mun75, p. 134–136]. If X is a topological space and $\sim$ is an equivalence relation on X, we define the *quotient topology* on the set $X/\sim$ of equivalence classes by decreeing that a set of classes is open if and only if the union of their inverse images in X is open.

Riemannian manifold, Riemannian metric [BG88, p. 126]. A *Riemannian metric* on a differentiable manifold X is a rule that gives for each point $p \in X$ an inner product on the tangent space T_pX, in such a way that the inner product varies smoothly with p. Giving an inner product also gives a notion of length for tangent vectors, and consequently for differentiable paths. A *Riemannian manifold* is a manifold with a Riemannian metric; such a manifold is a metric space in a natural way, the distance between two points being the infimum of the lengths of paths joining the two points.

semigroup. A *semigroup* is a set X with an associative multiplication law.

signature. A quadratic form Q on $\mathbf{R}^n$ is equivalent, up to a linear change of coordinates, to one of the form $x_1^2 + \ldots + x_r^2 - (x_{r+1}^2 + \ldots + x_{r+s}^2)$, with $r + s \leq n$. The *signature* of Q is the pair (r, s); it obviously characterizes the form Q up to change of coordinates. The *rank* of Q is $r + s$. A form is *positive definite* if $r = n$ (that is, if it evaluates to a positive number on any non-zero vector), and *positive semi-definite* if $s = 0$ (that is, if it evaluates to a non-negative number on any non-zero vector). We define *negative definite* and *semi-definite* forms in the same way.

similarity [Ber87, vol. I, p. 183]. A *similarity* of Euclidean space is a transformation that multiplies all distances by the same factor k.

simply connected. See †fundamental group.

smooth. See †differentiable.

stabilizer. See †isotropy subgroup.

total space. See †fiber bundle.

tiling [Ber87, vol. I, p. 11–31]. Roughly speaking, a *tiling* of a space X is a way to fill up X with copies of one or more standard *tiles*, or shapes. There are various ways to restrict and formalize this definition to make it more manageable. We can stick to the following definition: A *tiling* of X consists of a connected compact subset $P \subset X$ and a group G of isometries of X such that the interior of P is non-empty, the union of images of P under G is X, and two images of P coincide if their interiors intersect.

topological group. A *topological group* is a topological space with a group law such that multiplication is a continuous function of the two variables, and inversion is also continuous.

universal cover. See †covering.

Bibliography

[Abi80] William Abikoff. *The real analytic theory of Teichmüller space.* Springer, New York, 1980.

[Ahl60] Lars Ahlfors. The complex analytic structure of the space of closed riemann surfaces. In et. al. Rolf Nevanlinna, editor, *Analytic Functions*, pages 45–66. Princeton University Press, Princeton, 1960.

[Bae28] Reinhold Baer. Isotopien von Kurven auf orientierbaren, geschlossenen Flächen. *Journal für die Reine und Angewandte Mathematik*, 159:101–116, 1928.

[Bar94] W. Barlow. Über die geometrischen Eigenschaften homogener starrer Strukturen und ihre Anwendung auf Krystalle. *Zeitschrift für Krystallographie und Mineralogie*, 23:1–63, 1894.

[BCH84] E. D. Block, R. Connelly, and D. W. Henderson. The space of simplexwise linear homeomorphisms of a convex 2-disk. *Topology*, 23(2), 1984.

[Ber81] Lipman Bers. Finite dimensional Teichmüller spaces and generalizations. *Bull. AMS*, 5:131–172, 1981.

[Ber87] Marcel Berger. *Geometry (2 vols.).* Springer, Heidelberg, 1987.

[BG88] Marcel Berger and Bernard Gostiaux. *Differential Geometry: Manifolds, Curves and Surfaces.* Springer, New York, 1988.

[Bia92] Luigi Bianchi. Sui gruppi di sostituzioni lineari con coefficienti appartenenti a corpi quadratici immaginari. *Mathematische Annalen*, 40:332–412, 1892.

[Bie11] Bieberbach. Über die Bewegungsgruppen der euklidischen Räume I. *Mathematische Annalen*, 70:297–336, 1911.

[Bie12] Bieberbach. Über die Bewegungsgruppen der euklidischen Räume II. *Mathematische Annalen*, 72:400–412, 1912.

[Bin54] R. H. Bing. Locally tame sets are tame. *Annals of Math. (2)*, 59:145–158, 1954.

[Bin59] R. H. Bing. An alternative proof that 3-manifolds can be triangulated. *Annals of Math. (2)*, 69:37–65, 1959.

[Bra49] Auguste Bravais. Mémoire sur les poly'edres de forme symétrique. *Journal de Mathematiques Pures et Appliquées*, 14:141–180, 1849.

[Bro12] Luitzen Egbertus Jan Brouwer. Beweiss des ebenen Translationsatzes. *Mathematische Annalen*, 72:37–54, 1912.

[Bro60] E. J. Brody. The topological classification of lens spaces. *Annals of Math. (2)*, 71:163–184, 1960.

[Bus55] Herbert Busemann. *The Geometry of Geodesics*. Academic Press, New York, 1955.

[Can79] James W. Cannon. Shrinking cell-like decompositions of manifolds: Codimension three. *Annals of Math. (2)*, 110:83–112, 1979.

[Cha86] Leonard Charlap. *Bieberbach groups and flat manifolds*. Springer, New York, 1986.

[dC76] Manfredo P. do Carmo. *Differential Geometry of Curves and Surfaces*. Prentice-Hall, Englewood Cliffs, NJ, 1976.

[dC92] Manfredo P. do Carmo. *RiemannianGeometry*. Birkhäuser, Boston, 1992.

[DE86] A. Douady and C. J. Earle. Conformally natural extension of homeomorphisms of the circle. *Acta Mathematica*, 157:23–48, 1986.

[Deh38] Max Dehn. Die gruppe der abbildungsklassen. *Acta Mathematica*, 69:135–206, 1938. Translated and reprinted as pp. 256–362 in *Papers on Group Theory and Topology by Max Dehn*, New York, Springer-Verlag, 1987.

[Eps66] David Epstein. Curves on 2-manifolds and isotopies. *Acta Mathematica*, 115:83–107, 1966.

[Fed49] Evgraf Stepanovich Fedorov. *Simmetriia i struktura kristallov*. Akademia nauk SSSR, Moscow, 1949. Translated as *Symmetry of crystals*, American Crystallographic Association, New York, 1971.

[FK12] R. Fricke and F. Klein. *Vorlesungen über die Theorie der automorphen Funktionen*. Teubner, Stuttgart, I(1897), II(1912).

[Fra87] George Francis. *A Topological Picturebook*. Springer, New York, 1987.

[Gab83] David Gabai. Foliations and the topology of 3-manifolds i. *Journal of Differential Geometry*, 18:445–503, 1983.

[Gab87] David Gabai. Foliations and the topology of 3-manifolds ii,iii. *Journal of Differential Geometry*, 26:461–478,479–536, 1987.

[Gar87] Frederick P. Gardiner. *Teichmüller theory and quadratic differentials*. Wiley-Interscience, New York, 1987.

[GHL90] Sylvestre Gallot, Dominique Hulin, and Jacques Lafontaine. *Riemannian Geometry*. Springer, Berlin, second edition, 1990.

[GM91] Charlie Gunn and Delle Maxwell (directors). *Not Knot*. A K Peters, Wellesley, 1991. 16-minute video produced by The Geometry Center (University of Minnesota); see http://www.geom.umn.edu/video/NotKnot.

[Hae58] André Haefliger. Structures feuilletées et cohomologie 'a valeur dans un faisceau de groupoïdes. *Commentarii Mathematici Helvetici*, 32:248–329, 1958.

[Hat83a] Allen E. Hatcher. Hyperbolic structures of arithmetic type on some link complements. *J. London Math. Soc. (2)*, 27:345–355, 1983.

[Hat83b] Allen E. Hatcher. A proof of a Smale conjecture, $\mathrm{Diff}(S^3) \simeq O(4)$. *Annals of Mathematics (2)*, 117:553–607, 1983.

[Hes30] J. F. C. Hessel. Krystallometrie oder Krystallometrie und Krystallographie. In *Gehler's Phys. Wörterbuch, Band 5, Abtheilung 2*. Schwickert, Liepzig, 1830.

[Hic65] Noel Hicks. *Notes on Differential Geometry*. Van Nostrand, Princeton, NJ, 1965.

[Hil01] David Hilbert. Über Flächen von konstanter gausscher Krümmung. *Transactions of the American Mathematical Society*, pages 87–99, 1901.

[Hir76] Morris Hirsch. *Differential Topology*. Springer, New York, 1976.

[Hoc65] Gerhard Hochschild. *The Structure of Lie Groups*. Holden-Day, San Francisco, 1965.

[Hop26] Heinz Hopf. Zum Clifford-Kleinschen Raumproblem. *Mathematische Annalen*, 95:313–339, 1926.

[Hör90] Lars Hörmander. *An introduction to complex analysis in several variables*. North-Holland/Elsevier, Amsterdam and New York, third edition, 1990.

[HR31] H. Hopf and W. Rinow. Über den Begriff der vollständingen differentialgeometrischen Flächen. *Comm. Math. Helv.*, 3:209–225, 1931.

[HW35] W. Hantschze and H. Wendt. Dreidimensionale euklidische Raumforme. *Mathematische Annalen*, 110:593–611, 1935.

[IT92] Y. Imayoshi and M. Taniguchi. *An Introduction to Teichmüller Spaces*. Springer, Tokyo and New York, 1992.

[Jor77] Troels Jorgensen. Compact 3-manifolds of constant negative curvature fibering over the circle. *Annals of Math. (2)*, 106:61–72, 1977.

[KS69] R. Kirby and L. Siebenmann. On the triangulation of manifolds and the hauptvermutung. *Bull. Amer. Math. Soc.*, 75:742–749, 1969.

[Kui55] Nicolaas Kuiper. On c^1-isometric embeddings ii. *Nederl. Akad. Wetensch. Proc. Ser. A*, pages 683–689, 1955.

[Leh87] Olli Lehto. *Univalent Functions and Teichmüller theory.* Springer, New York, 1987.

[Lic62] W. B. R. Lickorish. A representation of orientable combinatorial 3-manifolds. *Annals of Math. (2)*, 76:531–538, 1962.

[Lic65] W. B. R. Lickorish. A foliation for 3-manifolds. *Annals of Math. (2)*, 82:414–420, 1965.

[Mag74] Wilhelm Magnus. *Non-Euclidean Tesselations and Their Groups.* Academic Press, New York, 1974.

[Mar71] Jean Martinet. Formes de contact sur les variétés de dimension 3. In C. T. C. Wall, editor, *Proc. Liverpool Singularities Symposium II*, Lecture Notes in Mathematics, 209, pages 142–163. Springer, Berlin, 1971.

[Mil56] John Milnor. On manifolds homeomorphic to the 7-sphere. *Annals of Math. (2)*, 64:399–405, 1956.

[Mil58] John Milnor. On the existence of a connection with curvature zero. *Commentarii Mathematici Helvetici*, 32:215–223, 1958.

[Mil61] John Milnor. Two complexes which are homeomorphic but combinatorially distinct. *Annals of Math. (2)*, 74:575–590, 1961.

[Mil65] John Milnor. *Topology from a Differentiable Viewpoint.* University of Virginia, Charlottesville, 1965.

[Mil72] Tilla Klotz Milnor. Efimov's theorem about complete immersed surfaces of negative curvature. *Advances in Math.*, 8:474–543, 1972.

[Moi52] Edwin E. Moise. Affine structures in 3-manifolds, v: the triangulation theorem and the Hauptvermutung. *Annals of Math. (2)*, 56:96–114, 1952.

[Mun60] James Munkres. Obstructions to the smoothing of piecewise differentiable homeomorphisms. *Annals of Math. (2)*, 72:521–554, 1960.

[Mun66] James Munkres. *Elementary Differential Topology.* Princeton University Press, Princeton, 1966.

[Mun75] James Munkres. *Topology: A First Course.* Prentice-Hall, Englewood Cliffs, NJ, 1975.

[Mun84] James Munkres. *Elements of Algebraic Topology.* Addison-Wesley, Reading, MA, 1984.

[MZ55] Deane Montgomery and Leo Zippin. *Topological Transformation Groups*. Interscience, New York, 1955.

[Nag88] Subhashi Nag. *The complex analytic theory of Teichmüller spaces*. Wiley-Interscience, New York, 1988.

[Nov65] S.P. Novikov. Topology of foliations. *Transactions of the Moscow Mathematical Society*, 14:268–305, 1965.

[Now34] Werner Nowacki. Die euklidischen, dreidimensionalen, geschlossenen und offenen Raumformen. *Commentarii Mathematici Helvetici*, 7:81–93, 1934.

[O'N66] Barrett O'Neill. *Elementary Differential Geometry*. Academic Press, New York, 1966.

[Pap43] Christos Papakyriakopoulos. A new proof of the invariance of homology groups of a complex. *Bull. Greek Math. Soc.*, 22:1–154, 1943.

[Rad25] Tibor Radó. Über den Begriff der Riemannschen Flächen. *Acta Litt. Sci. Szeged*, 2:101–121, 1925.

[Ree52] Georges Reeb. *Sur certaines propriétés topologiques des variétés feuilletées*. Hermann, Paris, 1952.

[Ril82] Robert Riley. Seven excellent knots. In R. Brown and T. L. Thickstun, editors, *Low-Dimensional Topology*, pages 81–151. Cambridge University Press, Cambridge, 1982.

[Riv] Igor Rivin. Some applications of the hyperbolic volume formula of lobachevsky and milnor. preprint.

[Roy71] H. L. Royden. Automorphisms and isometries of teichmüller space. In L. Ahlfors et al., editor, *Advances in the Theory of Riemann Surfaces*, pages 369–384. Princeton University Press, 1971.

[sC55] Shiing shen Chern. An elementary proof of the existence of isothermal parameters on a surface. *Proc. Amer. Math. Soc.*, 6:771–782, 1955.

[Sch91] A. Schoenfliess. *Krystallsysteme und Krystallstruktur*. Teubner Verlag, Liepzig, 1891.

[Sma59] Stephen Smale. Diffeomorphisms of the 2-sphere. *Proceedings of the American Mathematical Society*, 10:621–626, 1959.

[Soh79] L. Sohncke. *Entwickelung einer Theorie der Krystallstruktur*. Teubner Verlag, Liepzig, 1879.

[Spa66] Edwin H. Spanier. *Algebraic Topology*. McGraw-Hill, New York, 1966.

[ST30] Herbert Seifert and William Threlfall. Topologische Untersuchung der Diskontinuitätsbereiche des dreidimensionalen sphärischen Raumes. *Mathematische Annalen*, 104:1–70, 1930. Continued in **107** (1932), pp. 543–586.

[Sti87] John Stillwell. The dehn–nielsen theorem. In *Papers on group theory and topology by Max Dehn*, pages 363–396. Springer-Verlag, New York, 1987.

[Tei39] Oswald Teichmüller. Extremale quasikonforme Abbildungen und quadratische Differentiale. *Abh. Preuss. Akad. Wiss. math-naturw. Kl.*, 22:1–197, 1939. Reprinted as pp. 335–531 in *Collected Papers*, Springer, Berlin, 1982.

[Tei43] Oswald Teichmüller. Bestimmung der extremalen quasikonformen Abbildungen bei geschlossenen orientierten riemannschen Flächen. *Abh. Preuss. Akad. Wiss. math-naturw. Kl.*, 4:1–42, 1943. Reprinted as pp. 635–676 in *Collected Papers*, Springer, Berlin, 1982.

[Thu74] William P. Thurston. The theory of foliations of codimension greater then one. *Comm. Math. Helv.*, 49:214–231, 1974.

[Thu76] William P. Thurston. Existence of codimension-one foliations. *Annals of Math. (2)*, 104:249–268, 1976.

[Whi36] Hassler Whitney. Differentiable manifolds. *Annals of Math. (2)*, 37:645–680, 1936.

[Whi40] J. H. C. Whitehead. On C^1 complexes. *Annals of Math. (2)*, 41:809–824, 1940.

[Wol67] Joseph A. Wolf. *Spaces of Constant Curvature*. McGraw-Hill, New York, 1967.

[Woo69] John W. Wood. Foliations on 3-manifolds. *Annals of Math. (2)*, 89:336–358, 1969.

[Zas48] H. Zassenhaus. Über einen Algorithmus, zur Bestimmung der Raumgruppen. *Commentarii Mathematici Helvetici*, 21:117–141, 1948.

Index

Pages where a term is defined are shown in italics. A †indicates the term can be found in the glossary.

(connected sum), 28

^ (one-point compactification), 4

abelian subgroups, 218–219, 226
absolute value, *105*
action, 6, †
– discrete, effective, free, wandering, *153*
– of $GL(2, \mathbf{C})$ at infinity, 100
– with non-Hausdorff quotient, 155
Adams, Colin, 129
adjoint representation, *186*, 187
Adobe Illustrator, iv
affine
– hull, 119
– manifold, *126*
– map, 20
– nilgeometry, 283
– space, 119
– torus, 142
– transformations, 72
affinely independent, 118
Alexander trick, *124*
almost abelian, 217–218, 221, 226, 254, *278*
alternate universes, 3
ambient, 5, †
analytic continuation, *139*–140
angle, *see* trigonometry
annuli for making a pseudosphere, 50
antipodes, 33
area
– and contact transformations, 176
– in hyperbolic plane, 84–86
– in sphere, 85
– -preserving vector field, 115
– -preserving automorphisms, 170
associated bundle, *161*
associated, *160*
atlas, $\mathcal{G}$-atlas, *110–111*
automorphism, $\mathcal{G}$-automorphism, *111*
– , contact, 170
– , lattice, 228
autosum, *30*
axis of a transformation, *86*, 88–89

Baer, Reinhold 260
ball, 3
balloon, 40
banana angles, 55, 59
barber shop, 31
barycentric subdivision, 22, *119*, *123*, †
base space, *158*
Beltrami, Eugenio, 8, 65
bi-invariant metric, *217*
Bieberbach groups and theorem, *222*
biholomorphic map, *117*
binary group, *252*

blowing up, *30*
body-centered cubic lattice, 228, *229*
Bolyai, János, 8
Borromean rings, 131–132
boundary, *118*
branch points, 52
Bravais class, group, lattice, *240*
Brouwer fixed-point theorem, 97
bundle, *158*
– , frame 28–29, *161*
– isomorphism, *161*
– map, *160*
– , normal, 176
– projection, *158*
– , principal, *161*
– , product, *159*
– , tangent, *161*
– , tangent circle/sphere, *161*, 172
– , vector *160*
Busemann, Herbert, 94
butterfly operation, 73

CX (cone over X), 120
Canary, Dick, iii
Cannon, Jim, iii
canonical smoothing of a manifold, 207–208
cell, cell division, cell map *18*
center of mass, *96*
charges on a convex polyhedron, 20
chart, $\mathcal{G}$-chart, *110*
cheese wedges, 38
$\chi(S)$ (Euler number), 18
circle
– inversion in, 54
– , wild, 192, †
– and oval, 255
classification
– of $S^2 \times \mathbf{E}^1$-manifolds, 277
– of complete (G, X)-structures, 157, 222
– of contact flows, 175
– of elliptic manifolds, 249
– of Euclidean manifolds, 222
– of flat bundles, 163
– of isometries of $\mathbf{H}^n$, 95
– of measure stiffenings, 116
– of spherical manifold, 245
– of surfaces, 25
closed
– manifold, *118*
– orbits, 206
– subgroups, *210*
cocompact, *157*, 281–285
cocycle condition, *160*, 230
cofinite, *157*, 280, 283
combinatorial class, *123*
combinatorially equivalent, *122*
common perpendicular for lines in $\mathbf{H}^3$, 88
compact
– group has fixed point, 96
– manifolds have no cusps, 257
– stabilizers, 144, 157
compact-open topology, 153, †
compactification, *4*
compatibility condition, 123, 230
compatible charts, atlas, *111*
complete (G, X)-manifold, *142*, 157, 222
complete space, 51
completeness criteria, 144–145, 147–148, 152, 157
completion of a hyperbolic surface, 150
complex
– coordinates for $\mathbf{H}^3$, 98
– manifold, *117*
– numbers, 98, 104
– trigonometry, 101
computer algorithm, 18
cone on a space, 40, *120*, †
conformal, *10*
congruence of ideal triangles, 84
conjugate
– quaternion, *105*
– fixed points, 96, 255

connected sum, *28*
connection, *165*
constant curvature, 47
contact
– automorphism, diffeomorphism, *170*
– flow, 175
– perturbations, 175
– plane field, *170*
– pseudogroup, 175
– structure, *168*, *170*–175
, strict, 178
convex
– distance function, *90*
– function, *90*
– hull, 118, †
– polyhedron, *119*, *127*
Cooper, Daryl, 129
coordinate change, map, patch, 22, *111*
cosines, law of, *75*–76
countable basis, 111
covering group, map, space, transformation, 5–6, 159, †
crystallographic group, *222*–233
cubic lattice, 228, *229*, 240
curvature, *45*, 47, 55, 167
cusp, *150*, *256*–257, 267
cutting up surfaces, 269

D^n (disk), 28
$d_{\mathrm{qi}}(M, M')$ (quasi-isometry metric), 266
deck transformation, *see* covering transformation
definite, positive, 64, †
deformation retract, 203, †
Dehn twist, *275*
Dehn–Nielsen theorem, 276
derivative
– of distance function, 94
– of a piecewise linear map, 196
developing map, *139*–*140*, 165
diagonal groups, 246
dictionary, 9
Diff S, Diff$_+$ S, 262
diffeomorphic, *111*
– embeddings, 28, 207
– of the unit disk, 207
differentiable
– manifold, *19*, 111 †
– map, 19, †
– structure, 194
– surface, 20
– triangulation, *20*
– vector field, 24
differential geometry, 45, 83
dihedral angles, 35, 128
dilatation, 57
discrete group of isometries, *153*
– generated by small elements is nilpotent, 211
– , flowchart for, 280, 282
– Euclidean is almost abelian, 218, 221, 226
– is properly discontinuous, 156
– of $\mathbf{H}^2 \times \mathbf{E}^1$ and $\widetilde{\mathrm{PSL}}(2, \mathbf{R})$, 278
– of nilgeometry, 279
– of solvegeometry, 280
– of sphere is almost abelian, 217, 221
– , structure of, 209
distance function is convex, 91
distinguishing geometries of discrete groups, 280, 282
divergence, *182*
dodecahedral space, 34, 37, 124, 252
dodecahedron, 34, 37, 70
double suspension, *124*
dual
– basis, *74*
– hyperspace, *71*
– spherical law of cosines, *75*
duality between a hyperplane and a point, 71

$\mathbf{E}^n$, *see* Euclidean

$\mathbf{E}^{2,1}$, *see* Lorentz space
edge, *18*, 119
effective action, *153*
Eilenberg–MacLane space, *145*
elementary move, *26*
elliptic
– space, *33*
– manifold, *127*, 242, 243
– manifolds, classification of, 249
– structure, *27*
– transformation, *87*, 90
embedding, 5, 20, 29, †
empty word, *219*
ends, *267*
Epstein, David, iii
equidistant curve, *13*–14, 87
equilateral triangle, 51, 228
equivariant, *156*, 256
– closed curve, *269*
– edge, 48
Euclidean
– axioms, 8
– isometries, 6, 9, 54, 90, 218
– geometry, 44
– manifold, *125*, 132, 221
, examples of, 126, 236, 239
has flat tangent bundle, 165
s, classification of, 222
covered by $T^2 \times \mathbf{E}^1$, 233
– median, mean, *96*–97
– similarities, 60
– space $\mathbf{E}^3$, 74
– stereoscopic vision, 70
– structure, *6*
– triangle, 35
– transformations, 6, 9, 54, 90, 218
Euler number, *18*–19, 122
even covering, 145, †
exact lattice group, *227*, 240
exotic sphere, *112*
$\exp v$, 211
exponential map, 144, *211*
exposition, philosophy of, v
extrinsic properties, 45
face-centered cubic lattice, 228–*229*
facet, *119*, 122
face, *18*, 119, 122
faithful representation, *264*
Fenchel–Nielsen coordinates, *272*
Ferguson, Helaman, iv, 138
Ferguson, Jon, iv
fiber, *158*
– bundle, *103*, 158
– optics, 32
– product, *246*
fibered geometries, manifolds based on 277–287
fibration, 103
figure-eight knot, 40–41, 127, 157
finite
– group has a fixed point, 96
– subgroups of $\mathrm{GL}(n, \mathbf{Z})$, 226
– subgroups of $\mathrm{SL}(2, \mathbf{Z})$, 228
– subgroups of $\mathrm{SL}(3, \mathbf{Z})$, 238
– type, surface of *267*
– volume, 157, 258, 280
Finsler manifold, 94
fixed point, 63, 96–97, 218, 235, 255
flare, 267
flat
– manifold, *125*, *165*
– connection, G-bundle, *163*
floppy surfaces, 51
flow, 104, †
flowchart for discrete groups, 280, 282
Floyd, Bill, iii
focal distance, 33
foliation, 113, *114*–115, 160
fractional linear transformation, *99*
frame 28, *43*, †
– bundle 28, *161*
Francis, George, 135
free action, *153*–157, 242–245
freely homotopic, *142*
Frobenius' theorem, 177
Frobenius map, 252–*253*
fundamental domain, 96

$\mathcal{G}$-atlas, *110–111*
$\mathcal{G}$-automorphism, *111*
$\mathcal{G}$-compatible charts, *110*
$\mathcal{G}$-isomorphism, *111*
$\mathcal{G}$-manifold, *110*
$\mathcal{G}$-relaxation, *112*
(G, X)-bundle, *158*
(G, X)-foliation, 160
(G, X)-manifold, *125*
Gauss, Carl Friedrich, 8, 47, 83, 85
– 's Theorema Egregium, 47
Gauss–Bonnet theorem, 28, 83
Gaussian curvature, 6, *45*, 55
generated, *113*
genus, *26*
genus-two surface, *7*, 16
geodesic, 7–9, 59, 66–67, †
– flow, *174*
– triangle, 35
–s accumulate on the boundary, 151
–s in nilgeometry, 185
geometric structures, 125ff
Geometry Center, Geometry Supercomputer Project, iii–iv, 70
germ, 140
$GL(2, \mathbf{C})$, 99–100
$GL(n, \mathbf{Z})$, finite subgroups of, 226
glass spheres, 32
glide-reflections, 237
gluing, *17*, *120*, *123*
– , complete, 148
– and Euler number, 122
– examples, 18, 39, 123, 133
– , polyhedral, 123
– , rectilinear, *120*
Gram–Schmidt process, 204
Grassmannian, 215
group action, *see* action
group cohomology, 230
growth function, *220*
Gunn, Charlie, iv, 70

$\mathbf{H}^n$, *see* hyperbolic space
Haar measure, *96*
handlebody, 135
half-space, 53, 60
Hauptvermutung, *192*
Hausdorff
– distance, *209*
– property, absence of, 154
– topology, *209*
Heisenberg group, *185*–186
hemisphere 53
– model, *57*–58, 68
Hermitian form, 100
Hessian, 46
hexagon
– , regular, 4
– , right-angled, 82–83, 263
hexagonal law of sines, *82*
Hilbert's Theorem, 51
holomorphic, 117
holonomy, *141*, 166
– defines structure, 264
– group, *141*
characterizes manifold, 142, 157
– holonomy, *163*
homeomorphism, 6
– of an interval, 17
homogeneity, *43*
homogeneous coordinates, *98*
homology, 121
homothety, 11, †
homotopy, 24, 103, †
Hopf
– circles, fibration, *103*–104
– flow, 104, 106, 108, 174
– foliation, 106
horizontal contact perturbations, 175
horoball, horocycle, horosphere, *61*, 148
hyperbolic
– end, *267*
– isometry, 9, 12, *55*, 86, *87*, 90, 95–97
– law of cosines, 81

– law of sines, 82
– lighting, 71
– lines, 9
– manifold, 7, *127*
with geodesic boundary, 133
with flat sphere bundle, 165
– median, mean, *96*–97
– paper, 49
– paraboloid, 46
– Pythagorean theorem, 81
– ray, 56
– space, 8–9, 36, 53ff, 69
– structure, *16*, 260
– transformation, 9, 12, *55*, 86, *87*, 90, 95–97
hyperboloid
– model, *64*–67
– , parabolic, 46

icosahedron, 124
ideal
– boundary, *267*
– dodecahedron, *37*
– triangles, 84
– vertex, *37*, *148*, 151
immersion, 17, †
indefinite metric, 64
index of vector field, *23*–24
induced
– orientation, 17
– representation, *223*
– structure, 112
infinity, point at, 4, 98
infinitesimal injectivity, 93
informal pictures, 74
inherent symmetry, 3
inhomogeneous coordinate, *98*
injective derivative, 92
injectivity radius, *253*
inner product, 65, 77–78
inside view
– of S^3, 32
– of T^3, 31
– of tiled $\mathbf{H}^3$, 69

integer Heisenberg group, *185*
integrable, *167*
interior, *119*
intermediate value theorem, 15, †
intrinsic geometry, 45, 50
invariance of domain, 19, †
invariant
– metric, existence of, 144, 217
– translation subspace, 224–225
inversion, *10*, *53*–54, 57
inversive models, 53
inversor (mechanical), 13
irrational foliation on the torus, 114
isolated zeros, 24
isometric lattices, 228, *229*
isometry, *43*; *see also* under hyperbolic, Euclidean, etc.
isothermal coordinates, *117*
isotopy, 17, †
isotropy, *43*
– subgroup, *see* stabilizer

Jacobi identity, *167*
Jordan normal form, 101

$K(G,1)$ (classifying space), 145
Kerckhoff, Steve, iii
Klein, Felix, 8, 117
– bottle, 26
– model, *65*–68
knot, 39–42, 127, 147, 157
knotted Y, 135, 138

lattice *157*
– automorphism, 228
– group, *227*, 240
– , isometric 228, *229*
– , uniform, *157*
law of sines, hyperbolic, *82*
law of cosines, hyperbolic, *81*
leaf of foliation, *114*
Legendrian
– curves, lifts, *169*
– foliation, *170*, 174

length, hyperbolic and Euclidean, 15
lens space, 37–*38*, 251
Levi-Civita connection, *166*
Levy, Silvio, iv
Lie
– algebra, bracket, *167*
– derivative, *116*
– group, 156
light cone, *66*
light rays, 32
light-like, *64*
linear
– algebra, 86
– fractional transformation, *99*
– group, see GL(,)
– vector field, 24
linearization, *198*–199
lines
– and circles, 54
– of curvature, *52*
link, *40*, *120*
linkage, 12
lk σ, 120
Lipkin, Lippman, 12
Lobachevskii, Nikolai Ivanovich, 8
local
– condition for $\mathcal{G}$-structures, *110*, 113, 125
– coordinate system, *110*
– cross-section, *143*
– finiteness, 119
– homeomorphism, 30
– $\mathcal{G}$-isomorphism, *112*
– isotropy, 49
– simple connectedness, *121*
– trivialization, *158*
logarithm map, *211*
Lorentz
– group, *66*
– metric, 64
– space, *64*, 74, 76
– transformation, *66*, 73
loxodromic, 87
$\mathcal{M}S$ (moduli space of S), 260
manifold, $\mathcal{G}$-manifold, *109*–111, *125*; *see also* under adjectives
manifold-with-boundary, *118*, 132
map of the earth, 53
map on the torus, 5
mapping
– class group, *262*
– torus, *159*, 189, 237
Marden, Albert, iii
marked surface, markings, *261*, 276
Mathematica, iv
Mathfig, iv
maximal translation subspace, 223
maximally symmetric, *227*
Maxwell, Delle, iv, 70
mcg S, mcg$_+$ S, 262
mean, 96
measure, 96
– stiffenings, 116
measured manifold, *115*
mechanical linkage, 12
median, *97*
minimal hyperbolic properties, 63
minimum distance and perpendicularity, 79
Minsky, Yair, iii
mirrors, 31
Möb$_{n-1}$, 62
Möbius
– group, *62*
– strip, 17, 27, 30, 159
– transformation, *62*
model
– of hyperbolic space, 8, 53
– geometry, 179, *180*–181
modular group, *262*–266
moduli space, 113, *259*–270
modulus, *260*
moon, 53
multiplication, right and left, 108
multiplicity, 271
Munzner, Aribert, iv

navigation, 53
negative curvature, 45, *94*
Nielsen–Dehn's theorem, 276
nilgeometry, nilmanifold, *184*–185, 280, 284
nilpotent, 185, 212, †
– of class k, *210*
non-Hausdorff, 154–155
nonorientable genus, *26*
nondiscrete quotient with Hausdorff quotients, 156
normalized, *76*
norm, *105*, 219
normal bundle, 176
Not Knot, 70
nullspace, 100

$O(n+1)$, 64, 66
$O(n,1)$, 66
octagon, 7, 17
octahedral symmetry, lattices with, 229
octahedron, 131
Olson, Kenneth, iv
one-point compactification, *4*
open manifold, *118*
order, 17, †
orientability of gluings, 17
orientation-preserving, orientation-reversing, 17, 90, †
orthogonal
– complement, *65*
– families of circles, 11
– map, 14
– trajectories, 11
– transformation, *64*, 66; *see also* elliptic, spherical, hyperbolic transformations

pants, *262*, 269
paper, 45, 49
parabolic
– adjustment, 98
– end, 267
– isometry, *87*, 90
paraboloid
– model, *73*
– of revolution, 46
paracompact, 111
parallel translation, *166*
parallelism, 8, 72, *88*, 93, 173
Peaucellier, 12
pentagon, 72, 82, 252
perpendicular, common, 88
$\mathrm{PGL}(2,\mathbf{C}) = \mathrm{GL}(2,\mathbf{C})/\mathbf{C}^*$, 99
piecewise
– integral projective, *193*
– linear, *190*
– projective, *192*
– piecewise smooth, *194*
plane translation theorem, *154*
plants, 51
Poincaré, Henri, 9, 100, 117
– ball model, *36*, 54–57
– disk model, *9*, 12, 14, 53, 73
– dodecahedral space, 34–35, 124, 252
– index theorem, 25
Poincaré–Bendixson theorem, 206
point
– at infinity, *98*
– , fixed, 63, 96–97, 218, 255
– groups, *231*
polygon, 4
–s, space of shapes of, 72
polygonal
– helix, 87
– region, 17, †
polyhedral
– gluing, 123
– models of negative curvature, 50
polyhedron, *119*
positive
– curvature, 45
– definite, 64, †
– semidefinite, 79, †
– Dehn twist, *275*–276
power of a point with respect to a circle, *10*

Princeton University, iii
principal
- bundle, *161*
- directions, *52*
procrastination, iii
product
- bundle, *159*
- coverings, 238
- of Riemmanifold manifolds, *90*
projective
- model, *65*
- plane, 17, 26
- space is SO(3), 107
- transformation, *72*, 99
proper map, 54, 93, *154*, 156, 279, †
- discontinuity, *154*, 158
properties of inversions, 10, 54
property hierarchy for group actions, 154
pseudogroup, *110*, 113
pseudosphere, 47–48, 50, 62
PSL(2, $\mathbf{C}$), 99
PSL(2, $\mathbf{R}$) = Isom$^+$ $\mathbf{H}^2$, 101
pullback, *162*, 169
pure quaternions, *105*

quadratic form, 47, 51, 64
quadrilateral, 17
quasi-isometry, *266*
quasiconformal, *267*
quaternion, *104*
quotient, 6, †

reflection, 60, 237; *see also* hyperbolic isometry, etc.
real analytic, 113
- immersions, 52
- manifolds, *113*
real quaternion, *105*
reduced Teichmüller space, *268*
Reeb foliation, 114
regular neighborhood, 135
relativistic universe, 73
Renderman, iv
representation, induced, *223*
Riemann, Georg Friedrich, 8, 260
- mapping theorem, 117, 207
- sphere, *98*
- surface, *117*, 260
Riemannian contact structure, 178
- geometry, 94
- manifold, 55, 94, 174
- metric, 12, 14, 16, 43, †
exists on every manifold, 163
right-angled
- polygons, 82–83, 263
- dodecahedron, 70
rigid motions, 107
rotational part, *218*
rotation, *87*, 235, 237; *see also* hyperbolic, Euclidean isometry
rubber ball, 45

$\mathcal{S}S$ (space of hyperbolic structures on S), 260
S^n, *see* sphere
$S^2 \times \mathbf{E}^1$-manifolds, classification of, 277
S^{n-1}_∞ (sphere at infinity), 57
saddle, *23*, 45–46
scalar multiple, 99
scaling, 83
screw motion, *87*
section, *161*
sectional curvature, 43, *168*
Seifert's theorem, 247
Seifert fiber space, *248*
Seifert–Weber dodecahedral space, 36–37
self-intersecting surfaces, 17
semidefinite, positive, 79, †
semidirect, *229*, 230
semigroup, 29, †
ΣX (suspension of X), 124
similarity, 22, *60*, †
- manifold, structure, 143
simple transitivity, 106
simplex, simplicial, *118–119*, *194*
simply connected, 6, *121*

sines, law of, *75*–76
sink, *23*
sixfold symmetry, 4
skeleton, *119*
skew field, *105*
skirts, 50
SL(2, **C**), 99
SL(2, **Z**), finite subitem subgroups of, 228
SL(3, **Z**), finite subitem subgroups of, 238
small elements, *209*, 214, 221, 254
Smillie's procedure, 143
smooth
– manifolds, *111*
– triangulation, *194*
– welding, *199*
smoothing, *112*, 193–*194*, 201, 207–208
SO(2), 106
SO(3), 107; *see also* spherical
SO(4), 107
solvegeometry, *188*–189, 285
source, *23*
space-like, *64*
special
– relativity, 64
– group, see SL(,)
sphere, 4, 19, 64
– at infinity, 56, *57*
– , exotic *112*
– S^3, inside view, 32
– S^3, group structure of, 104–105
–s, tangent, 63
spherical
– area, 85
– geometry, 33, 44, 74
– law of cosines, *75*
– links, 120–121
– triangle inequality, 92
– triangles, 34, 74
– manifold, *127*
s, classification of, 245
square
– lattice, 228
– , gluings of, 18
– roots, 90
– torus, 4
– , vector field on, 206
square-of-distance mean, 97
st σ, 120
stabilizer, 143–144, †
standard
– gluing, 123
– piecewise linear sphere, *191*
– piecewise projective sphere, *192*
star, *22*, 120
star-shaped, *24*, 256
Steiner's porism, 63
stereographic projection, *57*
stiffening, *112*, 151
Stokes' theorem, 169
straight line, *see* geodesic
straightening closed curves, 269–271
strict contact structures, 178
strictly convex, *90*
strips, disk with, 27
structure group, *158*
subdivision, *119*
subsimplex, *119*
sum of dihedral angles, 128
surface
–s, classification of, 25
– semigroup, 29
– of revolution, 48–49
suspension, *124*

$\mathcal{T}S$ (Teichmüller space of S), 260
tangent
– bundle, *161*
– circle bundle, *161*, 172
– cones, *197*
– horosphere *61*
– plane field, *167*
– space, *161*
– space, piecewise linear, 197
– sphere bundle, *161*, 172

– spheres, 63
– vectors to hyperboloid, 65
tangentially (G, X) foliation, *160*
τ_γ (Dehn twist), 275
Teichmüller space, 258, *259*–271
– metric, 267
– of the torus, 265
– of compact surfaces, 271
– reduced, *268*
tetrahedron, 18, 34, 128, 134
Theorema Egregium, 47
thick part, *253*, 258
thick-thin decomposition, 150, 253–257
thin part, *254*
Thompson, Richard, 193
thrice-punctured sphere, 149, 264
tiling, 5, 6, 16, 37, 56, 60, †
time-like, *64*
Top, 110
topological
– group, 96, †
– manifold, *110*
torus, 4, 17, 26
– T^2, affine, 142
– T^3, inside view, 31
– T^3, foliation on, 114–115
total space, *158*
totally geodesic, *55*
trace, 101–102
tractrix, 47–48
transition map, *111*
transitivity of symmetries, 43
translation
– distance, *95*
– , hyperbolic *87*
– subgroup, *218*
– subspace, *223*
–s form lattice, 218
transverse, *21*, 22
trefoil, 41–42
triangulation, 22, 24, *119*, 190, 194, 202
trigonometry, 15, 74–82
tripus, 135, 138
trivial
– bundle, *161*
– pseudogroup, *110*
truncated
– cone, 49
– tetrahedron, 134
twist parameter, *272*, 274
twistedness, *176*

umbilic, *52*
uniform lattice, *157*
uniformization theorem, *117*, 260
unimodular, *187*
uniqueness
– of line by two points, 33–34
– of gluings, 124
– of smoothing, 201
unit
– quaternion, *105*
– tangent bundle, *161*
unitary transformations, *106*
universal cover, 6, †
unrolling, 139
up to $\mathcal{G}$-automorphisms, *112*
upper half-space, 53, 60
UTM, 161

vector
– bundle, *160*
– field, 19
vertex, *18*, 119, *122*
Villarceau circles, *104*
visual sphere, *45*, 56, 68

wandering, *153*, 154
Weeks, Jeff, 129
welding, *199*–200
Whitehead link, *129*–131
wild circle, 192
word, *219*

Xfig, iv